D0082978

INTRODUCTION TO SET THEORY

PURE AND APPLIED MATHEMATICS

A Program of Monographs, Textbooks, and Lecture Notes

EXECUTIVE EDITORS

Earl J. Taft
Rutgers University
New Brunswick, New Jersey

Zuhair Nashed
University of Delaware
Newark, Delaware

EDITORIAL BOARD

M. S. Baouendi
University of California, San Diego

Jane Cronin
Rutgers University

Jack K. Hale
Georgia Institute of Technology

S. Kobayashi
University of California, Berkeley

Marvin Marcus
University of California, Santa Barbara

W. S. Massey
Yale University

Anil Nerode
Cornell University

Donald Passman
University of Wisconsin, Madison

Fred S. Roberts
Rutgers University

Gian-Carlo Rota
Massachusetts Institute of Technology

David L. Russell
Virginia Polytechnic Institute and State University

Walter Schempp
Universität Siegen

Mark Teply
University of Wisconsin, Milwaukee

MONOGRAPHS AND TEXTBOOKS IN PURE AND APPLIED MATHEMATICS

1. *K. Yano,* Integral Formulas in Riemannian Geometry (1970)
2. *S. Kobayashi,* Hyperbolic Manifolds and Holomorphic Mappings (1970)
3. *V. S. Vladimirov,* Equations of Mathematical Physics (A. Jeffrey, ed.; A. Littlewood, trans.) (1970)
4. *B. N. Pshenichnyi,* Necessary Conditions for an Extremum (L. Neustadt, translation ed.; K. Makowski, trans.) (1971)
5. *L. Narici et al.,* Functional Analysis and Valuation Theory (1971)
6. *S. S. Passman,* Infinite Group Rings (1971)
7. *L. Dornhoff,* Group Representation Theory. Part A: Ordinary Representation Theory. Part B: Modular Representation Theory (1971, 1972)
8. *W. Boothby and G. L. Weiss, eds.,* Symmetric Spaces (1972)
9. *Y. Matsushima,* Differentiable Manifolds (E. T. Kobayashi, trans.) (1972)
10. *L. E. Ward, Jr.,* Topology (1972)
11. *A. Babakhanian,* Cohomological Methods in Group Theory (1972)
12. *R. Gilmer,* Multiplicative Ideal Theory (1972)
13. *J. Yeh,* Stochastic Processes and the Wiener Integral (1973)
14. *J. Barros-Neto,* Introduction to the Theory of Distributions (1973)
15. *R. Larsen,* Functional Analysis (1973)
16. *K. Yano and S. Ishihara,* Tangent and Cotangent Bundles (1973)
17. *C. Procesi,* Rings with Polynomial Identities (1973)
18. *R. Hermann,* Geometry, Physics, and Systems (1973)
19. *N. R. Wallach,* Harmonic Analysis on Homogeneous Spaces (1973)
20. *J. Dieudonné,* Introduction to the Theory of Formal Groups (1973)
21. *I. Vaisman,* Cohomology and Differential Forms (1973)
22. *B.-Y. Chen,* Geometry of Submanifolds (1973)
23. *M. Marcus,* Finite Dimensional Multilinear Algebra (in two parts) (1973, 1975)
24. *R. Larsen,* Banach Algebras (1973)
25. *R. O. Kujala and A. L. Vitter, eds.,* Value Distribution Theory: Part A; Part B: Deficit and Bezout Estimates by Wilhelm Stoll (1973)
26. *K. B. Stolarsky,* Algebraic Numbers and Diophantine Approximation (1974)
27. *A. R. Magid,* The Separable Galois Theory of Commutative Rings (1974)
28. *B. R. McDonald,* Finite Rings with Identity (1974)
29. *J. Satake,* Linear Algebra (S. Koh et al., trans.) (1975)
30. *J. S. Golan,* Localization of Noncommutative Rings (1975)
31. *G. Klambauer,* Mathematical Analysis (1975)
32. *M. K. Agoston,* Algebraic Topology (1976)
33. *K. R. Goodearl,* Ring Theory (1976)
34. *L. E. Mansfield,* Linear Algebra with Geometric Applications (1976)
35. *N. J. Pullman,* Matrix Theory and Its Applications (1976)
36. *B. R. McDonald,* Geometric Algebra Over Local Rings (1976)
37. *C. W. Groetsch,* Generalized Inverses of Linear Operators (1977)
38. *J. E. Kuczkowski and J. L. Gersting,* Abstract Algebra (1977)
39. *C. O. Christenson and W. L. Voxman,* Aspects of Topology (1977)
40. *M. Nagata,* Field Theory (1977)
41. *R. L. Long,* Algebraic Number Theory (1977)
42. *W. F. Pfeffer,* Integrals and Measures (1977)
43. *R. L. Wheeden and A. Zygmund,* Measure and Integral (1977)
44. *J. H. Curtiss,* Introduction to Functions of a Complex Variable (1978)
45. *K. Hrbacek and T. Jech,* Introduction to Set Theory (1978)
46. *W. S. Massey,* Homology and Cohomology Theory (1978)
47. *M. Marcus,* Introduction to Modern Algebra (1978)
48. *E. C. Young,* Vector and Tensor Analysis (1978)
49. *S. B. Nadler, Jr.,* Hyperspaces of Sets (1978)
50. *S. K. Segal,* Topics in Group Kings (1978)
51. *A. C. M. van Rooij,* Non-Archimedean Functional Analysis (1978)
52. *L. Corwin and R. Szczarba,* Calculus in Vector Spaces (1979)
53. *C. Sadosky,* Interpolation of Operators and Singular Integrals (1979)
54. *J. Cronin,* Differential Equations (1980)
55. *C. W. Groetsch,* Elements of Applicable Functional Analysis (1980)

56. *I. Vaisman,* Foundations of Three-Dimensional Euclidean Geometry (1980)
57. *H. I. Freedan,* Deterministic Mathematical Models in Population Ecology (1980)
58. *S. B. Chae,* Lebesgue Integration (1980)
59. *C. S. Rees et al.,* Theory and Applications of Fourier Analysis (1981)
60. *L. Nachbin,* Introduction to Functional Analysis (R. M. Aron, trans.) (1981)
61. *G. Orzech and M. Orzech,* Plane Algebraic Curves (1981)
62. *R. Johnsonbaugh and W. E. Pfaffenberger,* Foundations of Mathematical Analysis (1981)
63. *W. L. Voxman and R. H. Goetschel,* Advanced Calculus (1981)
64. *L. J. Corwin and R. H. Szczarba,* Multivariable Calculus (1982)
65. *V. I. Istrățescu,* Introduction to Linear Operator Theory (1981)
66. *R. D. Järvinen,* Finite and Infinite Dimensional Linear Spaces (1981)
67. *J. K. Beem and P. E. Ehrlich,* Global Lorentzian Geometry (1981)
68. *D. L. Armacost,* The Structure of Locally Compact Abelian Groups (1981)
69. *J. W. Brewer and M. K. Smith, eds.,* Emmy Noether: A Tribute (1981)
70. *K. H. Kim,* Boolean Matrix Theory and Applications (1982)
71. *T. W. Wieting,* The Mathematical Theory of Chromatic Plane Ornaments (1982)
72. *D. B. Gauld,* Differential Topology (1982)
73. *R. L. Faber,* Foundations of Euclidean and Non-Euclidean Geometry (1983)
74. *M. Carmeli,* Statistical Theory and Random Matrices (1983)
75. *J. H. Carruth et al.,* The Theory of Topological Semigroups (1983)
76. *R. L. Faber,* Differential Geometry and Relativity Theory (1983)
77. *S. Barnett,* Polynomials and Linear Control Systems (1983)
78. *G. Karpilovsky,* Commutative Group Algebras (1983)
79. *F. Van Oystaeyen and A. Verschoren,* Relative Invariants of Rings (1983)
80. *I. Vaisman,* A First Course in Differential Geometry (1984)
81. *G. W. Swan,* Applications of Optimal Control Theory in Biomedicine (1984)
82. *T. Petrie and J. D. Randall,* Transformation Groups on Manifolds (1984)
83. *K. Goebel and S. Reich,* Uniform Convexity, Hyperbolic Geometry, and Nonexpansive Mappings (1984)
84. *T. Albu and C. Năstăsescu,* Relative Finiteness in Module Theory (1984)
85. *K. Hrbacek and T. Jech,* Introduction to Set Theory: Second Edition (1984)
86. *F. Van Oystaeyen and A. Verschoren,* Relative Invariants of Rings (1984)
87. *B. R. McDonald,* Linear Algebra Over Commutative Rings (1984)
88. *M. Namba,* Geometry of Projective Algebraic Curves (1984)
89. *G. F. Webb,* Theory of Nonlinear Age-Dependent Population Dynamics (1985)
90. *M. R. Bremner et al.,* Tables of Dominant Weight Multiplicities for Representations of Simple Lie Algebras (1985)
91. *A. E. Fekete,* Real Linear Algebra (1985)
92. *S. B. Chae,* Holomorphy and Calculus in Normed Spaces (1985)
93. *A. J. Jerri,* Introduction to Integral Equations with Applications (1985)
94. *G. Karpilovsky,* Projective Representations of Finite Groups (1985)
95. *L. Narici and E. Beckenstein,* Topological Vector Spaces (1985)
96. *J. Weeks,* The Shape of Space (1985)
97. *P. R. Gribik and K. O. Kortanek,* Extremal Methods of Operations Research (1985)
98. *J.-A. Chao and W. A. Woyczynski, eds.,* Probability Theory and Harmonic Analysis (1986)
99. *G. D. Crown et al.,* Abstract Algebra (1986)
00. *J. H. Carruth et al.,* The Theory of Topological Semigroups, Volume 2 (1986)
01. *R. S. Doran and V. A. Belfi,* Characterizations of C*-Algebras (1986)
02. *M. W. Jeter,* Mathematical Programming (1986)
03. *M. Altman,* A Unified Theory of Nonlinear Operator and Evolution Equations with Applications (1986)
04. *A. Verschoren,* Relative Invariants of Sheaves (1987)
05. *R. A. Usmani,* Applied Linear Algebra (1987)
06. *P. Blass and J. Lang,* Zariski Surfaces and Differential Equations in Characteristic $p > 0$ (1987)
07. *J. A. Reneke et al.,* Structured Hereditary Systems (1987)
08. *H. Busemann and B. B. Phadke,* Spaces with Distinguished Geodesics (1987)
09. *R. Harte,* Invertibility and Singularity for Bounded Linear Operators (1988)
10. *G. S. Ladde et al.,* Oscillation Theory of Differential Equations with Deviating Arguments (1987)
11. *L. Dudkin et al.,* Iterative Aggregation Theory (1987)
12. *T. Okubo,* Differential Geometry (1987)
13. *D. L. Stancl and M. L. Stancl,* Real Analysis with Point-Set Topology (1987)

114. *T. C. Gard*, Introduction to Stochastic Differential Equations (1988)
115. *S. S. Abhyankar*, Enumerative Combinatorics of Young Tableaux (1988)
116. *H. Strade and R. Farnsteiner*, Modular Lie Algebras and Their Representations (1988)
117. *J. A. Huckaba*, Commutative Rings with Zero Divisors (1988)
118. *W. D. Wallis*, Combinatorial Designs (1988)
119. *W. Więsław*, Topological Fields (1988)
120. *G. Karpilovsky*, Field Theory (1988)
121. *S. Caenepeel and F. Van Oystaeyen*, Brauer Groups and the Cohomology of Graded Rings (1989)
122. *W. Kozlowski*, Modular Function Spaces (1988)
123. *E. Lowen-Colebunders*, Function Classes of Cauchy Continuous Maps (1989)
124. *M. Pavel*, Fundamentals of Pattern Recognition (1989)
125. *V. Lakshmikantham et al.*, Stability Analysis of Nonlinear Systems (1989)
126. *R. Sivaramakrishnan*, The Classical Theory of Arithmetic Functions (1989)
127. *N. A. Watson*, Parabolic Equations on an Infinite Strip (1989)
128. *K. J. Hastings*, Introduction to the Mathematics of Operations Research (1989)
129. *B. Fine*, Algebraic Theory of the Bianchi Groups (1989)
130. *D. N. Dikranjan et al.*, Topological Groups (1989)
131. *J. C. Morgan II*, Point Set Theory (1990)
132. *P. Biler and A. Witkowski*, Problems in Mathematical Analysis (1990)
133. *H. J. Sussmann*, Nonlinear Controllability and Optimal Control (1990)
134. *J.-P. Florens et al.*, Elements of Bayesian Statistics (1990)
135. *N. Shell*, Topological Fields and Near Valuations (1990)
136. *B. F. Doolin and C. F. Martin*, Introduction to Differential Geometry for Engineers (1990)
137. *S. S. Holland, Jr.*, Applied Analysis by the Hilbert Space Method (1990)
138. *J. Okníński*, Semigroup Algebras (1990)
139. *K. Zhu*, Operator Theory in Function Spaces (1990)
140. *G. B. Price*, An Introduction to Multicomplex Spaces and Functions (1991)
141. *R. B. Darst*, Introduction to Linear Programming (1991)
142. *P. L. Sachdev*, Nonlinear Ordinary Differential Equations and Their Applications (1991)
143. *T. Husain*, Orthogonal Schauder Bases (1991)
144. *J. Foran*, Fundamentals of Real Analysis (1991)
145. *W. C. Brown*, Matrices and Vector Spaces (1991)
146. *M. M. Rao and Z. D. Ren*, Theory of Orlicz Spaces (1991)
147. *J. S. Golan and T. Head*, Modules and the Structures of Rings (1991)
148. *C. Small*, Arithmetic of Finite Fields (1991)
149. *K. Yang*, Complex Algebraic Geometry (1991)
150. *D. G. Hoffman et al.*, Coding Theory (1991)
151. *M. O. González*, Classical Complex Analysis (1992)
152. *M. O. González*, Complex Analysis (1992)
153. *L. W. Baggett*, Functional Analysis (1992)
154. *M. Sniedovich*, Dynamic Programming (1992)
155. *R. P. Agarwal*, Difference Equations and Inequalities (1992)
156. *C. Brezinski*, Biorthogonality and Its Applications to Numerical Analysis (1992)
157. *C. Swartz*, An Introduction to Functional Analysis (1992)
158. *S. B. Nadler, Jr.*, Continuum Theory (1992)
159. *M. A. Al-Gwaiz*, Theory of Distributions (1992)
160. *E. Perry*, Geometry: Axiomatic Developments with Problem Solving (1992)
161. *E. Castillo and M. R. Ruiz-Cobo*, Functional Equations and Modelling in Science and Engineering (1992)
162. *A. J. Jerri*, Integral and Discrete Transforms with Applications and Error Analysis (1992)
163. *A. Charlier et al.*, Tensors and the Clifford Algebra (1992)
164. *P. Biler and T. Nadzieja*, Problems and Examples in Differential Equations (1992)
165. *E. Hansen*, Global Optimization Using Interval Analysis (1992)
166. *S. Guerre-Delabrière*, Classical Sequences in Banach Spaces (1992)
167. *Y. C. Wong*, Introductory Theory of Topological Vector Spaces (1992)
168. *S. H. Kulkarni and B. V. Limaye*, Real Function Algebras (1992)
169. *W. C. Brown*, Matrices Over Commutative Rings (1993)
170. *J. Loustau and M. Dillon*, Linear Geometry with Computer Graphics (1993)
171. *W. V. Petryshyn*, Approximation-Solvability of Nonlinear Functional and Differential Equations (1993)
172. *E. C. Young*, Vector and Tensor Analysis: Second Edition (1993)
173. *T. A. Bick*, Elementary Boundary Value Problems (1993)

174. *M. Pavel*, Fundamentals of Pattern Recognition: Second Edition (1993)
175. *S. A. Albeverio et al.*, Noncommutative Distributions (1993)
176. *W. Fulks*, Complex Variables (1993)
177. *M. M. Rao*, Conditional Measures and Applications (1993)
178. *A. Janicki and A. Weron*, Simulation and Chaotic Behavior of α-Stable Stochastic Processes (1994)
179. *P. Neittaanmäki and D. Tiba*, Optimal Control of Nonlinear Parabolic Systems (1994)
180. *J. Cronin*, Differential Equations: Introduction and Qualitative Theory, Second Edition (1994)
181. *S. Heikkilä and V. Lakshmikantham*, Monotone Iterative Techniques for Discontinuous Nonlinear Differential Equations (1994)
182. *X. Mao*, Exponential Stability of Stochastic Differential Equations (1994)
183. *B. S. Thomson*, Symmetric Properties of Real Functions (1994)
184. *J. E. Rubio*, Optimization and Nonstandard Analysis (1994)
185. *J. L. Bueso et al.*, Compatibility, Stability, and Sheaves (1995)
186. *A. N. Michel and K. Wang*, Qualitative Theory of Dynamical Systems (1995)
187. *M. R. Darnel*, Theory of Lattice-Ordered Groups (1995)
188. *Z. Naniewicz and P. D. Panagiotopoulos*, Mathematical Theory of Hemivariational Inequalities and Applications (1995)
189. *L. J. Corwin and R. H. Szczarba*, Calculus in Vector Spaces: Second Edition (1995)
190. *L. H. Erbe et al.*, Oscillation Theory for Functional Differential Equations (1995)
191. *S. Agaian et al.*, Binary Polynomial Transforms and Nonlinear Digital Filters (1995)
192. *M. I. Gil'*, Norm Estimations for Operation-Valued Functions and Applications (1995)
193. *P. A. Grillet*, Semigroups: An Introduction to the Structure Theory (1995)
194. *S. Kichenassamy*, Nonlinear Wave Equations (1996)
195. *V. F. Krotov*, Global Methods in Optimal Control Theory (1996)
196. *K. I. Beidar et al.*, Rings with Generalized Identities (1996)
197. *V. I. Arnautov et al.*, Introduction to the Theory of Topological Rings and Modules (1996)
198. *G. Sierksma*, Linear and Integer Programming (1996)
199. *R. Lasser*, Introduction to Fourier Series (1996)
200. *V. Sima*, Algorithms for Linear-Quadratic Optimization (1996)
201. *D. Redmond*, Number Theory (1996)
202. *J. K. Beem et al.*, Global Lorentzian Geometry: Second Edition (1996)
203. *M. Fontana et al.*, Prüfer Domains (1997)
204. *H. Tanabe*, Functional Analytic Methods for Partial Differential Equations (1997)
205. *C. Q. Zhang*, Integer Flows and Cycle Covers of Graphs (1997)
206. *E. Spiegel and C. J. O'Donnell*, Incidence Algebras (1997)
207. *B. Jakubczyk and W. Respondek*, Geometry of Feedback and Optimal Control (1998)
208. *T. W. Haynes et al.*, Fundamentals of Domination in Graphs (1998)
209. *T. W. Haynes et al.*, Domination in Graphs: Advanced Topics (1998)
210. *L. A. D'Alotto et al.*, A Unified Signal Algebra Approach to Two-Dimensional Parallel Digital Signal Processing (1998)
211. *F. Halter-Koch*, Ideal Systems (1998)
212. *N. K. Govil et al.*, Approximation Theory (1998)
213. *R. Cross*, Multivalued Linear Operators (1998)
214. *A. A. Martynyuk*, Stability by Liapunov's Matrix Function Method with Applications (1998)
215. *A. Favini and A. Yagi*, Degenerate Differential Equations in Banach Spaces (1999)
216. *A. Illanes and S. Nadler, Jr.*, Hyperspaces: Fundamentals and Recent Advances (1999)
217. *G. Kato and D. Struppa*, Fundamentals of Algebraic Microlocal Analysis (1999)
218. *G. X.-Z. Yuan*, KKM Theory and Applications in Nonlinear Analysis (1999)
219. *D. Motreanu and N. H. Pavel*, Tangency, Flow Invariance for Differential Equations, and Optimization Problems (1999)
220. *K. Hrbacek and T. Jech*, Introduction to Set Theory, Third Edition (1999)
221. *G. E. Kolosov*, Optimal Design of Control Systems (1999)
222. *A. I. Prilepko et al.*, Methods for Solving Inverse Problems in Mathematical Physics (1999)

Additional Volumes in Preparation

INTRODUCTION TO SET THEORY

Third Edition, Revised and Expanded

Karel Hrbacek
The City College of the City University of New York
New York, New York

Thomas Jech
The Pennsylvania State University
University Park, Pennsylvania

MARCEL DEKKER

MARCEL DEKKER, INC. NEW YORK • BASEL

Library of Congress Cataloging-in-Publication

Hrbacek, Karel.
Introduction to set theory / Karel Hrbacek, Thomas Jech. — 3rd ed., rev. and expanded.
p. cm. — (Monographs and textbooks in pure and applied mathematics; 220).
Includes index.
ISBN 0-8247-7915-0 (alk. paper)
1. Set theory. I. Jech, Thomas J. II. Title. III. Series.
QA248.H68 1999
511.3'22—dc21 99-15458
CIP

This book is printed on acid-free paper.

Headquarters
Marcel Dekker, Inc.
270 Madison Avenue, New York, NY 10016
tel: 212-696-9000; fax: 212-685-4540

Eastern Hemisphere Distribution
Marcel Dekker AG
Hutgasse 4, Postfach 812, CH-4001 Basel, Switzerland
tel: 41-61-261-8482; fax: 41-61-261-8896

World Wide Web
http://www.dekker.com

The publisher offers discounts on this book when ordered in bulk quantities. For more information, write to Special Sales/Professional Marketing at the headquarters address above.

Copyright © 1999 by Marcel Dekker, Inc. All Rights Reserved.

Neither this book nor any part may be reproduced or transmitted in any form or by any means, electronic or mechanical, including photocopying, microfilming, and recording, or by any information storage and retrieval system, without permission in writing from the publisher.

Current printing (last digit)
10 9 8 7 6 5 4 3 2 1

PRINTED IN THE UNITED STATES OF AMERICA

Preface to the 3rd Edition

Modern set theory has grown tremendously since the first edition of this book was published in 1978, and even since the second edition appeared in 1984. Moreover, many ideas that were then at the forefront of research have become, by now, important tools in other branches of mathematics. Thus, for example, combinatorial principles, such as Principle Diamond and Martin's Axiom, are indispensable tools in general topology and abstract algebra. Non-well-founded sets turned out to be a convenient setting for semantics of artificial as well as natural languages. Nonstandard Analysis, which is grounded on structures constructed with the help of ultrafilters, has developed into an independent methodology with many exciting applications. It seems appropriate to incorporate some of the underlying set-theoretic ideas into a textbook intended as a general introduction to the subject. We do that in the form of four new chapters (Chapters 11-14), expanding on the topics of Chapter 11 in the second edition, and containing largely new material. Chapter 11 presents filters and ultrafilters, develops the basic properties of closed unbounded and stationary sets, and culminates with the proof of Silver's Theorem. The first two sections of Chapter 12 provide an introduction to the partition calculus. The next two sections study trees and develop their relationship to Suslin's Problem. Section 5 of Chapter 12 is an introduction to combinatorial principles. Chapter 13 is devoted to the measure problem and measurable cardinals. The topic of Chapter 14 is a fairly detailed study of well-founded and non-well-founded sets.

Chapters 1–10 of the second edition have been thoroughly revised and reorganized. The material on rational and real numbers has been consolidated in Chapter 10, so that it does not interrupt the development of set theory proper. To preserve continuity, a section on Dedekind cuts has been added to Chapter 4. New material (on normal forms and Goodstein sequences) has also been added to Chapter 6.

A solid basic course in set theory should cover most of Chapters 1–9. This can be supplemented by additional material from Chapters 10–14, which are almost completely independent of each other (except that Section 5 in Chapter 12, and Chapter 13, refer to some concepts introduced in earlier chapters).

Karel Hrbacek
Thomas Jech

Preface to the 2nd Edition

The first version of this textbook was written in Czech in spring 1968 and accepted for publication by Academia, Prague under the title *Úvod do teorie množin*. However, we both left Czechoslovakia later that year, and the book never appeared. In the following years we taught introductory courses in set theory at various universities in the United States, and found it difficult to select a textbook for use in these courses. Some existing books are based on the "naive" approach to set theory rather than the axiomatic one. We consider some understanding of set-theoretic "paradoxes" and of undecidable propositions (such as the Continuum Hypothesis) one of the important goals of such a course, but neither topic can be treated honestly with a "naive" approach. Moreover, set theory is a natural choice of a field where students can first become acquainted with an axiomatic development of a mathematical discipline. On the other hand, all currently available texts presenting set theory from an axiomatic point of view heavily stress logic and logical formalism. Most of them begin with a virtual minicourse in logic. We found that students often take a course in set theory before taking one in logic. More importantly, the emphasis on formalization obscures the essence of the axiomatic method. We felt that there was a need for a book which would present axiomatic set theory more mathematically, at the level of rigor customary in other undergraduate courses for math majors. This led us to the decision to rewrite our Czech text. We kept the original general plan, but the requirements for a textbook suitable for American colleges resulted in the production of a completely new work.

We wish to stress the following features of the book:

1. Set theory is developed axiomatically. The reasons for adopting each axiom, as arising both from intuition and mathematical practice, are carefully pointed out. A detailed discussion is provided in "controversial" cases, such as the Axiom of Choice.

2. The treatment is not formal. Logical apparatus is kept to a minimum and logical formalism is completely avoided.

3. We show that axiomatic set theory is powerful enough to serve as an underlying framework for mathematics by developing the beginnings of the theory of natural, rational, and real numbers in it. However, we carry the development only as far as it is useful to illustrate the general idea and to motivate set-theoretic generalizations of some of these concepts (such as the ordinal numbers and operations on them). Dreary, repetitive details, such as the proofs of all

the usual arithmetic laws, are relegated into exercises.

4. A substantial part of the book is devoted to the study of ordinal and cardinal numbers.

5. Each section is accompanied by many exercises of varying difficulty.

6. The final chapter is an informal outline of some recent developments in set theory and their significance for other areas of mathematics: the Axiom of Constructibility, questions of consistency and independence, and large cardinals. No proofs are given, but the exposition is sufficiently detailed to give a nonspecialist some idea of the problems arising in the foundations of set theory, methods used for their solution, and their effects on mathematics in general.

The first edition of the book has been extensively used as a textbook in undergraduate and first-year graduate courses in set theory. Our own experience, that of our many colleagues, and suggestions and criticism by the reviewers led us to consider some changes and improvements for the present second edition. Those turned out to be much more extensive than we originally intended. As a result, the book has been substantially rewritten and expanded. We list the main new features below.

1. The development of natural numbers in Chapter 3 has been greatly simplified. It is now based on the definition of the set of all natural numbers as the least inductive set, and the Principle of Induction. The introduction of transitive sets and a characterization of natural numbers as those transitive sets that are well-ordered and inversely well-ordered by $\in$ is postponed until the chapter on ordinal numbers (Chapter 7).

2. The material on integers and rational numbers (a separate chapter in the first edition) has been condensed into a single section (Section 1 of Chapter 5). We feel that most students learn this subject in another course (Abstract Algebra), where it properly belongs.

3. A series of new sections (Section 3 in Chapter 5, 6, and 9) deals with the set-theoretic properties of sets of real numbers (open, closed, and perfect sets, etc.), and provides interesting applications of abstract set theory to real analysis.

4. A new chapter called "Uncountable Sets" (Chapter 11) has been added. It introduces some of the concepts that are fundamental to modern set theory: ultrafilters, closed unbounded sets, trees and partitions, and large cardinals. These topics can be used to enrich the usual one-semester course (which would ordinarily cover most of the material in Chapters 1–10).

5. The study of linear orderings has been expanded and concentrated in one place (Section 4 of Chapter 4).

6. Numerous other small additions, changes, and corrections have been made throughout.

7. Finally, the discussion of the present state of set theory in Section 3 of Chapter 12 has been updated.

Karel Hrbacek
Thomas Jech

Contents

INTRODUCTION TO SET THEORY

Chapter 1

Sets

1. Introduction to Sets

The central concept of this book, that of a *set*, is, at least on the surface, extremely simple. A set is any collection, group, or conglomerate. So we have the set of all students registered at the City University of New York in February 1998; the set of all even natural numbers, the set of all points in the plane π exactly 2 inches distant from a given point P, the set of all pink elephants.

Sets are not objects of the real world, like tables or stars; they are created by our mind, not by our hands. A heap of potatoes is not a set of potatoes, the set of all molecules in a drop of water is not the same object as that drop of water. The human mind possesses the ability to abstract, to think of a variety of different objects as being bound together by some common property, and thus to form a set of objects having that property. The property in question may be nothing more than the ability to think of these objects (as being) together. Thus there is a set consisting of exactly the numbers 2, 7, 12, 13, 29, 34, and 11,000, although it is hard to see what binds exactly those numbers together, besides the fact that we collected them together in our mind. Georg Cantor, a German mathematician who founded set theory in a series of papers published over the last three decades of the nineteenth century, expressed it as follows: "Unter einer Menge verstehen wir jede Zusammenfassung M von bestimmten wohlunterschiedenen Objekten in unserer Anschauung oder unseres Denkens (welche die Elemente von M genannt werden) zu einem ganzen." [A set is a collection into a whole of definite, distinct objects of our intuition or our thought. The objects are called elements (members) of the set.]

Objects from which a given set is composed are called *elements* or *members* of that set. We also say that they *belong* to that set.

In this book, we want to develop the theory of sets as a foundation for other mathematical disciplines. Therefore, we are not concerned with sets of people or molecules, but only with sets of *mathematical* objects, such as numbers, points of space, functions, or sets. Actually, the first three concepts can be defined in set theory as sets with particular properties, and we do that in the following chapters. So the only objects with which we are concerned from now on are sets. For purposes of illustration, we talk about sets of numbers or points even before these notions are exactly defined. We do that, however, only in examples, exercises and problems, not in the main body of theory. Sets of mathematical objects are, for example:

1.1 Example

(a) The set of all prime divisors of 324.
(b) The set of all numbers divisible by 0.
(c) The set of all continuous real-valued functions on the interval $[0, 1]$.
(d) The set of all ellipses with major axis 5 and eccentricity 3.
(e) The set of all sets whose elements are natural numbers less than 20.

Examination of these and many other similar examples reveals that sets with which mathematicians work are relatively simple. They include the set of natural numbers and its various subsets (such as the set of all prime numbers), as well as sets of pairs, triples, and in general n-tuples of natural numbers. Integers and rational numbers can be defined using only such sets. Real numbers can then be defined as sets or sequences of rational numbers. Mathematical analysis deals with sets of real numbers and functions on real numbers (sets of ordered pairs of real numbers), and in some investigations, sets of functions or even sets of sets of functions are considered. But a working mathematician rarely encounters objects more complicated than that. Perhaps it is not surprising that uncritical usage of "sets" remote from "everyday experience" may lead to contradictions.

Consider for example the "set" R of all those sets which are not elements of themselves. In other words, R is a set of all sets x such that $x \notin x$ ($\in$ reads "belongs to," $\notin$ reads "does not belong to"). Let us now ask whether $R \in R$. If $R \in R$, then R is not an element of itself (because no element of R belongs to itself), so $R \notin R$; a contradiction. Therefore, necessarily $R \notin R$. But then R is a set which is not an element of itself, and all such sets belong to R. We conclude that $R \in R$; again, a contradiction.

The argument can be briefly summarized as follows: Define R by: $x \in R$ if and only if $x \notin x$. Now consider $x = R$; by definition of R, $R \in R$ if and only if $R \notin R$; a contradiction.

A few comments on this argument (due to Bertrand Russell) might be helpful. First, there is nothing wrong with R being a set of sets. Many sets whose elements are again sets are legitimately employed in mathematics — see Example 1.1 — and do not lead to contradictions. Second, it is easy to give examples of elements of R; e.g., if x is the set of all natural numbers, then $x \notin x$ (the set of all natural numbers is not a natural number) and so $x \in R$. Third, it is not so easy to give examples of sets which do not belong to R, but this is irrelevant. The foregoing argument would result in a contradiction even if there were no sets which are elements of themselves. (A plausible candidate for a set which is an element of itself would be the "set of all sets" V; clearly $V \in V$. However, the "set of all sets" leads to contradictions of its own in a more subtle way — see Exercises 3.3 and 3.6.)

How can this contradiction be resolved? We assumed that we have a set R defined as the set of all sets which are not elements of themselves, and derived a contradiction as an immediate consequence of the definition of R. This can only mean that there is no set satisfying the definition of R. In other words, this argument proves that there exists no set whose members would be precisely the sets which are not elements of themselves. The lesson contained in Russell's Paradox and other similar examples is that by merely defining a set we do not prove its existence (similarly as by defining a unicorn we do not prove that unicorns exist). There are properties which do not define sets; that is, it is not possible to collect all objects with those properties into one set. This observation leaves set-theorists with a task of determining the properties which do define sets. Unfortunately, no way how to do this is known, and some results in logic (especially the so-called Incompleteness Theorems discovered by Kurt Gödel) seem to indicate that a complete answer is not even possible.

Therefore, we attempt a less lofty goal. We formulate some of the relatively simple properties of sets used by mathematicians as *axioms*, and then take care to check that all theorems follow logically from the axioms. Since the axioms are obviously true and the theorems logically follow from them, the theorems are also true (not necessarily obviously). We end up with a body of truths about sets which includes, among other things, the basic properties of natural, rational, and real numbers, functions, orderings, etc., but as far as is known, no contradictions. Experience has shown that practically all notions used in contemporary mathematics can be defined, and their mathematical properties derived, in this axiomatic system. In this sense, the axiomatic set theory serves as a satisfactory foundation for the other branches of mathematics.

On the other hand, we do not claim that every true fact about sets can be derived from the axioms we present. The axiomatic system is not complete in this sense, and we return to the discussion of the question of completeness in the last chapter.

2. Properties

In the preceding section we introduced sets as collections of objects having some property in common. The notion of property merits some analysis. Some properties commonly considered in everyday life are so vague that they can hardly be admitted in a mathematical theory. Consider, for example, the "set of all the great twentieth century American novels." Different persons' judgments as to what constitutes a great literary work differ so much that there is no generally accepted way how to decide whether a given book is or is not an element of the "set."

For an even more startling example, consider the "set of those natural numbers which could be written down in decimal notation" (by "could" we mean that someone could actually do it with paper and pencil). Clearly, 0 can be so written down. If number n can be written down, then surely number $n+1$ can also be written down (imagine another, somewhat faster writer, or the person capable of writing n working a little faster). Therefore, by the familiar principle of induction, every natural number n can be written down. But that is plainly absurd; to write down say $10^{10^{10}}$ in decimal notation would require to follow 1 by 10^{10} zeros, which would take over 300 years of continuous work at a rate of a zero per second.

The problem is caused by the vague meaning of "could." To avoid similar difficulties, we now describe explicitly what we mean by a property. Only clear, mathematical properties are allowed; fortunately, these properties are sufficient for expression of all mathematical facts.

Our exposition in this section is informal. Readers who would like to see how this topic can be studied from a more rigorous point of view can consult some book on mathematical logic.

The basic set-theoretic property is the *membership* property: " ... is an element of ... ," which we denote by $\in$. So "$X \in Y$" reads "X is an element of Y" or "X is a member of Y" or "X belongs to Y."

The letters X and Y in these expressions are *variables*; they stand for (denote) unspecified, arbitrary sets. The proposition "$X \in Y$" holds or does not hold depending on sets (denoted by) X and Y. We sometimes say "$X \in Y$" is a *property of* X *and* Y. The reader is surely familiar with this informal way of speech from other branches of mathematics. For example, "m is less than n" is a property of m and n. The letters m and n are variables denoting unspecified numbers. Some m and n have this property (for example, "2 is less than 4" is true) but others do not (for example, "3 is less than 2" is false).

All other set-theoretic properties can be stated in terms of membership with the help of logical means: identity, logical connectives, and quantifiers.

We often speak of one and the same set in different contexts and find it convenient to denote it by different variables. We use the identity sign "=" to express that two variables denote the same set. So we write $X = Y$ if X *is the same set as* Y [X *is identical with* Y or X *is equal to* Y].

In the next example, we list some obvious facts about identity:

2.1 Example

(a)	$X = X$.	(X is identical with X.)
(b)	If $X = Y$, then $Y = X$.	(If X and Y are identical, then Y and X are identical.)
(c)	If $X = Y$ and $Y = Z$, then $X = Z$.	(If X is identical with Y and Y is identical with Z, then X is identical with Z.)
(d)	If $X = Y$ and $X \in Z$, then $Y \in Z$.	(If X and Y are identical and X belongs to Z, then Y belongs to Z.)
(e)	If $X = Y$ and $Z \in X$, then $Z \in Y$.	(If X and Y are identical and Z belongs to X, then Z belongs to Y.)

Logical connectives can be used to construct more complicated properties from simpler ones. They are expressions like "not ... ," " ... and ... ," "if ... , then ... ," and " ... if and only if ... ."

2.2 Example

(a) "$X \in Y$ or $Y \in X$" is a property of X and Y.
(b) "Not $X \in Y$ and not $Y \in X$" or, in more idiomatic English, "X is not an element of Y and Y is not an element of X" is also a property of X and Y.
(c) "If $X = Y$, then $X \in Z$ if and only if $Y \in Z$" is a property of X, Y and Z.
(d) "X is not an element of X" (or: "not $X \in X$") is a property of X.

We write $X \notin Y$ instead of "not $X \in Y$" and $X \neq Y$ instead of "not $X = Y$."

Quantifiers "for all" ("for every") and "there is" ("there exists") provide additional logical means. Mathematical practice shows that all mathematical facts can be expressed in the very restricted language we just described, but that this language does not allow vague expressions like the ones at the beginning of the section.

Let us look at some examples of properties which involve quantifiers.

2.3 Example

(a) "There exists $Y \in X$."
(b) "For every $Y \in X$, there is Z such that $Z \in X$ and $Z \in Y$."
(c) "There exists Z such that $Z \in X$ and $Z \notin Y$."

Truth or falsity of (a) obviously depends on the set (denoted by the variable) X. For example, if X is the set of all American presidents after 1789, then (a) is true; if X is the set of all American presidents before 1789, (a) becomes false. [Generally, (a) is true if X has some element and false if X is empty.] We say that (a) is a property of X or that (a) *depends on the parameter* X. Similarly, (b) is a property of X, and (c) is a property of X and Y. Notice also that Y is not a parameter in (a) since it does not make sense to inquire whether (a) is true for some particular set Y; we use the letter Y in the quantifier only for convenience and could as well say, "There exists $W \in X$," or "There exists some element of X." Similarly, (b) is not a property of Y, or Z, and (c) is not a property of Z.

Although precise rules for determining parameters of a given property can easily be formulated, we rely on the reader's common sense, and limit ourselves to one last example.

2.4 Example

(a) "$Y \in X$."
(b) "There is $Y \in X$."
(c) "For every X, there is $Y \in X$."

Here (a) is a property of X and Y; it is true for some pairs of sets X, Y and false for others. (b) is a property of X (but not of Y), while (c) has no parameters. (c) is, therefore, either true or false (it is, in fact, false). Properties which have no parameters (and are, therefore, either true or false) are called *statements*; all mathematical theorems are (true) statements.

We sometimes wish to refer to an arbitrary, unspecified property. We use boldface capital letters to denote statements and properties and, if convenient, list some or all of their parameters in parentheses. So $\mathbf{A}(X)$ stands for any property of the parameter X, e.g., (a), (b), or (c) in Example 2.3. $\mathbf{E}(X, Y)$ is a property of parameters X and Y, e.g., (c) in Example 2.3 or (a) in Example 2.4 or

(d) "$X \in Y$ or $X = Y$ or $Y \in X$."

In general, $\mathbf{P}(X, Y, \ldots, Z)$ is a property whose truth or falsity depends on parameters $X, Y, \ldots, Z$ (and possibly others).

We said repeatedly that all set-theoretic properties can be expressed in our restricted language, consisting of membership property and logical means. However, as the development proceeds and more and more complicated theorems are proved, it is practical to give names to various particular properties, i.e., to *define* new properties. A new symbol is then introduced (defined) to denote the property in question; we can view it as a shorthand for the explicit formulation. For example, the property of being a *subset* is defined by

2.5 $X \subseteq Y$ if and only if every element of X is an element of Y.

"X is a subset of Y" ($X \subseteq Y$) is a property of X and Y. We can use it in more complicated formulations and, whenever desirable, replace $X \subseteq Y$ by its definition. For example, the explicit definition of

"If $X \subseteq Y$ and $Y \subseteq Z$, then $X \subseteq Z$."

would be

"If every element of X is an element of Y and every element of Y is an element of Z, then every element of X is an element of Z."

It is clear that mathematics without definitions would be possible, but exceedingly clumsy.

For another type of definition, consider the property $\mathbf{P}(X)$:

"There exists no $Y \in X$."

We prove in Section 3 that

(a) There exists a set X such that $\mathbf{P}(X)$ (there exists a set X with no elements).

(b) There exists at most one set X such that $\mathbf{P}(X)$, i.e., if $\mathbf{P}(X)$ and $\mathbf{P}(X')$, then $X = X'$ (if X has no elements and X' has no elements, then X and X' are identical).

(a) and (b) together express the fact that there is a unique set X with the property $\mathbf{P}(X)$. We can then give this set a name, say $\emptyset$ (the empty set), and use it in more complicated expressions.

The full meaning of "$\emptyset \subseteq Z$" is then "the set X which has no elements is a subset of Z." We occasionally refer to $\emptyset$ as the *constant* defined by the property **P**.

For our last example of a definition, consider the property $\mathbf{Q}(X, Y, Z)$ of X, Y, and Z:

"For every U, $U \in Z$ if and only if $U \in X$ and $U \in Y$."

We see in the next section that
(a) For every X and Y there is Z such that $\mathbf{Q}(X, Y, Z)$.
(b) For every X and Y, if $\mathbf{Q}(X, Y, Z)$ and $\mathbf{Q}(X, Y, Z')$, then $Z = Z'$. [For every X and Y, there exists at most one Z such that $\mathbf{Q}(X, Y, Z)$.]

Conditions (a) and (b) (which have to be proved whenever this type of definition is used) guarantee that for every X and Y there is a unique set Z such that $\mathbf{Q}(X, Y, Z)$. We can then introduce a name, say $X \cap Y$, for this unique set Z, and call $X \cap Y$ *the* intersection of X and Y. So $\mathbf{Q}(X, Y, X \cap Y)$ holds. We refer to $\cap$ as the *operation* defined by the property **Q**.

3. The Axioms

We now begin to set up our axiomatic system and try to make clear the intuitive meaning of each axiom.

The first principle we adopt postulates that our "universe of discourse" is not void, i.e., that some sets exist. To be concrete, we postulate the existence of a specific set, namely the empty set.

The Axiom of Existence There exists a set which has no elements.

A set with no elements can be variously described intuitively, e.g., as the set of all U.S. Presidents before 1789, the set of all real numbers x for which $x^2 = -1$, etc. All examples of this kind describe one and the same set, namely the empty, vacuous set. So, intuitively, there is only one empty set. But we cannot yet prove this assertion. We need another postulate to express the fact that each set is determined by its elements. Let us see another example:

X is the set consisting exactly of numbers 2, 3, and 5.
Y is the set of all prime numbers greater than 1 and less than 7.
Z is the set of all solutions of the equation $x^3 - 10x^2 + 31x - 30 = 0$.

Here $X = Y$, $X = Z$, and $Y = Z$, and we have three different descriptions of one and the same set. This leads to the Axiom of Extensionality.

The Axiom of Extensionality If every element of X is an element of Y and every element of Y is an element of X, then $X = Y$.

Briefly, if two sets have the same elements, then they are identical. We can now prove Lemma 3.1.

3.1 Lemma *There exists only one set with no elements.*

Proof. Assume that A and B are sets with no elements. Then every element of A is an element of B (since A has no elements, the statement "$a \in A$ implies $a \in B$" is an implication with a false antecedent, and thus automatically true). Similarly, every element of B is an element of A (since B has no elements). Therefore, $A = B$, by the Axiom of Extensionality. □

3.2 Definition The (unique) set with no elements is called the *empty set* and is denoted $\emptyset$.

Notice that the definition of the constant $\emptyset$ is justified by the Axiom of Existence and Lemma 3.1.

Intuitively, sets are collections of objects sharing some common property, so we expect to have axioms expressing this fact. But, as demonstrated by the paradoxes in Section 1, not every property describes a set; properties "$X \notin X$" or "$X = X$" are typical examples.

In both cases, the problem seems to be that in order to collect all objects having such a property into a set, we already have to be able to perceive all sets. The difficulty is avoided if we postulate the existence of a set of all objects with a given property only if there already exists some set to which they all belong.

The Axiom Schema of Comprehension Let $\mathbf{P}(x)$ be a property of x. For any set A, there is a set B such that $x \in B$ if and only if $x \in A$ and $\mathbf{P}(x)$.

This is a schema of axioms, i.e., for each property $\mathbf{P}$, we have one axiom. For example, if $\mathbf{P}(x)$ is "$x = x$," the axiom says:

> For any set A, there is a set B such that $x \in B$ if and only if $x \in A$ and $x = x$. (In this case, $B = A$.)

If $\mathbf{P}(x)$ is "$x \notin x$", the axiom postulates:

> For any set A, there is a set B such that $x \in B$ if and only if $x \in A$ and $x \notin x$.

Although the supply of axioms is unlimited, this causes no problems, since it is easy to recognize whether a particular statement is or is not an axiom and since every proof uses only finitely many axioms.

The property $\mathbf{P}(x)$ can depend on other parameters $p, \dots, q$; the corresponding axiom then postulates that for any sets $p, \dots, q$ and any A, there is a set B (depending on $p, \dots, q$ and, of course, on A) consisting exactly of all those $x \in A$ for which $\mathbf{P}(x, p, \dots, q)$.

3.3 Example If P and Q are sets, then there is a set R such that $x \in R$ if and only if $x \in P$ and $x \in Q$.

Proof. Consider the property $\mathbf{P}(x, Q)$ of x and Q: "$x \in Q$." Then, by the Comprehension Schema, for every Q and for every P there is a set R such that $x \in R$ if and only if $x \in P$ and $\mathbf{P}(x, Q)$, i.e., if and only if $x \in P$ and $x \in Q$. (P plays the role of A, Q is a parameter.) □

3.4 Lemma *For every A, there is only one set B such that $x \in B$ if and only if $x \in A$ and $\mathbf{P}(x)$.*

Proof. If B' is another set such that $x \in B'$ if and only if $x \in A$ and $\mathbf{P}(x)$, then $x \in B$ if and only if $x \in B'$, so $B = B'$, by the Axiom of Extensionality. □

We are now justified to introduce a name for the uniquely determined set B.

3.5 Definition $\{x \in A \mid \mathbf{P}(x)\}$ *is the set of all $x \in A$ with the property $\mathbf{P}(x)$.*

3.6 Example The set from Example 3.3 could be denoted $\{x \in P \mid x \in Q\}$.

Our axiomatic system is not yet very powerful; the only set we proved to exist is the empty set, and applications of the Comprehension Schema to the empty set produce again the empty set: $\{x \in \emptyset \mid \mathbf{P}(x)\} = \emptyset$ no matter what property $\mathbf{P}$ we take. (Prove it.) The next three principles postulate that some of the constructions frequently used in mathematics yield sets.

The Axiom of Pair For any A and B, there is a set C such that $x \in C$ if and only if $x = A$ or $x = B$.

So $A \in C$ and $B \in C$, and there are no other elements of C. The set C is unique (prove it); therefore, we define the *unordered pair* of A and B as the set having exactly A and B as its elements and introduce notation $\{A, B\}$ for the unordered pair of A and B. In particular, if $A = B$, we write $\{A\}$ instead of $\{A, A\}$.

3.7 Example

(a) Set $A = \emptyset$ and $B = \emptyset$; then $\{\emptyset\} = \{\emptyset, \emptyset\}$ is a set for which $\emptyset \in \{\emptyset\}$, and if $x \in \{\emptyset\}$, then $x = \emptyset$. So $\{\emptyset\}$ has a unique element $\emptyset$. Notice that $\{\emptyset\} \neq \emptyset$, since $\emptyset \in \{\emptyset\}$, but $\emptyset \notin \emptyset$.

(b) Let $A = \emptyset$ and $B = \{\emptyset\}$; then $\emptyset \in \{\emptyset, \{\emptyset\}\}$ and $\{\emptyset\} \in \{\emptyset, \{\emptyset\}\}$, and $\emptyset$ and $\{\emptyset\}$ are the only elements of $\{\emptyset, \{\emptyset\}\}$.

Note that $\emptyset \neq \{\emptyset, \{\emptyset\}\}$, $\{\emptyset\} \neq \{\emptyset, \{\emptyset\}\}$.

The Axiom of Union For any set S, there exists a set U such that $x \in U$ if and only if $x \in A$ for some $A \in S$.

Again, the set U is unique (prove it); it is called the *union* of S and denoted by $\bigcup S$. We say that S is a *system of sets* or a *collection of sets* when we want to stress that elements of S are sets (of course, this is always true — all our objects are sets — and thus the expressions "set" and "system of sets" have the same meaning). The union of a system of sets S is then a set of precisely those x which belong to some set from the system S.

3.8 Example
(a) Let $S = \{\emptyset, \{\emptyset\}\}$; $x \in \bigcup S$ if and only if $x \in A$ for some $A \in S$, i.e., if and only if $x \in \emptyset$ or $x \in \{\emptyset\}$. Therefore, $x \in \bigcup S$ if and only if $x = \emptyset$; $\bigcup S = \{\emptyset\}$.
(b) $\bigcup \emptyset = \emptyset$.
(c) Let M and N be sets; $x \in \bigcup\{M, N\}$ if and only if $x \in M$ or $x \in N$.
The set $\bigcup\{M, N\}$ is called the *union* of M and N and is denoted $M \cup N$.

So we finally introduced one of the simple set-theoretic operations with which the reader is surely familiar. The Axiom of Pair and the Axiom of Union are necessary to define union of two sets (and the Axiom of Extensionality is needed to guarantee that it is unique). The union of two sets has the usual meaning: $x \in M \cup N$ if and only if $x \in M$ or $x \in N$.

3.9 Example $\{\{\emptyset\}\} \cup \{\emptyset, \{\emptyset\}\} = \{\emptyset, \{\emptyset\}\}$.

The Axiom of Union is, of course, much more powerful; it enables us to form unions of not just two, but of any, possibly infinite, collection of sets.

If A, B, and C are sets, we can now prove the existence and uniqueness of the set P whose elements are exactly A, B, and C (see Exercise 3.5). P is denoted $\{A, B, C\}$ and is called an *unordered triple* of A, B, and C. Analogously, we could define an unordered quadruple or 17-tuple.

Before introducing the last axiom of this section, we define another simple concept.

3.10 Definition A is a *subset* of B if and only if every element of A belongs to B. In other words, A is a subset of B if, for every x, $x \in A$ implies $x \in B$.

We write $A \subseteq B$ to denote that A is a subset of B.

3.11 Example
(a) $\{\emptyset\} \subseteq \{\emptyset, \{\emptyset\}\}$ and $\{\{\emptyset\}\} \subseteq \{\emptyset, \{\emptyset\}\}$.
(b) $\emptyset \subseteq A$ and $A \subseteq A$ for every set A.
(c) $\{x \in A \mid \mathbf{P}(x)\} \subseteq A$.
(d) If $A \in S$, then $A \subseteq \bigcup S$.

The next axiom postulates that all subsets of a given set can be collected into one set.

The Axiom of Power Set For any set S, there exists a set P such that $X \in P$ if and only if $X \subseteq S$.

Since the set P is again uniquely determined, we call the set of all subsets of S the *power set* of S and denote it by $\mathcal{P}(S)$.

3.12 Example

(a) $\mathcal{P}(\emptyset) = \{\emptyset\}$.
(b) $\mathcal{P}(\{a\}) = \{\emptyset, \{a\}\}$.
(c) The elements of $\mathcal{P}(\{a, b\})$ are $\emptyset$, $\{a\}$, $\{b\}$, and $\{a, b\}$.

We conclude this section with another notational convention. Let $\mathbf{P}(x)$ be a property of x (and, possibly, of other parameters).

If there is a set A such that, for all x, $\mathbf{P}(x)$ implies $x \in A$, then $\{x \in A \mid \mathbf{P}(x)\}$ exists, and, moreover, does not depend on A. That means that if A' is another set such that for all x, $\mathbf{P}(x)$ implies $x \in A'$, then $\{x \in A' \mid \mathbf{P}(x)\} = \{x \in A \mid \mathbf{P}(x)\}$. (Prove it.)

We can now define $\{x \mid \mathbf{P}(x)\}$ to be the set $\{x \in A \mid \mathbf{P}(x)\}$, where A is any set for which $\mathbf{P}(x)$ implies $x \in A$ (since it does not matter which such set A we use). $\{x \mid \mathbf{P}(x)\}$ is the *set of all x with the property* $\mathbf{P}(x)$. We stress once again that this notation can be used only after it has been proved that some A contains all x with the property $\mathbf{P}$.

3.13 Example

(a) $\{x \mid x \in P \text{ and } x \in Q\}$ exists.

> *Proof.* $\mathbf{P}(x, P, Q)$ is the property "$x \in P$ and $x \in Q$"; let $A = P$. Then $\mathbf{P}(x, P, Q)$ implies $x \in A$. Therefore, $\{x \mid x \in P \text{ and } x \in Q\} = \{x \in P \mid x \in P \text{ and } x \in Q\} = \{x \in P \mid x \in Q\}$ is the set R from Example 3.3. □

(b) $\{x \mid x = a \text{ or } x = b\}$ exists; for a proof put $A = \{a, b\}$; also show that $\{x \mid x = a \text{ or } x = b\} = \{a, b\}$.
(c) $\{x \mid x \notin x\}$ does not exist (because of Russell's Paradox); thus in this instance the notation $\{x \mid \mathbf{P}(x)\}$ is inadmissible.

Although our list of axioms is not complete, we postpone the introduction of the remaining postulates until the need for them arises. Quite a few concepts can be introduced and some theorems proved from the postulates we now have available. The reader may have noticed that we did not guarantee existence of any infinite sets. This deficiency is removed in Chapter 3. Other axioms are introduced in Chapters 6 and 8. The complete list of axioms can be found in Section 1 of Chapter 15. This axiomatic system was essentially formulated by Ernst Zermelo in 1908 and is often referred to as the *Zermelo-Fraenkel axiomatic system* for set theory.

Exercises

3.1 Show that the set of all x such that $x \in A$ and $x \notin B$ exists.

3.2 Replace the Axiom of Existence by the following weaker postulate:
Weak Axiom of Existence Some set exists.
Prove the Axiom of Existence using the Weak Axiom of Existence and the Comprehension Schema. [*Hint*: Let A be a set known to exist; consider $\{x \in A \mid x \neq x\}$.]

3.3 (a) Prove that a "set of all sets" does not exist. [*Hint*: if V is a set of all sets, consider $\{x \in V \mid x \notin x\}$.]
(b) Prove that for any set A there is some $x \notin A$.

3.4 Let A and B be sets. Show that there exists a unique set C such that $x \in C$ if and only if either $x \in A$ and $x \notin B$ or $x \in B$ and $x \notin A$.

3.5 (a) Given A, B, and C, there is a set P such that $x \in P$ if and only if $x = A$ or $x = B$ or $x = C$.
(b) Generalize to four elements.

3.6 Show that $\mathcal{P}(X) \subseteq X$ is false for any X. In particular, $\mathcal{P}(X) \neq X$ for any X. This proves again that a "set of all sets" does not exist. [*Hint*: Let $Y = \{u \in X \mid u \notin u\}$; $Y \in \mathcal{P}(X)$ but $Y \notin X$.]

3.7 The Axiom of Pair, the Axiom of Union, and the Axiom of Power Set can be replaced by the following weaker versions.
Weak Axiom of Pair For any A and B, there is a set C such that $A \in C$ and $B \in C$.
Weak Axiom of Union For any S, there exists U such that if $X \in A$ and $A \in S$, then $X \in U$.
Weak Axiom of Power Set For any set S, there exists P such that $X \subseteq S$ implies $X \in P$.
Prove the Axiom of Pair, the Axiom of Union, and the Axiom of Power Set using these weaker versions. [*Hint*: Use also the Comprehension Schema.]

4. Elementary Operations on Sets

The purpose of this section is to elaborate somewhat on the notions introduced in the preceding section. In particular, we introduce simple set-theoretic operations (union, intersection, difference, etc.) and prove some of their basic properties. The reader is certainly familiar with them to some extent and we leave out most of the details.

Definition 3.10 tells us what it means that A is a subset of B (*included* in B), $A \subseteq B$. The property $\subseteq$ is called *inclusion*. It is easy to prove that, for any sets A, B, and C,
(a) $A \subseteq A$.
(b) If $A \subseteq B$ and $B \subseteq A$, then $A = B$.
(c) If $A \subseteq B$ and $B \subseteq C$, then $A \subseteq C$.
For example, to verify (c) we have to prove: If $x \in A$, then $x \in C$. But if $x \in A$, then $x \in B$, since $A \subseteq B$. Now, $x \in B$ implies $x \in C$, since $B \subseteq C$. So $x \in A$ implies $x \in C$.

If $A \subseteq B$ and $A \neq B$, we say that A is a *proper subset* of B (A is *properly contained* in B) and write $A \subset B$. We also write $B \supseteq A$ instead of $A \subseteq B$ and $B \supset A$ instead of $A \subset B$.

Most of the forthcoming set-theoretic operations have been mentioned before. The reader probably knows how they can be visualized using Venn diagrams (see Figure 1).

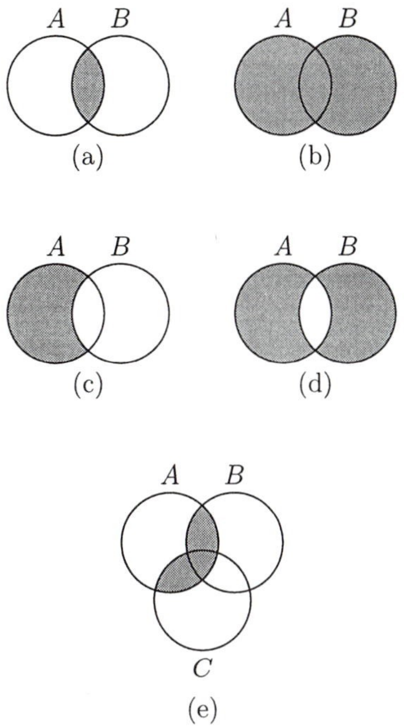

Figure 1: Venn diagrams. (a) Intersection: The shaded part is $A \cap B$. (b) Union: The shaded part is $A \cup B$. (c) Difference: The shaded part is $A - B$. (d) Symmetric difference: The shaded part is $A \triangle B$. (e) Distributive law: The shaded part obviously represents both $A \cap (B \cup C)$ and $(A \cap B) \cup (A \cap C)$.

4.1 Definition The *intersection* of A and B, $A \cap B$, is the set of all x which belong to both A and B. The *union* of A and B, $A \cup B$, is the set of all x which belong in either A or B (or both). The *difference* of A and B, $A - B$, is the set of all $x \in A$ which do not belong to B. The *symmetric difference* of A and B, $A \triangle B$, is defined by $A \triangle B = (A - B) \cup (B - A)$. (See Examples 3.3 and 3.8 and Exercises 3.1 and 3.4 for proofs of existence and uniqueness.)

As an exercise, the reader can work out proofs of some simple properties of these operations.

Commutativity
$$A \cap B = B \cap A$$
$$A \cup B = B \cup A$$

Associativity
$$(A \cap B) \cap C = A \cap (B \cap C)$$
$$(A \cup B) \cup C = A \cup (B \cup C)$$

So forgetting the parentheses we can write simply $A \cap B \cap C$ for the intersection of sets A, B, and C. Similarly, we do not need parentheses for the union and for more than three sets.

$$\text{Distributivity} \quad A \cap (B \cup C) = (A \cap B) \cup (A \cap C)$$
$$A \cup (B \cap C) = (A \cup B) \cap (A \cup C)$$

$$\text{DeMorganLaws} \quad C - (A \cap B) = (C - A) \cup (C - B)$$
$$C - (A \cup B) = (C - A) \cap (C - B)$$

Some of the properties of the difference and the symmetric difference are

$$A \cap (B - C) = (A \cap B) - C$$
$$A - B = \emptyset \text{ if and only if } A \subseteq B$$
$$A \triangle A = \emptyset$$
$$A \triangle B = B \triangle A$$
$$(A \triangle B) \triangle C = A \triangle (B \triangle C)$$

Drawing Venn diagrams often helps one discover and prove these and similar relationships. For example, Figure 1(e) illustrates the distributive law $A \cap (B \cup C) = (A \cap B) \cup (A \cap C)$. The rigorous proof proceeds as follows: We have to prove that the sets $A \cap (B \cup C)$ and $(A \cap B) \cup (A \cap C)$ have the same elements. That requires us to show two facts:

(a) Every element of $A \cap (B \cup C)$ belongs to $(A \cap B) \cup (A \cap C)$.
(b) Every element of $(A \cap B) \cup (A \cap C)$ belongs to $A \cap (B \cup C)$.

To prove (a), let $a \in A \cap (B \cup C)$. Then $a \in A$ and also $a \in B \cup C$. Therefore, either $a \in B$ or $a \in C$. So $a \in A$ and $a \in B$ or $a \in A$ and $a \in C$. This means that $a \in A \cap B$ or $a \in A \cap C$; hence, finally, $a \in (A \cap B) \cup (A \cap C)$.

To prove (b), let $a \in (A \cap B) \cup (A \cap C)$. Then $a \in A \cap B$ or $a \in A \cap C$. In the first case, $a \in A$ and $a \in B$, so $a \in A$ and $a \in B \cup C$, and $a \in A \cap (B \cup C)$. In the second case, $a \in A$ and $a \in C$, so again $a \in A$ and $a \in B \cup C$, and finally, $a \in A \cap (B \cup C)$. □

The exercises should provide sufficient material for practicing similar elementary arguments about sets.

The union of a system of sets S was defined in the preceding section. We now define the *intersection* $\bigcap S$ of a nonempty system of sets S: $x \in \bigcap S$ if and only if $x \in A$ for all $A \in S$. Then intersection of two sets is again a special case of the more general operation: $A \cap B = \bigcap\{A, B\}$. Notice that we do not define $\bigcap \emptyset$; the reason is that every x belongs to all $A \in \emptyset$ (since there is no such A), so $\bigcap \emptyset$ would have to be a set of all sets. We postpone more detailed investigation of general unions and intersections until Chapter 2, where a more wieldy notation becomes available.

Finally, we say that sets A and B are *disjoint* if $A \cap B = \emptyset$. More generally, S is a *system of mutually disjoint sets* if $A \cap B = \emptyset$ for all $A, B \in S$ such that $A \neq B$.

Exercises

4.1 Prove all the displayed formulas in this section and visualize them using Venn diagrams.

4.2 Prove:

(a) $A \subseteq B$ if and only if $A \cap B = A$ if and only if $A \cup B = B$ if and only if $A - B = \emptyset$.

(b) $A \subseteq B \cap C$ if and only if $A \subseteq B$ and $A \subseteq C$.

(c) $B \cup C \subseteq A$ if and only if $B \subseteq A$ and $C \subseteq A$.

(d) $A - B = (A \cup B) - B = A - (A \cap B)$.

(e) $A \cap B = A - (A - B)$.

(f) $A - (B - C) = (A - B) \cup (A \cap C)$.

(g) $A = B$ if and only if $A \triangle B = \emptyset$.

4.3 For each of the following (false) statements draw a Venn diagram in which it fails:

(a) $A - B = B - A$.

(b) $A \cap B \subset A$.

(c) $A \subseteq B \cup C$ implies $A \subseteq B$ or $A \subseteq C$.

(d) $B \cap C \subseteq A$ implies $B \subseteq A$ or $C \subseteq A$.

4.4 Let A be a set; show that a "complement" of A does not exist. (The "complement" of A is the set of all $x \notin A$.)

4.5 Let $S \neq \emptyset$ and A be sets.

(a) Set $T_1 = \{Y \in \mathcal{P}(A) \mid Y = A \cap X \text{ for some } X \in S\}$, and prove $A \cap \bigcup S = \bigcup T_1$ (generalized distributive law).

(b) Set $T_2 = \{Y \in \mathcal{P}(A) \mid Y = A - X \text{ for some } X \in S\}$, and prove

$$A - \bigcup S = \bigcap T_2$$
$$A - \bigcap S = \bigcup T_2$$

(generalized De Morgan laws).

4.6 Prove that $\bigcap S$ exists for all $S \neq \emptyset$. Where is the assumption $S \neq \emptyset$ used in the proof?

Chapter 2

Relations, Functions, and Orderings

1. Ordered Pairs

In this chapter we begin our program of developing set theory as a foundation for mathematics by showing how various general mathematical concepts, such as relations, functions, and orderings can be represented by sets.

We begin by introducing the notion of the *ordered pair*. If a and b are sets, then the *unordered pair* $\{a, b\}$ is a set whose elements are exactly a and b. The "order" in which a and b are put together plays no role; $\{a, b\} = \{b, a\}$. For many applications, we need to pair a and b in a way making possible to "read off" which set comes "first" and which comes "second." We denote this *ordered pair* of a and b by (a, b); a is the *first coordinate* of the pair (a, b), b is the *second coordinate*.

As any object of our study, the ordered pair has to be a set. It should be defined in such a way that two ordered pairs are equal if and only if their first coordinates are equal and their second coordinates are equal. This guarantees in particular that $(a, b) \neq (b, a)$ if $a \neq b$. (See Exercise 1.3.)

There are many ways how to define (a, b) so that the foregoing condition is satisfied. We give one such definition and refer the reader to Exercise 1.6 for an alternative approach.

1.1 Definition $(a, b) = \{\{a\}, \{a, b\}\}$.

If $a \neq b$, (a, b) has two elements, a singleton $\{a\}$ and an unordered pair $\{a, b\}$. We find the first coordinate by looking at the element of $\{a\}$. The second coordinate is then the other element of $\{a, b\}$. If $a = b$, then $(a, a) = \{\{a\}, \{a, a\}\} = \{\{a\}\}$ has only one element. In any case, it seems obvious that both coordinates can be uniquely "read off" from the set (a, b). We make this statement precise in the following theorem.

1.2 Theorem *$(a,b) = (a',b')$ if and only if $a = a'$ and $b = b'$.*

Proof. If $a = a'$ and $b = b'$, then, of course, $(a,b) = \{\{a\},\{a,b\}\} = \{\{a'\},\{a',b'\}\} = (a',b')$. The other implication is more intricate. Let us assume that $\{\{a\},\{a,b\}\} = \{\{a'\},\{a',b'\}\}$. If $a \neq b$, $\{a\} = \{a'\}$ and $\{a,b\} = \{a',b'\}$. So, first, $a = a'$ and then $\{a,b\} = \{a,b'\}$ implies $b = b'$. If $a = b$, $\{\{a\},\{a,a\}\} = \{\{a\}\}$. So $\{a\} = \{a'\}$, $\{a\} = \{a',b'\}$, and we get $a = a' = b'$, so $a = a'$ and $b = b'$ holds in this case, too. □

With ordered pairs at our disposal, we can define *ordered triples*

$$(a,b,c) = ((a,b),c),$$

ordered quadruples

$$(a,b,c,d) = ((a,b,c),d),$$

and so on. Also, we define ordered "*one-tuples*"

$$(a) = a.$$

However, the general definition of ordered n-tuples has to be postponed until Chapter 3, where natural numbers are defined.

Exercises

1.1 Prove that $(a,b) \in \mathcal{P}(\mathcal{P}(\{a,b\}))$ and $a, b \in \bigcup(a,b)$. More generally, if $a \in A$ and $b \in A$, then $(a,b) \in \mathcal{P}(\mathcal{P}(A))$.

1.2 Prove that (a,b), (a,b,c), and (a,b,c,d) exist for all a, b, c, and d.

1.3 Prove: If $(a,b) = (b,a)$, then $a = b$.

1.4 Prove that $(a,b,c) = (a',b',c')$ implies $a = a'$, $b = b'$, and $c = c'$. State and prove an analogous property of quadruples.

1.5 Find a, b, and c such that $((a,b),c) \neq (a,(b,c))$. Of course, we could use the second set to define ordered triples, with equal success.

1.6 To give an alternative definition of ordered pairs, choose two different sets $\Box$ and $\triangle$ (for example, $\Box = \emptyset$, $\triangle = \{\emptyset\}$) and define

$$\langle a,b \rangle = \{\{a,\Box\},\{b,\triangle\}\}.$$

State and prove an analogue of Theorem 1.2 for this notion of ordered pairs. Define ordered triples and quadruples.

2. Relations

Mathematicians often study relations between mathematical objects. Relations between objects of two sorts occur most frequently; we call them *binary* relations. For example, let us say that a line l is in relation R_1 with a point P if l passes through P. Then R_1 is a binary relation between objects called lines and objects called points. Similarly, we define a binary relation R_2 between positive integers and positive integers by saying that a positive integer m is in relation R_2 with a positive integer n if m divides n (without remainder).

Let us now consider the relation R_1' between lines and points such that a line l is in relation R_1' with a point P if P lies on l. Obviously, a line l is in relation R_1 to a point P exactly when l is in relation R_1' to P. Although different properties were used to describe R_1 and R_1', we would ordinarily consider R_1 and R_1' to be one and the same relation, i.e., $R_1 = R_1'$. Similarly, let a positive integer m be in relation R_2' with a positive integer n if n is a multiple of m. Again, the same ordered pairs (m, n) are related in R_2 as in R_2', and we consider R_2 and R_2' to be the same relation.

A binary relation is, therefore, determined by specifying all ordered pairs of objects in that relation; it does not matter by what property the set of these ordered pairs is described. We are led to the following definition.

2.1 Definition A set R is a *binary relation* if all elements of R are ordered pairs, i.e., if for any $z \in R$ there exist x and y such that $z = (x, y)$.

2.2 Example The relation R_2 is simply the set $\{z \mid$ there exist positive integers m and n such that $z = (m, n)$ and m divides $n\}$. Elements of R_2 are ordered pairs

$$\begin{array}{l}(1,1),(1,2),(1,3),\ldots\\(2,2),(2,4),(2,6),\ldots\\(3,3),(3,6),(3,9),\ldots\\\ldots\end{array}$$

It is customary to write xRy instead of $(x, y) \in R$. We say that *x is in relation R with y* if xRy holds.

We now introduce some terminology associated with relations.

2.3 Definition Let R be a binary relation.

(a) The set of all x which are in relation R with some y is called the *domain* of R and denoted by $\operatorname{dom} R$. So $\operatorname{dom} R = \{x \mid$ there exists y such that $xRy\}$. $\operatorname{dom} R$ is the set of all first coordinates of ordered pairs in R.

(b) The set of all y such that, for some x, x is in relation R with y is called the *range* of R, denoted by $\operatorname{ran} R$. So $\operatorname{ran} R = \{y \mid$ there exists x such that $xRy\}$. $\operatorname{ran} R$ is the set of all second coordinates of ordered pairs in R. Both $\operatorname{dom} R$ and $\operatorname{ran} R$ exist for any relation R. (Prove it. See Exercise 2.1).

(c) The set $\operatorname{dom} R \cup \operatorname{ran} R$ is called the *field* of R and is denoted by $\operatorname{field} R$.

(d) If $\operatorname{field} R \subseteq X$, we say that R is a *relation in* X or that R is a relation *between* elements of X.

2.4 Example Let R_2 be the relation from Example 2.2.

$$\begin{aligned}\operatorname{dom} R_2 &= \{m \mid \text{there exists } n \text{ such that } m \text{ divides } n\}\\ &= \text{the set of all positive integers}\end{aligned}$$

because each positive integer m divides some n, e.g., $n = m$;

$$\begin{aligned}\operatorname{ran} R_2 &= \{n \mid \text{there exists } m \text{ such that } m \text{ divides } n\} \\ &= \text{the set of all positive integers}\end{aligned}$$

because each positive integer n is divided by some m, e.g., by $m = n$;

$$\text{field } R_2 = \operatorname{dom} R_2 \cup \operatorname{ran} R_2 = \text{the set of all positive integers;}$$

R_2 is a relation between positive integers.

We next generalize Definition 2.3.

2.5 Definition

(a) The *image* of A under R is the set of all y from the range of R related in R to some element of A; it is denoted by $R[A]$. So

$$R[A] = \{y \in \operatorname{ran} R \mid \text{there exists } x \in A \text{ for which } xRy\}.$$

(b) The *inverse image* of B under R is the set of all x from the domain of R related in R to some element of B; it is denoted $R^{-1}[B]$. So

$$R^{-1}[B] = \{x \in \operatorname{dom} R \mid \text{there exists } y \in B \text{ for which } xRy\}.$$

2.6 Example $R_2^{-1}[\{3, 8, 9, 12\}] = \{1, 2, 3, 4, 6, 8, 9, 12\}$; $R_2[\{2\}] =$ the set of all even positive integers.

2.7 Definition Let R be a binary relation. The *inverse* of R is the set

$$R^{-1} = \{z \mid z = (x, y) \text{ for some } x \text{ and } y \text{ such that } (y, x) \in R\}.$$

2.8 Example Again let

$$\begin{aligned}R_2 &= \{z \mid z = (m, n),\ m \text{ and } n \text{ are positive integers, and } m \text{ divides } n\}; \text{then} \\ R_2^{-1} &= \{w \mid w = (n, m), \text{ and } (m, n) \in R_2\} \\ &= \{w \mid w = (n, m),\ m \text{ and } n \text{ are positive integers, and } m \text{ divides } n\}.\end{aligned}$$

In our description of R_2, we use variable m for the first coordinate and variable n for the second coordinate; we also state the property describing R_2 so that the variable m is mentioned first. It is a customary (though not necessary) practice to describe R_2^{-1} in the same way. All we have to do is use letter m instead of n, letter n instead of m, and change the wording:

$$\begin{aligned}R_2^{-1} &= \{w \mid w = (m, n),\ n\ ,\ m \text{ are positive integers, and } n \text{ divides } m\} \\ &= \{w \mid w = (m, n),\ m\ ,\ n \text{ are positive integers, and } m \text{ is a multiple of } n\}.\end{aligned}$$

Now R_2 and R_2^{-1} are described in a parallel way. In this sense, the inverse of the relation "divides" is the relation "is a multiple."

The reader may notice that the symbol $R^{-1}[B]$ in Definition 2.5(b) for the inverse image of B under R now also denotes the image of B under R^{-1}. Fortunately, these two sets are equal.

2.9 Lemma *The inverse image of B under R is equal to the image of B under R^{-1}.*

Proof. Notice first that $\operatorname{dom} R = \operatorname{ran} R^{-1}$ (see Exercise 2.4). Now, $x \in \operatorname{dom} R$ belongs to the inverse image of B under R if and only if for some $y \in B$, $(x, y) \in R$. But $(x, y) \in R$ if and only if $(y, x) \in R^{-1}$. Therefore, x belongs to the inverse image of B under R if and only if for some $y \in B$, $(y, x) \in R^{-1}$, i.e., if and only if x belongs to the image of B under R^{-1}. □

In the rest of the book we often define various relations, i.e., sets of ordered pairs having some particular property. To simplify our notation, we introduce the following conventions. Instead of

$$\{w \mid w = (x, y) \text{ for some } x \text{ and } y \text{ such that } \mathbf{P}(x, y)\},$$

we simply write

$$\{(x, y) \mid \mathbf{P}(x, y)\}.$$

For example, the inverse of R could be described in this notation as $\{(x, y) \mid (y, x) \in R\}$. [As in the general case, use of this notation is admissible only if we prove that there exists a set A such that, for all x and y, $\mathbf{P}(x, y)$ implies $(x, y) \in A$.]

2.10 Definition Let R and S be binary relations. The *composition* of R and S is the relation

$$S \circ R = \{(x, z) \mid \text{there exists } y \text{ for which } (x, y) \in R \text{ and } (y, z) \in S\}.$$

So $(x, z) \in S \circ R$ means that for some y, xRy and ySz. To find objects related to x in $S \circ R$, we first find objects y related to x in R, and then objects related to those y in S. Notice that R is performed first and S second, but notation $S \circ R$ is customary (at least in the case of functions; see Section 3).

Several types of relations are of special interest. We introduce some of them in this section and others in the rest of the chapter.

2.11 Definition The *membership relation on A* is defined by

$$\in_A = \{(a, b) \mid a \in A,\ b \in A, \text{ and } a \in b\}.$$

The *identity relation on A* is defined by

$$\mathrm{Id}_A = \{(a, b) \mid a \in A,\ b \in A, \text{ and } a = b\}.$$

2.12 Definition Let A and B be sets. The set of all ordered pairs whose first coordinate is from A and whose second coordinate is from B is called the *cartesian product* of A and B and denoted $A \times B$. In other words,

$$A \times B = \{(a, b) \mid a \in A \text{ and } b \in B\}.$$

Thus $A \times B$ is a relation in which every element of A is related to every element of B.

It is not completely trivial to show that the set $A \times B$ exists. However, one gets from Exercise 1.1 that if $a \in A$ and $b \in B$, then $(a, b) \in \mathcal{P}(\mathcal{P}(A \cup B))$. Therefore,

$$A \times B = \{(a, b) \in \mathcal{P}(\mathcal{P}(A \cup B)) \mid a \in A \text{ and } b \in B\}.$$

Since $\mathcal{P}(\mathcal{P}(A \cup B))$ was proved to exist, the existence of $A \times B$ follows from the Axiom Schema of Comprehension. [To be completely explicit, we can write,

$$A \times B = \{w \in \mathcal{P}(\mathcal{P}(A \cup B)) \mid w = (a, b) \text{ for some } a \in A \text{ and } b \in B\}.]$$

We denote $A \times A$ by A^2. The cartesian product of three sets can be introduced readily:

$$A \times B \times C = (A \times B) \times C.$$

Notice that

$$A \times B \times C = \{(a, b, c) \mid a \in A,\ b \in B, \text{ and } c \in C\}$$

(using an obvious extension of our notational convention). $A \times A \times A$ is usually denoted A^3.

We can also define *ternary relations.*

2.13 Definition A *ternary relation* is a set of unordered triples. More explicitly, S is a ternary relation if for every $u \in S$, there exist x, y, and z such that $u = (x, y, z)$. If $S \subseteq A^3$, we say that S is a ternary relation *in* A. (Note that a binary relation R is in A if and only if $R \subseteq A^2$.)

We could extend the concepts of this section to ternary relations and also define 4-ary or 17-ary relations. We postpone these matters until Section 5 in Chapter 3, where natural numbers become available, and we are able to define n-ary relations in general. At this stage we only define, for technical reasons, *unary relations* by specifying that a unary relation is any set. A *unary relation in* A is any subset of A. This agrees with the general conception that a unary relation in A should be a set of 1-tuples of elements of A and with the definition of $(x) = x$ in Section 1.

Exercises

2.1 Let R be a binary relation; let $A = \bigcup(\bigcup R)$. Prove that $(x, y) \in R$ implies $x \in A$ and $y \in A$. Conclude from this that $\operatorname{dom} R$ and $\operatorname{ran} R$ exist.

2.2 (a) Show that R^{-1} and $S \circ R$ exist. [*Hint*: $R^{-1} \subseteq (\operatorname{ran} R) \times (\operatorname{dom} R)$, $S \circ R \subseteq (\operatorname{dom} R) \times (\operatorname{ran} S)$.]
(b) Show that $A \times B \times C$ exists.

2.3 Let R be a binary relation and A and B sets. Prove:
(a) $R[A \cup B] = R[A] \cup R[B]$.
(b) $R[A \cap B] \subseteq R[A] \cap R[B]$.
(c) $R[A - B] \supseteq R[A] - R[B]$.

(d) Show by an example that $\subseteq$ and $\supseteq$ in parts (b) and (c) cannot be replaced by $=$.
(e) Prove parts (a)–(d) with R^{-1} instead of R.
(f) $R^{-1}[R[A]] \supseteq A \cap \operatorname{dom} R$ and $R[R^{-1}[B]] \supseteq B \cap \operatorname{ran} R$; give examples where equality does not hold.

2.4 Let $R \subseteq X \times Y$. Prove:
(a) $R[X] = \operatorname{ran} R$ and $R^{-1}[Y] = \operatorname{dom} R$.
(b) If $a \notin \operatorname{dom} R$, $R[\{a\}] = \emptyset$; if $b \notin \operatorname{ran} R$, $R^{-1}[\{b\}] = \emptyset$.
(c) $\operatorname{dom} R = \operatorname{ran} R^{-1}$; $\operatorname{ran} R = \operatorname{dom} R^{-1}$.
(d) $(R^{-1})^{-1} = R$.
(e) $R^{-1} \circ R \supseteq \operatorname{Id}_{\operatorname{dom} R}$; $R \circ R^{-1} \supseteq \operatorname{Id}_{\operatorname{ran} R}$.

2.5 Let $X = \{\emptyset, \{\emptyset\}\}$, $Y = \mathcal{P}(X)$. Describe
(a) $\in_Y$,
(b) Id_Y.
Determine the domain, range, and field of both relations.

2.6 Prove that for any three binary relations R, S, and T

$$T \circ (S \circ R) = (T \circ S) \circ R.$$

(The operation $\circ$ is associative.)

2.7 Give examples of sets X, Y, and Z such that
(a) $X \times Y \neq Y \times X$.
(b) $X \times (Y \times Z) \neq (X \times Y) \times Z$.
(c) $X^3 \neq X \times X^2$ [i.e., $(X \times X) \times X \neq X \times (X \times X)$].
[*Hint for part* (c): $X = \{a\}$.]

2.8 Prove:
(a) $A \times B = \emptyset$ if and only if $A = \emptyset$ or $B = \emptyset$.
(b) $(A_1 \cup A_2) \times B = (A_1 \times B) \cup (A_2 \times B)$;
$A \times (B_1 \cup B_2) = (A \times B_1) \cup (A \times B_2)$.
(c) Same as part (b), with $\cup$ replaced by $\cap$, $-$, and $\triangle$.

3. Functions

Function, as understood in mathematics, is a procedure, a rule, assigning to any object a from the domain of the function a unique object b, the value of the function at a. A function, therefore, represents a special type of relation, a relation where every object a from the domain is related to precisely one object in the range, namely, to the value of the function at a.

3.1 Definition A binary relation F is called a *function* (or *mapping*, *correspondence*) if aFb_1 and aFb_2 imply $b_1 = b_2$ for any a, b_1, and b_2. In other words, a binary relation F is a function if and only if for every a from $\operatorname{dom} F$ there is exactly one b such that aFb. This unique b is called the *value of F at a* and is denoted $F(a)$ or F_a. [$F(a)$ is not defined if $a \notin \operatorname{dom} F$.] If F is a function with $\operatorname{dom} F = A$ and $\operatorname{ran} F \subseteq B$, it is customary to use the notations

$F : A \to B$, $\langle F(a) \mid a \in A\rangle$, $\langle F_a \mid a \in A\rangle$, $\langle F_a\rangle_{a\in A}$ for the function F. The range of the function F can then be denoted $\{F(a) \mid a \in A\}$ or $\{F_a\}_{a\in A}$.

The Axiom of Extensionality can be applied to functions as follows.

3.2 Lemma *Let F and G be functions. $F = G$ if and only if $\operatorname{dom} F = \operatorname{dom} G$ and $F(x) = G(x)$ for all $x \in \operatorname{dom} F$.*

We leave the proof to the reader. □

Since functions are binary relations, concepts of domain, range, image, inverse image, inverse, and composition can be applied to them. We introduce several additional definitions.

3.3 Definition Let F be a function and A and B sets.
(a) F is a function *on* A if $\operatorname{dom} F = A$.
(b) F is a function *into* B if $\operatorname{ran} F \subseteq B$.
(c) F is a function *onto* B if $\operatorname{ran} F = B$.
(d) The *restriction* of the function F *to* A is the function

$$F \restriction A = \{(a, b) \in F \mid a \in A\}.$$

If G is a restriction of F to some A, we say that F is an *extension* of G.

3.4 Example Let $F = \{(x, 1/x^2) \mid x \neq 0,\ x \text{ is a real number}\}$. F is a function: If aFb_1 and aFb_2, $b_1 = 1/a^2$ and $b_2 = 1/a^2$, so $b_1 = b_2$.

Slightly stretching our notational conventions, we can also write $F = \langle 1/x^2 \mid x \text{ is a real number}, x \neq 0\rangle$. The value of F at x, $F(x)$, equals $1/x^2$. It is a function on A, where $A = \{x \mid x \text{ is a real number and } x \neq 0\} = \operatorname{dom} F$. F is a function into the set of all real numbers, but not onto the set of all real numbers. If $B = \{x \mid x \text{ is a real number and } x > 0\}$, then F is onto B. If $C = \{x \mid 0 < x \leq 1\}$, then $f[C] = \{x \mid x \geq 1\}$ and $f^{-1}[C] = \{x \mid x \leq -1 \text{ or } x \geq 1\}$.

Let us find the composition $f \circ f$:

$$\begin{aligned} f \circ f &= \{(x, z) \mid \text{there is } y \text{ for which } (x, y) \in f \text{ and } (y, z) \in f\} \\ &= \{(x, z) \mid \text{there is } y \text{ for which } x \neq 0,\ y = 1/x^2, \text{ and } y \neq 0,\ z = 1/y^2\} \\ &= \{(x, z) \mid x \neq 0 \text{ and } z = x^4\} \\ &= \langle x^4 \mid x \neq 0\rangle. \end{aligned}$$

Notice that $f \circ f$ is a function. This is not an accident.

3.5 Theorem *Let f and g be functions. Then $g \circ f$ is a function. $g \circ f$ is defined at x if and only if f is defined at x and g is defined at $f(x)$, i.e.,*

$$\operatorname{dom}(g \circ f) = \operatorname{dom} f \cap f^{-1}[\operatorname{dom} g].$$

Also, $(g \circ f)(x) = g(f(x))$ for all $x \in \operatorname{dom}(g \circ f)$.

Proof. We prove first that $g \circ f$ is a function. If $x(g \circ f)z_1$ and $x(g \circ f)z_2$, there exist y_1 and y_2 such that xfy_1, y_1gz_1, and xfy_2, y_2gz_2. Since f is a function, $y_1 = y_2$. So we get y_1gz_1, y_1gz_2, and $z_1 = z_2$, because g is also a function.

Now we investigate the domain of $g \circ f$. $x \in \operatorname{dom}(g \circ f)$ if and only if there is some z such that $x(g \circ f)z$, i.e., if and only if there is some z and some y such that xfy and ygz. But this happens if and only if $x \in \operatorname{dom} f$ and $y = f(x) \in \operatorname{dom} g$. The last statement can be equivalently expressed as $x \in \operatorname{dom} f$ and $x \in f^{-1}[\operatorname{dom} g]$. □

This theorem is used in calculus to find domains of compositions of functions. Let us give one typical example.

3.6 Example Let $f = \langle x^2 - 1 \mid x \text{ real} \rangle$, $g = \langle \sqrt{x} \mid x \geq 0 \rangle$. Find the composition $g \circ f$.

We determine the domain of $g \circ f$ first. $\operatorname{dom} f$ is the set of all real numbers and $\operatorname{dom} g = \{x \mid x \geq 0\}$. We find $f^{-1}[\operatorname{dom} g] = \{x \mid f(x) \in \operatorname{dom} g\} = \{x \mid x^2 - 1 \geq 0\} = \{x \mid x \geq 1 \text{ or } x \leq -1\}$. Therefore, $\operatorname{dom}(g \circ f) = (\operatorname{dom} f) \cap f^{-1}[\operatorname{dom} g] = \{x \mid x \geq 1 \text{ or } x \leq -1\}$ and $g \circ f = \{(x, z) \mid x \geq 1 \text{ or } x \leq -1 \text{ and, for some } y,\ x^2 - 1 = y \text{ and } \sqrt{y} = z\} = \langle \sqrt{x^2 - 1} \mid x \geq 1 \text{ or } x \leq -1 \rangle$.

If f is a function, its inverse f^{-1} is a relation, but it may not be a function. We say that a function f is *invertible* if f^{-1} is a function.

It is important to find necessary and sufficient conditions for a function to be invertible.

3.7 Definition A function f is called *one-to-one* or *injective* if $a_1 \in \operatorname{dom} f$, $a_2 \in \operatorname{dom} f$, and $a_1 \neq a_2$ implies $f(a_1) \neq f(a_2)$. In other words if $a_1 \in \operatorname{dom} f$, $a_2 \in \operatorname{dom} f$, and $f(a_1) = f(a_2)$, then $a_1 = a_2$. Thus a one-to-one function attains different values for different elements from its domain.

3.8 Theorem *A function is invertible if and only if it is one-to-one. If f is invertible, then f^{-1} is also invertible and $(f^{-1})^{-1} = f$.*

Proof.

(a) Let f be invertible; then f^{-1} is a function. It follows that $f^{-1}(f(a)) = a$ for all $a \in \operatorname{dom} f$. If $a_1, a_2 \in \operatorname{dom} f$ and $f(a_1) = f(a_2)$, we get $f^{-1}(f(a_1)) = f^{-1}(f(a_2))$ and $a_1 = a_2$. So f is one-to-one.

(b) Let f be one-to-one. If $af^{-1}b_1$ and $af^{-1}b_2$, we have b_1fa and b_2fa. Therefore, $b_1 = b_2$, and we have proved that f^{-1} is a function.

(c) We know that $(f^{-1})^{-1} = f$ by Exercise 2.4(d), so f^{-1} is also invertible (consequently, f^{-1} is also one-to-one).

□

3.9 Example

(a) Let $f = \langle 1/x^2 \mid x \neq 0 \rangle$; find f^{-1}. As $f = \{(x, 1/x^2) \mid x \neq 0\}$, we get $f^{-1} = \{(1/x^2, x) \mid x \neq 0\}$. f^{-1} is not a function since $(1, -1) \in f^{-1}$, $(1, 1) \in f^{-1}$. Therefore, f is not one-to-one; $(1, 1) \in f$, $(-1, 1) \in f$.

(b) Let $g = \langle 2x - 1 \mid x \text{ real} \rangle$; find g^{-1}. g is one-to-one: If $2x_1 - 1 = 2x_2 - 1$, then $2x_1 = 2x_2$ and $x_1 = x_2$. $g = \{(x, y) \mid y = 2x - 1, x \text{ real}\}$, therefore, $g^{-1} = \{(y, x) \mid y = 2x - 1, x \text{ real}\}$.

As customary when describing functions, we express the second coordinate (value) in terms of the first:

$$g^{-1} = \left\{ (y, x) \;\middle|\; x = \frac{y+1}{2},\ y \text{ real} \right\}.$$

Finally, it is usual to denote the first ("independent") variable x and the second ("dependent") variable y. So we change notation:

$$g^{-1} = \left\{ (x, y) \;\middle|\; y = \frac{x+1}{2},\ x \text{ real} \right\} = \left\langle \frac{x+1}{2} \;\middle|\; x \text{ real} \right\rangle.$$

3.10 Definition

(a) Functions f and g are called *compatible* if $f(x) = g(x)$ for all $x \in \operatorname{dom} f \cap \operatorname{dom} g$.

(b) A set of functions F is called a *compatible system* of functions if any two functions f and g from F are compatible.

3.11 Lemma

(a) Functions f and g are compatible if and only if $f \cup g$ is a function.

(b) Functions f and g are compatible if and only if $f \restriction (\operatorname{dom} f \cap \operatorname{dom} g) = g \restriction (\operatorname{dom} f \cap \operatorname{dom} g)$.

We leave the easy proof to the reader, but we prove the following.

3.12 Theorem *If F is a compatible system of functions, then $\bigcup F$ is a function with $\operatorname{dom}(\bigcup F) = \bigcup\{\operatorname{dom} f \mid f \in F\}$. The function $\bigcup F$ extends all $f \in F$.*

Functions from a compatible system can be pieced together to form a single function which extends them all.

Proof. Clearly, $\bigcup F$ is a relation; we now prove that it is a function. If $(a, b_1) \in \bigcup F$ and $(a, b_2) \in \bigcup F$, there are functions $f_1, f_2 \in F$ such that $(a, b_1) \in f_1$ and $(a, b_2) \in f_2$. But f_1 and f_2 are compatible, and $a \in \operatorname{dom} f_1 \cap \operatorname{dom} f_2$; so $b_1 = f(a_1) = f(a_2) = b_2$.

It is trivial to show that $x \in \operatorname{dom}(\bigcup F)$ if and only if $x \in \operatorname{dom} f$ for some $f \in F$. □

3.13 Definition Let A and B be sets. The set of all functions on A into B is denoted B^A. Of course, we have to show that B^A exists; this is done in Exercise 3.9.

It is useful to define a more general notion of product of sets in terms of functions.

Let $S = \langle S_i \mid i \in I \rangle$ be a function with domain I. The reader is probably familiar mainly with functions possessing numerical values; but for us, the values S_i are arbitrary sets. We call the function $\langle S_i \mid i \in I \rangle$ an *indexed system of sets*, whenever we wish to stress that the values of S are sets.

Now let $S = \langle S_i \mid i \in I \rangle$ be an indexed system of sets. We define the *product* of the indexed system S as the set

$$\prod S = \{f \mid f \text{ is a function on } I \text{ and } f_i \in S_i \text{ for all } i \in I\}.$$

Other notations we occasionally use are

$$\prod \langle S(i) \mid i \in I \rangle, \quad \prod_{i \in I} S(i), \quad \prod_{i \in I} S_i.$$

The existence of the product of any indexed system is proved in Exercise 3.9.

The reader is probably curious to know how this product is related to the previously defined notions $A \times B$ and $A \times B \times C$. We return to this technical problem in Section 5 of Chapter 3. At this time, we notice only that if the indexed system S is such that $S_i = B$ for all $i \in I$, then

$$\prod_{i \in I} S_i = B^I.$$

The "exponentiation" of sets is related to "multiplication" of sets in the same way as similar operations on numbers are related.

We conclude this section with two remarks concerning notation.

$\bigcup A$ and $\bigcap A$ were defined for any system of sets A ($A \neq \emptyset$ in case of intersection). Often the system A is given as a range of some function, i.e., of some indexed system. (See Exercise 3.8 for proof that any system A can be so presented, if desired.)

We say that A is *indexed* by S if

$$A = \{S_i \mid i \in I\} = \operatorname{ran} S,$$

where S is a function on I. It is then customary to write

$$\bigcup A = \bigcup \{S_i \mid i \in I\} = \bigcup_{i \in I} S_i,$$

and similarly for intersections. In the future, we use this more descriptive notation.

Let f be a function on a subset of the product $A \times B$. It is customary to denote the value of f at $(x, y) \in A \times B$ by $f(x, y)$ rather than $f((x, y))$ and regard f as a function of two variables x and y.

Exercises

3.1 Prove: If $\operatorname{ran} f \subseteq \operatorname{dom} g$, then $\operatorname{dom}(g \circ f) = \operatorname{dom} f$.

3.2 The functions $f_i : i = 1, 2, 3$ are defined as follows:

$$f_1 = \langle 2x - 1 \mid x \text{ real} \rangle,$$
$$f_2 = \langle \sqrt{x} \mid x > 0 \rangle,$$
$$f_3 = \langle 1/x \mid x \text{ real}, x \neq 0 \rangle.$$

Describe each of the following functions, and determine their domains and ranges: $f_2 \circ f_1$, $f_1 \circ f_2$, $f_3 \circ f_1$, $f_1 \circ f_3$.

3.3 Prove that the functions f_1, f_2, f_3 from Exercise 3.2 are one-to-one, and find the inverse functions. In each case, verify that $\operatorname{dom} f_i = \operatorname{ran}(f_i^{-1})$, $\operatorname{ran} f_i = \operatorname{dom}(f_i^{-1})$.

3.4 Prove:

(a) If f is invertible, $f^{-1} \circ f = \mathrm{Id}_{\operatorname{dom} f}$, $f \circ f^{-1} = \mathrm{Id}_{\operatorname{ran} f}$.

(b) Let f be a function. If there exists a function g such that $g \circ f = \mathrm{Id}_{\operatorname{dom} f}$ then f is invertible and $f^{-1} = g \restriction \operatorname{ran} f$. If there exists a function h such that $f \circ h = \mathrm{Id}_{\operatorname{ran} f}$ then f may fail to be invertible.

3.5 Prove: If f and g are one-to-one functions, $g \circ f$ is also a one-to-one function, and $(g \circ f)^{-1} = f^{-1} \circ g^{-1}$.

3.6 The images and inverse images of sets by functions have the properties exhibited in Exercise 2.3, but some of the inequalities can now be replaced by equalities. Prove:

(a) If f is a function, $f^{-1}[A \cap B] = f^{-1}[A] \cap f^{-1}[B]$.

(b) If f is a function, $f^{-1}[A - B] = f^{-1}[A] - f^{-1}[B]$.

3.7 Give an example of a function f and a set A such that $f \cap A^2 \neq f \restriction A$.

3.8 Show that every system of sets A can be indexed by a function. [*Hint*: Take $I = A$ and set $S_i = i$ for all $i \in A$.]

3.9 (a) Show that the set B^A exists. [*Hint*: $B^A \subseteq \mathcal{P}(A \times B)$.]

(b) Let $\langle S_i \mid i \in I \rangle$ be an indexed system of sets; show that $\prod_{i \in I} S_i$ exists. [*Hint*: $\prod_{i \in I} S_i \subseteq \mathcal{P}(I \times \bigcup_{i \in I} S_i)$.]

3.10 Show that unions and intersections satisfy the following general form of the associative law:

$$\bigcup_{a \in \bigcup S} F_a = \bigcup_{C \in S} \Big(\bigcup_{a \in C} F_a \Big),$$

$$\bigcap_{a \in \bigcup S} F_a = \bigcap_{C \in S} \Big(\bigcap_{a \in C} F_a \Big),$$

if S is a nonempty system of nonempty sets.

3.11 Other properties of unions and intersections can be generalized similarly. De Morgan Laws:

$$B - \bigcup_{a \in A} F_a = \bigcap_{a \in A} (B - F_a),$$

$$B - \bigcap_{a \in A} F_a = \bigcup_{a \in A} (B - F_a).$$

Distributive Laws:

$$\Big(\bigcup_{a\in A} F_a\Big)\cap\Big(\bigcup_{b\in B} G_b\Big) = \bigcup_{(a,b)\in A\times B} (F_a\cap G_b),$$

$$\Big(\bigcap_{a\in A} F_a\Big)\cup\Big(\bigcap_{b\in B} G_b\Big) = \bigcap_{(a,b)\in A\times B} (F_a\cup G_b).$$

3.12 Let f be a function. Then

$$f\Big[\bigcup_{a\in A} F_a\Big] = \bigcup_{a\in A} f[F_a],$$

$$f^{-1}\Big[\bigcup_{a\in A} F_a\Big] = \bigcup_{a\in A} f^{-1}[F_a],$$

$$f\Big[\bigcap_{a\in A} F_a\Big] \subseteq \bigcap_{a\in A} f[F_a],$$

$$f^{-1}\Big[\bigcap_{a\in A} F_a\Big] = \bigcap_{a\in A} f^{-1}[F_a].$$

If f is one-to-one, then $\subseteq$ in the third formula can be replaced by $=$.

3.13 Prove the following form of the distributive law:

$$\bigcap_{a\in A}\Big(\bigcup_{b\in B} F_{a,b}\Big) = \bigcup_{f\in B^A}\Big(\bigcap_{a\in A} F_{a,f(a)}\Big),$$

assuming that $F_{a,b_1}\cap F_{a,b_2} = \emptyset$ for all $a\in A$ and $b_1, b_2\in B$, $b_1\neq b_2$. [*Hint*: Let L be the set on the left and R the set on the right. $F_{a,f(a)}\subseteq \bigcup_{b\in B} F_{a,b}$; hence $\bigcap_{a\in A} F_{a,f(a)} \subseteq \bigcap_{a\in A}(\bigcup_{b\in B} F_{a,b}) = L$, so finally, $R\subseteq L$. To prove that $L\subseteq R$, take any $x\in L$. Put $(a,b)\in f$ if and only if $x\in F_{a,b}$, and prove that f is a function on A into B for which $x\in\bigcap_{a\in A} F_{a,f(a)}$; so $x\in R$.]

4. Equivalences and Partitions

Binary relations of several special types enter our considerations more frequently.

4.1 Definition Let R be a binary relation in A.

(a) R is called *reflexive in A* if for all $a\in A$, aRa.

(b) R is called *symmetric in A* if for all $a, b\in A$, aRb implies bRa.

(c) R is called *transitive in A* if for all $a, b, c\in A$, aRb and bRc imply aRc.

(d) R is called an *equivalence on A* if it is reflexive, symmetric, and transitive in A.

4.2 Example

(a) Let P be the set of all people living on Earth. We say that a person p is equivalent to a person q ($p \equiv q$) if p and q both live in the same country. Trivially, $\equiv$ is reflexive, symmetric, and transitive in P. Notice that the set P can be broken into classes of mutually equivalent elements; all people living in the United States form one class, all people living in France are another class, etc. All members of the same class are mutually equivalent; members of different classes are never equivalent. The equivalence classes correspond exactly to different countries.

(b) Define an equivalence E on the set Z of all integers as follows: xEy if and only if $x - y$ is divisible by 2. (Two numbers are equivalent if their difference is even.) The reader should verify 4.1(a)–(c). Again, the set Z can be divided into equivalence classes under (or, as is customary to say, *modulo*) the equivalence E. In this case, there are two equivalence classes: the set of even integers and the set of odd integers. Any two even integers are equivalent; so are any two odd integers. But an even integer cannot be equivalent to an odd one.

The situation encountered in the previous examples is quite general. Any equivalence on A partitions A into equivalence classes; conversely, given a suitable partition of A, there is an equivalence on A determined by it. The following definitions and theorems establish this correspondence.

4.3 Definition Let E be an equivalence on A and let $a \in A$. The *equivalence class of a modulo E* is the set

$$[a]_E = \{x \in A \mid xEa\}.$$

4.4 Lemma *Let $a, b \in A$.*

(a) a is equivalent to b modulo E if and only if $[a]_E = [b]_E$.

(b) a is not equivalent to b modulo E if and only if $[a]_E \cap [b]_E = \emptyset$.

Proof.

(a) (1) Assume that aEb. Let $x \in [a]_E$, i.e., xEa. By transitivity, xEa and aEb imply xEb, i.e., $x \in [b]_E$. Similarly, $x \in [b]_E$ implies $x \in [a]_E$ (bEa is true because E is symmetric). So $[a]_E = [b]_E$.

(2) Assume that $[a]_E = [b]_E$. Since E is reflexive, aEa, so $a \in [a]_E$. But then $a \in [b]_E$, that is, aEb.

(b) (1) Assume aEb is not true; we have to prove $[a]_E \cap [b]_E = \emptyset$. If not, there is $x \in [a]_E \cap [b]_E$; so xEa and xEb. But then, using first symmetry and then transitivity, aEx and xEb, so aEb, a contradiction.

(2) Assume finally that $[a]_E \cap [b]_E = \emptyset$. If a and b were equivalent modulo E, aEb would hold, so $a \in [b]_E$. But also $a \in [a]_E$, implying $[a]_E \cap [b]_E \neq \emptyset$, a contradiction.

□

4.5 Definition A system S of nonempty sets is called a *partition* of A if

(a) S is a system of mutually disjoint sets, i.e., if $C \in S$, $D \in S$, and $C \neq D$, then $C \cap D = \emptyset$,

(b) the union of S is the whole set A, i.e., $\bigcup S = A$.

4.6 Definition Let E be an equivalence on A. The system of all equivalence classes modulo E is denoted by A/E; so $A/E = \{[a]_E \mid a \in A\}$.

4.7 Theorem *Let E be an equivalence on A; then A/E is a partition of A.*

Proof. Property (a) follows from Lemma 4.4: If $[a]_E \neq [b]_E$, then a and b are not E-equivalent, so $[a]_E \cap [b]_E = \emptyset$. To prove (b), notice that $\bigcup A/E = A$ because $a \in [a]_E$. Notice also that no equivalence class is empty; surely at least $a \in [a]_E$. □

We now show that, conversely, for each partition there is a corresponding equivalence relation. For example, the partition of people by their country of residence yields the equivalence from (a) of Example 4.2.

4.8 Definition Let S be a partition of A. The relation E_S in A is defined by

$$E_S = \{(a, b) \in A \times A \mid \text{there is } C \in S \text{ such that } a \in C \text{ and } b \in C\}.$$

Objects a and b are related by E_S if and only if they belong to the same set from the partition S.

4.9 Theorem *Let S be a partition of A; then E_S is an equivalence on A.*

Proof.

(a) *Reflexivity.* Let $a \in A$; since $A = \bigcup S$, there is $C \in S$ for which $a \in C$, so $(a, a) \in E_S$.

(b) *Symmetry.* Assume aE_Sb; then there is $C \in S$ for which $a \in C$ and $b \in C$. Then, of course, $b \in C$ and $a \in C$, so bE_Sa.

(c) *Transitivity.* Assume aE_Sb and bE_Sc; then there are $C \in S$ and $D \in S$ such that $a \in C$ and $b \in C$ and $b \in D$ and $c \in D$. We see that $b \in C \cap D$, so $C \cap D \neq \emptyset$. But S is a system of mutually disjoint sets, so $C = D$. Now we have $a \in C$, $c \in C$, and so aE_Sc.

□

The next theorem should further clarify the relationship between equivalences and partitions. We leave its proof to the reader.

4.10 Theorem

(a) If E is an equivalence on A and $S = A/E$, then $E_S = E$.

(b) If S is a partition of A and E_S is the corresponding equivalence, then $A/E_S = S$.

So equivalence relations and partitions are two different descriptions of the same "mathematical reality." Every equivalence E determines a partition $S = A/E$. The equivalence E_S determined by this partition S is identical with the original E. Conversely, each partition S determines an equivalence E_S; when we form equivalence classes modulo E_S, we recover the original partition S.

When working with equivalences or partitions, it is often very convenient to have a set which contains exactly one "representative" from each equivalence class.

4.11 Definition A set $X \subseteq A$ is called a *set of representatives* for the equivalence E_S (or for the partition S of A) if for every $C \in S$, $X \cap C = \{a\}$ for some $a \in C$.

Returning to Example 4.2 we see that in (a) the set X of all Heads of State is a set of representatives for the partition by country of residence. The set $X = \{0, 1\}$ could serve as a set of representatives for the partition of integers into even and odd ones.

Does every partition have some set of representatives? Intuitively, the answer may seem to be yes, but it is not possible to prove existence of some such set for every partition on the basis of our axioms. We return to this problem when we discuss the Axiom of Choice. For the moment we just say that for many mathematically interesting equivalences a choice of a natural set of representatives is both possible and useful.

Exercises

4.1 For each of the following relations, determine whether they are reflexive, symmetric, or transitive:
  (a) Integer x is greater than integer y.
  (b) Integer n divides integer m.
  (c) $x \neq y$ in the set of all natural numbers.
  (d) $\subseteq$ and $\subset$ in $\mathcal{P}(A)$.
  (e) $\emptyset$ in $\emptyset$.
  (f) $\emptyset$ in a nonempty set A.

4.2 Let f be a function on A onto B. Define a relation E in A by: aEb if and only if $f(a) = f(b)$.
  (a) Show that E is an equivalence relation on A.
  (b) Define a function φ on A/E onto B by $\varphi([a]_E) = f(a)$ (verify that $\varphi([a]_E) = \varphi([a']_E)$ if $[a]_E = [a']_E$).
  (c) Let j be the function on A onto A/E given by $j(a) = [a]_E$. Show that $\varphi \circ j = f$.

4.3 Let $P = \{(r, \gamma) \in R \times R \mid r > 0\}$, where R is the set of all real numbers. View elements of P as polar coordinates of points in the plane, and define a relation on P by: $(r, \gamma) \sim (r', \gamma')$ if and only if $r = r'$ and $\gamma - \gamma'$ is an integer multiple of 2π. Show that $\sim$ is an equivalence relation on P. Show that each equivalence class contains a unique pair (r, γ) with $0 \leq \gamma < 2\pi$. The set of all such pairs is therefore a set of representatives for $\sim$.

5. Orderings

Orderings are another frequently encountered type of relation.

5.1 Definition A binary relation R in A is *antisymmetric* if for all $a, b \in A$, aRb and bRa imply $a = b$.

5.2 Definition A binary relation R in A which is reflexive, antisymmetric, and transitive is called a *(partial) ordering* of A. The pair (A, R) is called an *ordered set.*

aRb can be read as "a is less than or equal to b" or "b is greater than or equal to a" (in the ordering R). So, every element of A is less than or equal to itself. If a is less than or equal to b, and, at the same time, b is less than or equal to a, then $a = b$. Finally, if a is less than or equal to b and b is less than or equal to c, a has to be less than or equal to c.

5.3 Example
(a) $\leq$ is an ordering on the set of all (natural, rational, real) numbers.
(b) Define the relation $\subseteq_A$ in A as follows: $x \subseteq_A y$ if and only if $x \subseteq y$ and $x, y \in A$. Then $\subseteq_A$ is an ordering of the set A.
(c) Define the relation $\supseteq_A$ in A as follows: $x \supseteq_A y$ if and only if $x \supseteq y$ and $x, y \in A$. Then $\supseteq_A$ is also an ordering of the set A.
(d) The relation $\mid$ defined by: $n \mid m$ if and only if n divides m is an ordering of the set of all positive integers.
(e) The relation Id_A is an ordering of A.

The symbols $\leq$ or $\preccurlyeq$ are often used to denote orderings.

A different description of orderings is sometimes convenient. For example, instead of the relation $\leq$ between numbers, we might prefer to use the relation $<$ (strictly less). Similarly, we might use $\subset_A$ (proper subset) instead of $\subseteq_A$. Any ordering can be described in either one of these two mutually interchangeable ways.

5.4 Definition A relation S in A is *asymmetric* if aSb implies that bSa does not hold (for any $a, b \in A$). That is, aSb and bSa can never both be true.

5.5 Definition A relation S in A is a *strict ordering* if it is asymmetric and transitive.

We now establish relationships between orderings and strict orderings.

5.6 Theorem
(a) Let R be an ordering of A; then the relation S defined in A by

$$aSb \quad \text{if and only if} \quad aRb \text{ and } a \neq b$$

is a strict ordering of A.

(b) *Let S be a strict ordering of A; then the relation R defined in A by*

$$aRb \quad \textit{if and only if} \quad aSb \text{ or } a = b$$

is an ordering of A.

We say that the strict ordering S *corresponds to the ordering* R and vice versa.

Proof.

(a) Let us show that S is asymmetric: Assume that both aSb and bSa hold for some $a, b \in A$. Then also aRb and bRa, so $a = b$ (because R is antisymmetric). That contradicts the definition of aSb. We leave the verification of transitivity of S to the reader.

(b) Let us show that R is antisymmetric: Assume that aRb and bRa. Because aSb and bSa cannot hold simultaneously (S is asymmetric), we conclude that $a = b$. Reflexivity and transitivity of R are verified similarly.

□

5.7 Definition Let $a, b \in A$, and let $\leq$ be an ordering of A. We say that a and b are *comparable* in the ordering $\leq$ if $a \leq b$ or $b \leq a$. We say that a and b are *incomparable* if they are not comparable (i.e., if neither $a \leq b$ nor $b \leq a$ holds). Both definitions can be stated equivalently in terms of the corresponding strict ordering $<$; for example, a and b are incomparable in $<$ if $a \neq b$ and neither $a < b$ nor $b < a$ holds.

5.8 Example

(a) Any two real numbers are comparable in the ordering $\leq$.

(b) 2 and 3 are incomparable in the ordering $|$.

(c) Any two distinct $a, b \in A$ are incomparable in Id_A.

(d) If the set A has at least two elements, then there are incomparable elements in the ordered set $(\mathcal{P}(A), \subseteq_{\mathcal{P}(A)})$.

5.9 Definition An ordering $\leq$ (or $<$) of A is called *linear* or *total* if any two elements of A are comparable. The pair $(A, \leq)$ is then called a *linearly ordered set.*

So the ordering $\leq$ of positive integers is total, while $|$ is not.

5.10 Definition Let $B \subseteq A$, where A is ordered by $\leq$. B is a *chain in A* if any two elements of B are comparable.

For example, the set of all powers of 2 (i.e., $\{2^0, 2^1, 2^2, 2^3, \dots\}$) is a chain in the set of all positive integers ordered by $|$.

A problem that arises quite often is to find a least or greatest element among certain elements of an ordered set. Closer scrutiny reveals that there are several different notions of "least" and "greatest."

5.11 Definition Let $\leq$ be an ordering of A, and let $B \subseteq A$.

(a) $b \in B$ is the *least* element of B in the ordering $\leq$ if $b \leq x$ for every $x \in B$.

(b) $b \in B$ is a *minimal* element of B in the ordering $\leq$ if there exists no $x \in B$ such that $x \leq b$ and $x \neq b$.

(a′) Similarly, $b \in B$ is the *greatest* element of B in the ordering $\leq$ if, for every $x \in B$, $x \leq b$.

(b′) $b \in B$ is a *maximal* element of B in the ordering $\leq$ if there exists no $x \in B$ such that $b \leq x$ and $x \neq b$.

5.12 Example Let N be the set of positive integers ordered by divisibility relation $|$. Then 1 is the least element of N, but N has no greatest element. Let B be the set of all positive integers greater (in magnitude) than 1, $B = \{2, 3, 4, \ldots\}$. Then B does not have a least element in $|$ (e.g., 2 is not the least element because $2 \mid 3$ fails), but it has (infinitely) many minimal elements: numbers 2, 3, 5, etc. (exactly all prime numbers) are minimal. B has neither greatest nor maximal elements.

We list some of the properties of least and minimal elements in Theorem 5.13. The proof is left as an exercise.

5.13 Theorem *Let A be ordered by $\leq$, and let $B \subseteq A$.*

(a) B has at most one least element.

(b) The least element of B (if it exists) is also minimal.

(c) If B is a chain, then every minimal element of B is also least.

The theorem remains true if the words "least" and "minimal" are replaced by "greatest" and "maximal", respectively.

5.14 Definition Let $\leq$ be an ordering of A, and let $B \subseteq A$.

(a) $a \in A$ is a *lower bound* of B in the ordered set $(A, \leq)$ if $a \leq x$ for all $x \in B$.

(b) $a \in A$ is called an *infimum* of B in $(A, \leq)$ [or the *greatest lower bound* of B in $(A, \leq)$] if it is the greatest element of the set of all lower bounds of B in $(A, \leq)$.

Similarly,

(a′) $a \in A$ is an *upper bound* of B in the ordered set $(A, \leq)$ if $x \leq a$ for all $x \in B$.

(b′) $a \in A$ is called a *supremum* of B in $(A, \leq)$ [or the *least upper bound* of B in $(A, \leq)$] if it is the least element of the set of all upper bounds of B in $(A, \leq)$.

Note that the difference between the least element of B and a lower bound of B is that the second notion does not require $b \in B$. A set can have many lower bounds. But the set of all lower bounds of B can have at most one greatest element, so B can have at most one infimum.

We now summarize some properties of suprema and infima.

5.15 Theorem *Let $(A, \leq)$ be an ordered set and let $B \subseteq A$.*
(a) B has at most one infimum.
(b) If b is the least element of B, then b is the infimum of B.
(c) If b is the infimum of B and $b \in B$, then b is the least element of B.
(d) $b \in A$ is an infimum of B in $(A, \leq)$ if and only if

(i) $b \leq x$ for all $x \in B$.

(ii) If $b' \leq x$ for all $x \in B$, then $b' \leq b$.

The theorem remains true if the words "least" and "infimum" are replaced by the words "greatest" and "supremum" and "$\leq$" is replaced by "$\geq$" in (i) and (ii).

Proof.
(a) We proved this in the remark preceding the theorem.
(b) The least element b of B is certainly a lower bound of B. If b' is any lower bound of B, $b' \leq b$ because $b \in B$. So b is the greatest element of the set of all lower bounds of B.
(c) This is obvious.
(d) This is only a reformulation of the definition of infimum.

□

We use notations $\inf(B)$ and $\sup(B)$ for the infimum of B and the supremum of B, if they exist. If B is linearly ordered, we also use $\min(B)$ and $\max(B)$ to denote the minimal (least) and the maximal (greatest) elements of B, if they exist.

5.16 Example Let $\leq$ be the usual ordering of the set of real numbers; let $B_1 = \{x \mid 0 < x < 1\}$, $B_2 = \{x \mid 0 \leq x < 1\}$, $B_3 = \{x \mid x > 0\}$, and $B_4 = \{x \mid x < 0\}$. Then B_1 has no least element and no greatest element, but any $b \leq 0$ is a lower bound of B_1, so 0 is the greatest lower bound of B_1; i.e., $0 = \inf(B)$. Similarly, any $b \geq 1$ is an upper bound of B_1, so $1 = \sup(B_1)$. The set B_2 has a least element; so $0 = \min(B_2) = \inf(B_2)$; it does not have a greatest element. Nevertheless, $\sup(B_2) = 1$. The set B_3 has neither a greatest element nor a supremum (actually B_3 has no upper bound in $\leq$); of course, $\inf(B_3) = 0$. Similarly, B_4 has no lower bounds, hence no infimum.

5.17 Definition An *isomorphism* between two ordered sets $(P, <)$ and $(Q, \prec)$ is a one-to-one function h with domain P and range Q such that for all $p_1, p_2 \in P$

$$p_1 < p_2 \quad \text{if and only if} \quad h(p_1) \prec h(p_2).$$

If an isomorphism exists between $(P, <)$ and $(Q, \prec)$, then $(P, <)$ and $(Q, \prec)$ are *isomorphic*.

We study isomorphisms in a more general setting in Chapter 3. At this point, let us make the following observation.

5.18 Lemma *Let $(P, <)$ and $(Q, \prec)$ be linearly ordered sets, and let h be a one-to-one function with domain P and range Q such that $h(p_1) \prec h(p_2)$ whenever $p_1 < p_2$. Then h is an isomorphism between $(P, <)$ and $(Q, \prec)$.*

Proof. We have to verify that if $p_1, p_2 \in P$ are such that $h(p_1) \prec h(p_2)$, then $p_1 < p_2$. But if p_1 is not less than p_2, then, because $<$ is a linear ordering of P, either $p_1 = p_2$ or $p_2 < p_1$. If $p_1 = p_2$, then $h(p_1) = h(p_2)$, and if $p_2 < p_1$, then $h(p_2) \prec h(p_1)$, by the assumption. Either case contradicts $h(p_1) \prec h(p_2)$. □

Exercises

5.1 (a) Let R be an ordering of A, S be the corresponding strict ordering of A, and R^* be the ordering corresponding to S. Show that $R^* = R$.
(b) Let S be a strict ordering of A, R be the corresponding ordering, and S^* be the strict ordering corresponding to R. Then $S^* = S$.

5.2 State the definitions of incomparable elements, maximal, minimal, greatest, and least elements and suprema and infima in terms of strict orderings.

5.3 Let R be an ordering of A. Prove that R^{-1} is also an ordering of A, and for $B \subseteq A$,
(a) a is the least element of B in R^{-1} if and only if a is the greatest element of B in R;
(b) similarly for (minimal and maximal) and (supremum and infimum).

5.4 Let R be an ordering of A and let $B \subseteq A$. Show that $R \cap B^2$ is an ordering of B.

5.5 Give examples of a finite ordered set $(A, \leq)$ and a subset B of A so that
(a) B has no greatest element.
(b) B has no least element.
(c) B has no greatest element, but B has a supremum.
(d) B has no supremum.

5.6 (a) Let $(A, <)$ be a strictly ordered set and $b \notin A$. Define a relation $\prec$ in $B = A \cup \{b\}$ as follows:

$$x \prec y \quad \text{if and only if} \quad (x, y \in A \text{ and } x < y) \text{ or } (x \in A \text{ and } y = b).$$

Show that $\prec$ is a strict ordering of B and $\prec \cap A^2 = <$. (Intuitively, $\prec$ keeps A ordered in the same way as $<$ and makes b greater than every element of A.)
(b) Generalize part (a): Let $(A_1, <_1)$ and $(A_2, <_2)$ be strict orderings, $A_1 \cap A_2 = \emptyset$. Define a relation $\prec$ on $B = A_1 \cup A_2$ as follows:

$$\begin{aligned} x \prec y \quad \text{if and only if} \quad & x, y \in A_1 \text{ and } x <_1 y \\ \text{or} \quad & x, y \in A_2 \text{ and } x <_2 y \\ \text{or} \quad & x \in A_1 \text{ and } y \in A_2. \end{aligned}$$

Show that $\prec$ is a strict ordering of B and $\prec \cap A_1^2 = <_1$, $\prec \cap A_2^2 = <_2$. (Intuitively, $\prec$ puts every element of A_1 before every element of A_2 and coincides with the original orderings of A_1 and A_2.)

5.7 Let R be a reflexive and transitive relation in A (R is called a *preordering* of A). Define E in A by

$$aEb \quad \text{if and only if} \quad aRb \text{ and } bRa.$$

Show that E is an equivalence relation on A. Define the relation R/E in A/E by

$$[a]_E \, R/E \, [b]_E \quad \text{if and only if} \quad aRb.$$

Show that the definition does not depend on the choice of representatives for $[a]_E$ and $[b]_E$. Prove that R/E is an ordering of A/E.

5.8 Let $A = \mathcal{P}(X)$, $X \neq \emptyset$. Prove:
(a) Any $S \subseteq A$ has a supremum in the ordering $\subseteq_A$; $\sup S = \bigcup S$.
(b) Any $S \subseteq A$ has an infimum in $\subseteq_A$; $\inf S = \bigcap S$ if $S \neq \emptyset$; $\inf \emptyset = X$.

5.9 Let $\mathrm{Fn}(X, Y)$ be the set of all functions mapping a subset of X into Y [i.e., $\mathrm{Fn}(X, Y) = \bigcup_{Z \subseteq X} Y^Z$]. Define a relation $\leq$ in $\mathrm{Fn}(X, Y)$ by

$$f \leq g \quad \text{if and only if} \quad f \subseteq g.$$

(a) Prove that $\leq$ is an ordering of $\mathrm{Fn}(X, Y)$.
(b) Let $F \subseteq \mathrm{Fn}(X, Y)$. Show that $\sup F$ exists if and only if F is a compatible system of functions; then $\sup F = \bigcup F$.

5.10 Let $A \neq \emptyset$; let $\mathrm{Pt}(A)$ be the set of all partitions of A. Define a relation $\preccurlyeq$ in $\mathrm{Pt}(A)$ by

$$S_1 \preccurlyeq S_2 \text{ if and only if for every } C \in S_1 \text{ there is } D \in S_2 \text{ such that } C \subseteq D.$$

(We say that the partition S_1 is a *refinement* of the partition S_2 if $S_1 \preccurlyeq S_2$ holds.)
(a) Show that $\preccurlyeq$ is an ordering.
(b) Let $S_1, S_2 \in \mathrm{Pt}(A)$. Show that $\{S_1, S_2\}$ has an infimum. [*Hint*: Define $S = \{C \cap D \mid C \in S_1 \text{ and } D \in S_2\}$.] How is the equivalence relation E_S related to the equivalences E_{S_1} and E_{S_2}?
(c) Let $T \subseteq \mathrm{Pt}(A)$. Show that $\inf T$ exists.
(d) Let $T \subseteq \mathrm{Pt}(A)$. Show that $\sup T$ exists. [*Hint*: Let T' be the set of all partitions S with the property that every partition from T is a refinement of S. Show that $\sup T' = \inf T$.]

5.11 Show that if $(P, <)$ and $(Q, \prec)$ are isomorphic strictly ordered sets and $<$ is a linear ordering, then $\prec$ is a linear ordering.

5.12 The identity function on P is an isomorphism between $(P, <)$ and $(P, <)$.

5.13 If h is isomorphism between $(P, <)$ and $(Q, \prec)$, then h^{-1} is an isomorphism between $(Q, \prec)$ and $(P, <)$.

5.14 If f is an isomorphism between $(P_1, <_1)$ and $(P_2, <_2)$, and if g is an isomorphism between $(P_2, <_2)$ and $(P_3, <_3)$, then $g \circ f$ is an isomorphism between $(P_1, <_1)$ and $(P_3, <_3)$.

Chapter 3

Natural Numbers

1. Introduction to Natural Numbers

In order to develop mathematics within the framework of the axiomatic set theory, it is necessary to define natural numbers. We all know natural numbers intuitively: 0, 1, 2, 3, ... , 17, ... , 324, etc., and we can easily give examples of sets having zero, one, two, or three elements:

$\emptyset$ has 0 elements.

$\{\emptyset\}$ or, in general, $\{a\}$ for any a, has one element.

$\{\emptyset, \{\emptyset\}\}$, or $\{\{\{\emptyset\}\}, \{\{\{\emptyset\}\}\}\}$, or, in general, $\{a, b\}$ where $a \neq b$, has two elements, etc.

The purpose of the investigations in this section is to supplement this intuitive understanding by a rigorous definition.

To define number 0, we choose a representative of all sets having no elements. But this is easy, since there is only one such set. We define $0 = \emptyset$. Let us proceed to sets having one element (singletons): $\{\emptyset\}$, $\{\{\emptyset\}\}$, $\{\{\emptyset, \{\emptyset\}\}\}$; in general, $\{x\}$. How should we choose a representative? Since we already defined one particular object, namely 0, a natural choice is $\{0\}$. So we define

$$1 = \{0\} = \{\emptyset\}.$$

Next we consider sets with two elements: $\{\emptyset, \{\emptyset\}\}$, $\{\{\emptyset\}, \{\emptyset, \{\emptyset\}\}\}$, $\{\{\emptyset\}, \{\{\emptyset\}\}\}$, etc. By now, we have defined 0 and 1, and $0 \neq 1$. We single out a particular two-element set, the set whose elements are the previously defined numbers 0 and 1:

$$2 = \{0, 1\} = \{\emptyset, \{\emptyset\}\}.$$

It should begin to be obvious how the process continues:

$$3 = \{0, 1, 2\} = \{\emptyset, \{\emptyset\}, \{\emptyset, \{\emptyset\}\}\}$$
$$4 = \{0, 1, 2, 3\} = \{\emptyset, \{\emptyset\}, \{\emptyset, \{\emptyset\}\}, \{\emptyset, \{\emptyset\}, \{\emptyset, \{\emptyset\}\}\}\}$$
$$5 = \{0, 1, 2, 3, 4\} \text{ etc.}$$

The idea is simply to define a natural number n as the set of all smaller natural numbers: $\{0, 1, \ldots, n-1\}$. In this way, n is a particular set of n elements.

This idea still has a fundamental deficiency. We have defined 0, 1, 2, 3, 4, and 5 and could easily define 17 and—not so easily—324. But no list of such definitions tells us what a natural number is in general. We need a statement of the form: A set n is a natural number if .... We cannot just say that a set n is a natural number if its elements are all the smaller natural numbers, because such a "definition" would involve the very concept being defined.

Let us observe the construction of the first few numbers again. We defined $2 = \{0, 1\}$. To get 3, we had to adjoin a third element to 2, namely, 2 itself:

$$3 = 2 \cup \{2\} = \{0, 1\} \cup \{2\}.$$

Similarly,

$$4 = 3 \cup \{3\} = \{0, 1, 2\} \cup \{3\},$$
$$5 = 4 \cup \{4\}, \text{ etc.}$$

Given a natural number n, we get the "next" number by adjoining one more element to n, namely, n itself. The procedure works even for 1 and 2: $1 = 0 \cup \{0\}$, $2 = 1 \cup \{1\}$, but, of course, not for 0, the least natural number.

These considerations suggest the following.

1.1 Definition The *successor* of a set x is the set $S(x) = x \cup \{x\}$.

Intuitively, the successor $S(n)$ of a natural number n is the "one bigger" number $n+1$. We use the more suggestive notation $n+1$ for $S(n)$ in what follows. We later define addition of natural numbers (using the notion of successor) in such a way that $n + 1$ indeed equals the sum of n and 1. Until then, it is just a notation, and no properties of addition are assumed or implied by it.

We can now summarize the intuitive understanding of natural numbers as follows:

(a) 0 is a natural number.
(b) If n is a natural number, then its successor $n + 1$ is also a natural number.
(c) All natural numbers are obtained by application of (a) and (b), i.e., by starting with 0 and repeatedly applying the successor operation: 0, $0 + 1 = 1$, $1 + 1 = 2$, $2 + 1 = 3$, $3 + 1 = 4$, $4 + 1 = 5$, ... etc.

1.2 Definition A set I is called *inductive* if
(a) $0 \in I$.
(b) If $n \in I$, then $(n + 1) \in I$.

An inductive set contains 0 and, with each element, also its successor. According to (c), an inductive set should contain all natural numbers. The precise meaning of (c) is that the set of natural numbers is an inductive set which contains no other elements but natural numbers, i.e., it is the *smallest* inductive set. This leads to the following definition.

1.3 Definition *The set of all natural numbers* is the set

$$\boldsymbol{N} = \{x \mid x \in I \text{ for every inductive set } I\}.$$

The elements of $\boldsymbol{N}$ are called *natural numbers.* Thus a set x is a natural number if and only if it belongs to every inductive set.

We have to justify the existence of $\boldsymbol{N}$ on the basis of the Axiom of Comprehension (see the remarks following Example 3.12 in Chapter 1), but that is easy. Let A be any particular inductive set; then clearly

$$\boldsymbol{N} = \{x \in A \mid x \in I \text{ for every inductive set } I\}.$$

The only remaining question is whether there are any inductive sets at all. The intuitive answer is, of course, yes: the set of natural numbers is a prime example. But a careful look at the axioms we have adopted so far indicates that existence of infinite sets (such as $\boldsymbol{N}$) cannot be proved from them. Roughly, the reason is that these axioms have a general form:

"For every set X, there exists a set Y such that ... ,"

where, if the set X is finite, the set Y is also finite. Since the only set whose existence we postulated outright is $\emptyset$, which is finite, all the other sets whose existence is required by the axioms are also finite. (See Section 2 of Chapter 4 for a more rigorous elaboration of these remarks.) The point is that we need another axiom.

The Axiom of Infinity. An inductive set exists.

Some mathematicians object to the Axiom of Infinity on the grounds that a collection of objects produced by an infinite process (such as $\boldsymbol{N}$) should not be treated as a completed entity. However, most people with some mathematical training have no difficulty visualizing the collection of natural numbers in that way. Infinite sets are basic tools of modern mathematics and the essence of set theory. No contradiction resulting from their use has ever been discovered in spite of the enormous body of research founded on them. Therefore, we treat the Axiom of Infinity on a par with our other axioms.

We now have the set of natural numbers $\boldsymbol{N}$ at our disposal. Before proceeding further, let us check that the set $\boldsymbol{N}$ is indeed inductive.

1.4 Lemma *$\boldsymbol{N}$ is inductive. If I is any inductive set, then $\boldsymbol{N} \subseteq I$.*

Proof. $0 \in \boldsymbol{N}$ because $0 \in I$ for any inductive I.

If $n \in \boldsymbol{N}$, then $n \in I$ for any inductive I, so $(n+1) \in I$ for any inductive I, and consequently $(n+1) \in \boldsymbol{N}$. This shows that $\boldsymbol{N}$ is inductive. The second part of the Lemma follows immediately from the definition of $\boldsymbol{N}$. □

The next step is to define the ordering of natural numbers by size. As it is our guiding idea to define each natural number as a set of smaller natural numbers, we are led to the following.

1.5 Definition The relation $<$ on $\boldsymbol{N}$ is defined by: $m < n$ if and only if $m \in n$.

It is of course necessary to prove that $<$ is indeed a linear ordering and that the ordered set $(\boldsymbol{N}, <)$ actually has the properties that we expect natural numbers to have. The needed theory is developed in the rest of this chapter.

Exercises

1.1 $x \subseteq S(x)$ and there is no z such that $x \subset z \subset S(x)$.

2. Properties of Natural Numbers

In the preceding section we defined the set $\boldsymbol{N}$ of natural numbers to be the least set such that (a) $0 \in \boldsymbol{N}$ and (b) if $n \in \boldsymbol{N}$, then $(n+1) \in \boldsymbol{N}$. We also defined $m < n$ to mean $m \in n$. It is the goal of this section to show that the above-mentioned concepts really behave in the familiar ways.

We begin with a fundamental tool for study of natural numbers, the well-known principle of proof by mathematical induction.

The Induction Principle. Let $\mathbf{P}(x)$ be a property (possibly with parameters). Assume that
(a) $\mathbf{P}(0)$ holds.
(b) For all $n \in \boldsymbol{N}$, $\mathbf{P}(n)$ implies $\mathbf{P}(n+1)$.
Then $\mathbf{P}$ holds for all natural numbers n.

Proof. This is an immediate consequence of our definition of $\boldsymbol{N}$. The assumptions (a) and (b) simply say that the set $A = \{n \in \boldsymbol{N} \mid \mathbf{P}(n)\}$ is inductive. $\boldsymbol{N} \subseteq A$ follows. □

The following lemma establishes two simple properties of natural numbers and gives an example of a simple proof by induction.

2.1 Lemma
(i) $0 \leq n$ *for all* $n \in \boldsymbol{N}$.
(ii) For all $k, n \in \boldsymbol{N}$, $k < n+1$ *if and only if* $k < n$ *or* $k = n$.

Proof. (i) We let $\mathbf{P}(x)$ to be the property "$0 \leq x$" and proceed to establish the assumptions of the Induction Principle.
(a) $\mathbf{P}(0)$ holds. $\mathbf{P}(0)$ is the statement "$0 \leq 0$," which is certainly true $(0 = 0)$.
(b) $\mathbf{P}(n)$ implies $\mathbf{P}(n+1)$. Let us assume that $\mathbf{P}(n)$ holds, i.e., $0 \leq n$. This means, by definition of $<$, that $0 = n$ or $0 \in n$. In either case, $0 \in n \cup \{n\} = n+1$, so $0 < (n+1)$ and $\mathbf{P}(n+1)$ holds.

Having proved (a) and (b) we use the Induction Principle to conclude that $\mathbf{P}(n)$ holds for all $n \in \boldsymbol{N}$, i.e., $0 \leq n$ for all $n \in \boldsymbol{N}$.

(ii) This part does not require induction. It suffices to observe that $k \in n \cup \{n\}$ if and only if $k \in n$ or $k = n$. □

The proof of the next theorem provides several other, somewhat more complicated examples of inductive proofs.

2.2 Theorem *$(\boldsymbol{N}, <)$ is a linearly ordered set.*

Proof.

(i) *The relation $<$ is transitive on $\boldsymbol{N}$.*

We have to prove that for all $k, m, n \in \boldsymbol{N}$, $k < m$ and $m < n$ imply $k < n$. We proceed by induction on n; i.e., we use as $\mathbf{P}(x)$ the property "for all $k, m \in \boldsymbol{N}$, if $k < m$ and $m < x$, then $k < x$."

(a) $\mathbf{P}(0)$ holds. $\mathbf{P}(0)$ asserts: for all $k, m \in \boldsymbol{N}$, if $k < m$ and $m < 0$, then $k < 0$. By Lemma 2.1(i), there is no $m \in \boldsymbol{N}$ such that $m < 0$, so $\mathbf{P}(0)$ is trivially true.

(b) Assume $\mathbf{P}(n)$, i.e., assume that for all $k, m \in \boldsymbol{N}$, if $k < m$ and $m < n$, then $k < n$. We have to prove $\mathbf{P}(n+1)$, i.e., we have to show that $k < m$ and $m < (n+1)$ imply $k < (n+1)$. But if $k < m$ and $m < (n+1)$, then by Lemma 2.1(ii) $m < n$ or $m = n$. If $m < n$, we get $k < n$ by the inductive assumption $\mathbf{P}(n)$. If $m = n$, we have $k < n$ from $k < m$. In either case, $k < n+1$ by Lemma 2.1(ii). This establishes $\mathbf{P}(n+1)$.

The Induction Principle now asserts the validity of $\mathbf{P}(n)$ for all $n \in \boldsymbol{N}$; this is precisely the statement of transitivity of $(\boldsymbol{N}, <)$.

(ii) *The relation $<$ is asymmetric on $\boldsymbol{N}$.*

Assume that $n < k$ and $k < n$. By transitivity, this implies $n < n$. So we only have to show that the latter is impossible. We proceed by induction. Clearly, $0 < 0$ is impossible (it would mean that $\emptyset \in \emptyset$). Let us assume that $n < n$ is false and prove that $(n+1) < (n+1)$ is false. If $(n+1) < (n+1)$ were true, we would have either $n+1 < n$ or $n+1 = n$ [Lemma 2.1(ii)]. Since $n < n+1$ holds by Lemma 2.1(ii) and we have proved transitivity of $<$ previously, we conclude that $n < n$, thus contradicting our inductive assumption (to wit, that $n < n$ is false). We have now established both (a) and (b) in the Induction Principle [with $\mathbf{P}(x)$ being "$x < x$ is false"]. We can conclude that $n < n$ is impossible for any $n \in \boldsymbol{N}$. We now know that $<$ is a (strict) ordering of $\boldsymbol{N}$.

It remains to prove

(iii) *$<$ is a linear ordering of $\boldsymbol{N}$.*

We have to prove that for all $m, n \in \boldsymbol{N}$ either $m < n$ or $m = n$ or $n < m$. We proceed by induction on n.

(a) For all $m \in \boldsymbol{N}$, either $m < 0$ or $m = 0$ or $0 < m$. This follows immediately from Lemma 2.1(i).

(b) Assume that for all $m \in \boldsymbol{N}$, either $m < n$ or $m = n$ or $n < m$. We have to prove an analogous statement with $(n+1)$ in place of n. If $m < n$, then $m < (n+1)$ by Lemma 2.1(ii) and transitivity. Similarly, if $m = n$ then $m < (n+1)$. Finally, if $n < m$, we would like to conclude that $n+1 \leq m$. This would show that, for all $m \in \boldsymbol{N}$, either $m < (n+1)$ or $m = (n+1)$ or $(n+1) < m$, establishing (b), and completing the proof. So we prove that if $n < m$, then $(n+1) \leq m$ holds for all $m \in \boldsymbol{N}$ by induction on m [n is a parameter; that is, we are going to apply the Induction Principle to the property $\mathbf{P}(x)$: "If $n < x$, then $n + 1 \leq x$"]. If $m = 0$, the statement "if $n < 0$, then $n+1 \leq 0$" is true (since its assumption must be false). Assume $\mathbf{P}(m)$, i.e., if $n < m$, then $(n + 1) \leq m$. To prove $\mathbf{P}(m + 1)$, assume that $n < m + 1$; then $n < m$ or $n = m$. If $n < m$, $n + 1 \leq m$ by the inductive assumption, and so $n + 1 < m + 1$. If $n = m$ then of course $n + 1 = m + 1$. In either case $\mathbf{P}(m + 1)$ is proved. We conclude that $\mathbf{P}(m)$ holds for all $m \in \boldsymbol{N}$, as needed.

Finally, we finish the proof of (iii) by observing that the assumptions (a) and (b) of the Induction Principle have now been established. □

The reader should study the preceding proof carefully for its varied applications of the Induction Principle. In particular, the proof of part (iii) is an example of "double induction": In order to prove a statement depending on two variables, m and n, we proceed by induction on one of them (n); the proof of the induction assumption (b) then in itself requires induction on the other variable m (for fixed n). See Exercise 2.13.

Before proceeding further, we state and prove another version of the Induction Principle that is often more convenient.

The Induction Principle, Second Version. Let $\mathbf{P}(x)$ be a property (possibly with parameters). Assume that, for all $n \in \boldsymbol{N}$,

(*) If $\mathbf{P}(k)$ holds for all $k < n$, then $\mathbf{P}(n)$.

Then $\mathbf{P}$ holds for all natural numbers n.

In other words, in order to prove $\mathbf{P}(n)$ for all $n \in \boldsymbol{N}$, it suffices to prove $\mathbf{P}(n)$ (for all $n \in \boldsymbol{N}$) under the assumption that it holds for all smaller natural numbers.

Proof. Assume that (*) is true. Consider the property $\mathbf{Q}(n)$: $\mathbf{P}(k)$ holds for all $k < n$. Clearly $\mathbf{Q}(0)$ is true (there are no $k < 0$). If $\mathbf{Q}(n)$ holds, then $\mathbf{Q}(n + 1)$ holds: If $\mathbf{Q}(n)$ holds, then $\mathbf{P}(k)$ holds for all $k < n$, and consequently also for $k = n$ [by (*)]. Lemma 2.1(ii) enables us to conclude that $\mathbf{P}(k)$ holds for all $k < n + 1$, and therefore $\mathbf{Q}(n + 1)$ holds. By the Induction Principle, $\mathbf{Q}(n)$ is true for all $n \in \boldsymbol{N}$; since for $k \in \boldsymbol{N}$ there is some $n > k$ (e.g., $n = k+1$), we have $\mathbf{P}(k)$ true for all $k \in \boldsymbol{N}$, as desired. □

The ordering of natural numbers by size has an additional important property that distinguishes it from, say, ordering of integers or rational numbers by size.

2.3 Definition A linear ordering $\prec$ of a set A is a *well-ordering* if every nonempty subset of A has a least element. The ordered set $(A, \prec)$ is then called a *well-ordered set*.

Well-ordered sets form a backbone of set theory and we study them extensively in Chapter 6. At this point, we prove only:

2.4 Theorem $(\boldsymbol{N}, <)$ *is a well-ordered set.*

Proof. Let X be a nonempty subset of $\boldsymbol{N}$; we have to show that X has a least element. So let us assume that X does not have a least element and let us consider $\boldsymbol{N} - X$. The crucial step is to observe that if $k \in \boldsymbol{N} - X$ for all $k < n$, then $n \in \boldsymbol{N} - X$: otherwise, n would be the least element of X. By the second version of the Induction Principle we conclude that $n \in \boldsymbol{N} - X$ holds for all natural numbers n [$\mathbf{P}(x)$ is the property "$x \in \boldsymbol{N} - X$"] and therefore that $X = \emptyset$, contradicting our initial assumption. □

We conclude this section with another property of the ordering $<$.

2.5 Theorem *If a nonempty set of natural numbers has an upper bound in the ordering $<$, then it has a greatest element.*

Proof. Let $A \subseteq \boldsymbol{N}$, $A \neq \emptyset$ be given; let $B = \{k \in \boldsymbol{N} | k$ is an upper bound of $A\}$; we assume that $B \neq \emptyset$.

By Theorem 2.4, B has a least element n, so $n = \sup(A)$. The proof is completed by showing that $n \in A$. [See Theorem 5.15(c) in Chapter 2, with "supremum" in place of "infimum," etc.] Trivial induction proves that either $n = 0$ or $n = k + 1$ for some $k \in \boldsymbol{N}$ (see Exercise 2.4). Assume that $n \notin A$; we then have $n > m$ for all $m \in A$. Since $A \neq \emptyset$, it means that $n \neq 0$. Therefore, $n = k + 1$ for some $k \in \boldsymbol{N}$, which gives $k \geq m$ for all $m \in A$ [Lemma 2.1(ii) again!]. Thus k is an upper bound of A and $k < n$, a contradiction. □

Exercises

2.1 Let $n \in \boldsymbol{N}$. Prove that there is no $k \in \boldsymbol{N}$ such that $n < k < n + 1$.

2.2 Use Exercise 2.1 to prove for all $m, n \in \boldsymbol{N}$: if $m < n$, then $m + 1 \leq n$. Conclude that $m < n$ implies $m + 1 < n + 1$ and that therefore the successor $S(n) = n + 1$ defines a one-to-one function on $\boldsymbol{N}$.

2.3 Prove that there is a one-to-one mapping of $\boldsymbol{N}$ onto a proper subset of $\boldsymbol{N}$. [*Hint*: Use Exercise 2.2.]

2.4 For every $n \in \boldsymbol{N}$, $n \neq 0$, there is a unique $k \in \boldsymbol{N}$ such that $n = k + 1$.

2.5 For every $n \in \boldsymbol{N}$, $n \neq 0, 1$, there is a unique $k \in \boldsymbol{N}$ such that $n = (k + 1) + 1$.

2.6 Prove that each natural number is the set of all smaller natural numbers, i.e.,

$$n = \{m \in \boldsymbol{N} \mid m < n\}.$$

[*Hint*: Use induction to prove that all elements of a natural number are natural numbers.]

2.7 For all $m, n \in \boldsymbol{N}$

$$m < n \quad \text{if and only if} \quad m \subset n.$$

2.8 Prove that there is no function $f : N \to N$ such that for all $n \in N$, $f(n) > f(n+1)$. (There is no infinite decreasing sequence of natural numbers.)

2.9 If $X \subseteq N$, then $\langle X, < \cap X^2 \rangle$ is well-ordered.

2.10 In Exercise 5.6 of Chapter 2, let $A = N$, $b = N$. Prove that $\prec$ as defined there is a well-ordering of $B = N \cup \{N\}$. Notice that $x \prec y$ if and only if $x \in y$ holds for all $x, y \in B$.

2.11 Let $\mathbf{P}(x)$ be a property. Assume that $k \in N$ and

(a) $\mathbf{P}(k)$ holds.

(b) For all $n \geq k$, if $\mathbf{P}(n)$ then $\mathbf{P}(n+1)$.

Then $\mathbf{P}(n)$ holds for all $n \geq k$.

2.12 (Finite Induction Principle) Let $\mathbf{P}(x)$ be a property. Assume that $k \in N$ and

(a) $\mathbf{P}(0)$.

(b) For all $n < k$, $\mathbf{P}(n)$ implies $\mathbf{P}(n+1)$.

Then $\mathbf{P}(n)$ holds for all $n \leq k$.

2.13 (Double Induction) Let $\mathbf{P}(x, y)$ be a property. Assume

(**) If $\mathbf{P}(k, l)$ holds for all $k, l \in N$ such that $k < m$ or ($k = m$ and $l < n$), then $\mathbf{P}(m, n)$ holds.

Conclude that $\mathbf{P}(m, n)$ holds for all $m, n \in N$.

3. The Recursion Theorem

Our next task is to show how to define addition, multiplication, and other familiar operations of arithmetic. To facilitate this, we develop an important general method for defining functions on N.

We begin with some new terminology. A *sequence* is a function whose domain is either a natural number or N. A sequence whose domain is some natural number $n \in N$ is called a *finite sequence of length* n and is denoted

$$\langle a_i \mid i < n \rangle \quad \text{or} \quad \langle a_i \mid i = 0, 1, \ldots, n-1 \rangle \quad \text{or} \quad \langle a_0, a_1, \ldots, a_{n-1} \rangle.$$

In particular, $\langle \rangle (= \emptyset)$ is the unique sequence of length 0, the *empty sequence.* $\operatorname{Seq}(A) = \bigcup_{n \in N} A^n$ denotes the set of all finite sequences of elements of A. (Prove that it exists!) If the domain of a sequence is N, we call it an *infinite sequence* and denote it

$$\langle a_i \mid i \in N \rangle \quad \text{or} \quad \langle a_i \mid i = 0, 1, 2 \ldots \rangle \quad \text{or} \quad \langle a_i \rangle_{i=0}^{\infty}.$$

So infinite sequences of elements of A are just members of A^N. The notation simply specifies a function with an appropriate domain, whose value at i is a_i. We also use the notation $\{a_i \mid i \in N\}$, $\{a_i\}_{i=0}^{\infty}$, etc., for the range of the sequence $\langle a_i \mid i \in N \rangle$. Similarly, $\{a_i \mid i < n\}$ or $\{a_0, a_1, \ldots, a_{n-1}\}$ denotes the range of $\langle a_i \mid i < n \rangle$.

Let us now consider two examples of infinite sequences.

(a) The sequence $s : \boldsymbol{N} \to \boldsymbol{N}$ is defined by

$$s_0 = 1;$$
$$s_{n+1} = n^2 \quad \text{for all } n \in \boldsymbol{N}.$$

(b) The sequence $f : \boldsymbol{N} \to \boldsymbol{N}$ is defined by

$$f_0 = 1;$$
$$f_{n+1} = f_n \times (n+1) \quad \text{for all } n \in \boldsymbol{N}.$$

The two definitions, in spite of superficial similarity, exhibit a crucial difference. The definition of s gives *explicit* instructions how to compute s_x for any $x \in \boldsymbol{N}$. More precisely, it enables us to formulate a property $\mathbf{P}$ such that

$$s_x = y \quad \text{if and only if} \quad \mathbf{P}(x, y);$$

namely, let $\mathbf{P}$ be "either $x = 0$ and $y = 1$ or, for some $n \in \boldsymbol{N}$, $x = n+1$ and $y = n^2$." The existence and uniqueness of a sequence s satisfying (a) then immediately follows from our axioms:

$$s = \{(x, y) \in \boldsymbol{N} \times \boldsymbol{N} \mid \mathbf{P}(x, y)\}.$$

In contrast, the instructions supplied by the definition of f tell us only how to compute f_x *provided* that the value of f for some smaller number (namely, $x - 1$) was already computed. It is not immediately obvious how to formulate a property $\mathbf{P}$, not involving the function f being defined, such that

$$f_x = y \quad \text{if and only if} \quad \mathbf{P}(x, y).$$

We might view the definition (b) as giving *conditions* the sequence f ought to satisfy: "f is a function on $\boldsymbol{N}$ to $\boldsymbol{N}$ which satisfies the 'initial condition': $f_0 = 1$, and the 'recursive condition': for all $n \in \boldsymbol{N}$, $f_{n+1} = f_n \times (n+1)$."

Such definitions are widely used in mathematics; the reader might wish to draw a parallel, e.g., with the implicit definitions of functions in calculus. However, a definition of this kind is justified only if it is possible to show that there exists some function satisfying the required conditions, and that there do not exist two or more such functions. In calculus, this is provided for by the Implicit Function Theorem. We now state and prove an analogous result for our situation.

The Recursion Theorem For any set A, any $a \in A$, and any function $g : A \times \boldsymbol{N} \to A$, there exists a unique infinite sequence $f : \boldsymbol{N} \to A$ such that
(a) $f_0 = a$;
(b) $f_{n+1} = g(f_n, n)$ for all $n \in \boldsymbol{N}$.

In Example (b), we had $A = \boldsymbol{N}$, $a = 1$, and $g(u, v) = u \times (v + 1)$. The set a is the "initial value" of f. The role of g is to provide instructions for computing f_{n+1} assuming f_n has been already computed.

The proof of the Recursion Theorem consists of devising an explicit definition of f. Consider again Example (b); then f_n is the n-factorial; and an explicit definition of f can be readily written down:

$$f_0 = 1 \quad \text{and} \quad f_m = 1 \times 2 \times \cdots \times (m-1) \times m \text{ if } m \neq 0 \text{ and } m \in \boldsymbol{N}.$$

The problem is in making "$\cdots$" precise. It can be resolved by stating that f_m is the *result of a computation*

$$\begin{array}{l} 1 \\ 1 \times 1 \\ [1 \times 1] \times 2 \\ [1 \times 1 \times 2] \times 3 \\ \quad\vdots \\ [1 \times 1 \times 2 \times \cdots \times (m-1)] \times m. \end{array}$$

A computation is a finite sequence starting with the "initial value" of f and repeatedly applying g. In the example above, the *m-step computation* t is a finite sequence that is of length $m+1$ where $t_0 = 1$ and $t_{k+1} = t_k \times (k+1) = g(t_k, k)$ for all $k < m$, $k \geq 0$. The rigorous *explicit* definition of f then is:

$$f_m = t_m \quad \text{where } t \text{ is an } m\text{-step computation (based on } a = 1 \text{ and } g).$$

The problem of the existence and uniqueness of f is reduced to the problem of showing that there is precisely one m-step computation for each $m \in \boldsymbol{N}$.

We now proceed with the formal proof of the Recursion Theorem. As this theorem and its generalizations are among the most important methods of set theory, the reader should study the preceding intuitive example and the proof itself carefully.

Proof. *The existence of f.* A function $t : (m+1) \to A$ is called an *m-step computation based on a and g* if $t_0 = a$, and, for all k such that $0 \leq k < m$, $t_{k+1} = g(t_k, k)$. Notice that $t \subseteq \boldsymbol{N} \times A$. Let

$$F = \{t \in \mathcal{P}(\boldsymbol{N} \times A) \mid t \text{ is an } m\text{-step computation for some natural number } m\}.$$

Let $f = \bigcup F$.

3.1 Claim *f is a function.*

It suffices to show that the system of functions F is compatible—see Theorem 3.12 in Chapter 2. So let $t, u \in F$, $\operatorname{dom} t = n \in \boldsymbol{N}$, $\operatorname{dom} u = m \in \boldsymbol{N}$. Assume, e.g., $n \leq m$; then $n \subseteq m$, and it suffices to show that $t_k = u_k$ for all $k < n$. This can be done by induction (in the form stated in Exercise 2.12). Surely, $t_0 = a = u_0$. Next let k be such that $k+1 < n$, and assume $t_k = u_k$. Then $t_{k+1} = g(t_k, k) = g(u_k, k) = u_{k+1}$. Thus $t_k = u_k$ for all $k < n$.

3.2 Claim $\operatorname{dom} f = \mathbf{N}$; $\operatorname{ran} f \subseteq A$.

We know immediately that $\operatorname{dom} f \subseteq \mathbf{N}$ and $\operatorname{ran} f \subseteq A$. To show that $\operatorname{dom} f = \mathbf{N}$, it suffices to prove that for each $n \in \mathbf{N}$ there is an n-step computation t. We use the Induction Principle. Clearly, $t = \{(0, a)\}$ is a 0-step computation.

Assume that t is an n-step computation. Then the following function t^+ on $(n+1)+1$ is an $(n+1)$-step computation:

$$t_k^+ = t_k \quad \text{if } k \leq n;$$
$$t_{n+1}^+ = g(t_n, n).$$

We conclude that each $n \in \mathbf{N}$ is in the domain of some computation $t \in F$, so $\mathbf{N} \subseteq \bigcup_{t \in F} \operatorname{dom} t = \operatorname{dom} f$.

3.3 Claim *f satisfies conditions (a) and (b).*

Clearly, $f_0 = a$ since $t_0 = a$ for all $t \in F$. To show that $f_{n+1} = g(f_n, n)$ for any $n \in \mathbf{N}$, let t be an $(n+1)$-step computation; then $t_k = f_k$ for all $k \in \operatorname{dom} t$, so $f_{n+1} = t_{n+1} = g(t_n, n) = g(f_n, n)$.

The existence of a function f with properties required by the Recursion Theorem follows from Claims 3.1, 3.2, and 3.3.

The uniqueness of f. Let $h : \mathbf{N} \to A$ be such that
(a') $h_0 = a$;
(b') $h_{n+1} = g(h_n, n)$ for all $n \in \mathbf{N}$.
We show that $f_n = h_n$ for all $n \in \mathbf{N}$, again using induction. Certainly $f_0 = a = h_0$. If $f_n = h_n$, then $f_{n+1} = g(f_n, n) = g(h_n, n) = h_{n+1}$; therefore, $f = h$, as claimed. □

As a typical example of the use of the Recursion Theorem, we prove that the properties of the ordering of natural numbers by size established in the previous section uniquely characterize the ordered set $(\mathbf{N}, <)$.

3.4 Theorem *Let $(A, \prec)$ be a nonempty linearly ordered set with the properties:*
(a) For every $p \in A$, there is $q \in A$ such that $q \succ p$.
(b) Every nonempty subset of A has a $\prec$-least element.
(c) Every nonempty subset of A that has an upper bound has a $\prec$-greatest element.
Then $(A, \prec)$ is isomorphic to $(\mathbf{N}, <)$.

Proof. We construct the isomorphism f using the Recursion Theorem. Let a be the least element of A and let $g(x, n)$ be the least element of A greater than x (for any n). Then $a \in A$ and g is a function on $A \times \mathbf{N}$ into A; notice that $g(x, n)$ is defined for any $x \in A$ because of assumptions (a) and (b) in Theorem 3.4, and does not depend on n. The Recursion Theorem guarantees the existence of a function $f : \mathbf{N} \to A$ such that

(i) $f_0 = a =$ the least element of A.
(ii) $f_{n+1} = g(f_n, n) =$ the least element of A greater than f_n.

It is obvious that $f_n \prec f_{n+1}$ for each $n \in \boldsymbol{N}$. By induction, we get $f_n \prec f_m$ whenever $n < m$ (see Exercise 3.1). Consequently, f is a one-to-one function. It remains to show that the range of f is A.

If not, $A - \operatorname{ran} f \neq \emptyset$; let p be the least element of $A - \operatorname{ran} f$. The set $B = \{q \in A \mid q \prec p\}$ has an upper bound p, and is nonempty (otherwise, p would be the least element of A, but then $p = f_0$). Let q be the greatest element of B [it exists by assumption (c) in Theorem 3.4]. Since $q \prec p$, we have $q = f_m$ for some $m \in \boldsymbol{N}$. However, it is now easily seen that p is the least element of A greater than q. Therefore, $p = f_{m+1}$ by the recursive condition (ii). Consequently, $p \in \operatorname{ran} f$, a contradiction. □

In some recursive definitions, the value of f_{n+1} depends not only on f_n, but also on f_k for other $k < n$. A typical example is the Fibonacci sequence:

$$1, 1, 2, 3, 5, 8, 13, 21, \ldots .$$

Here $f_0 = 1$, $f_1 = 1$, and $f_{n+1} = f_n + f_{n-1}$ for $n > 0$. The following theorem formalizes this apparently more general recursive construction.

3.5 Theorem *For any set S and any function $g : \operatorname{Seq}(S) \to S$ there exists a unique sequence $f : \boldsymbol{N} \to S$ such that*

$$f_n = g(f \restriction n) = g(\langle f_0, \ldots, f_{n-1} \rangle) \quad \textit{for all } n \in \boldsymbol{N}.$$

Notice that, in particular, $f_0 = g(f \restriction 0) = g(\langle\rangle) = g(\emptyset)$. To obtain the Fibonacci sequence, we let

$$g(t) = \begin{cases} 1 & \text{if } t \text{ is a finite sequence of length 0 or 1;} \\ t_{n-1} + t_{n-2} & \text{if } t \text{ is a finite sequence of length } n > 1. \end{cases}$$

Proof. The idea is to use the Recursion Theorem to define the sequence $\langle F_n \mid n \in \boldsymbol{N} \rangle = \langle f \restriction n \mid n \in \boldsymbol{N} \rangle$.

So let us define

$$F_0 = \langle\rangle;$$
$$F_{n+1} = F_n \cup \{\langle n, g(F_n) \rangle\} \quad \text{for all } n \in \boldsymbol{N}.$$

The existence of the sequence $\langle F_n \mid n \in \boldsymbol{N} \rangle$ follows from the Recursion Theorem with $A = \operatorname{Seq}(S)$, $a = \langle\rangle$, and $G : A \times \boldsymbol{N} \to A$ defined by

$$G(t, n) = \begin{cases} t \cup \{\langle n, g(t) \rangle\} & \text{if } t \text{ is a sequence of length } n; \\ \langle\rangle & \text{otherwise.} \end{cases}$$

It is easy to prove by induction that each F_n belongs to S^n and that $F_n \subseteq F_{n+1}$ for all $n \in \boldsymbol{N}$. Therefore (see Exercise 3.1), $\{F_n \mid n \in \boldsymbol{N}\}$ is a compatible system of functions. Let $f = \bigcup_{n \in \mathbf{N}} F_n$; then clearly $f : \mathbf{N} \to S$ and $f \restriction n = F_n$ for all $n \in \mathbf{N}$; we conclude that $f_n = F_{n+1}(n) = g(F_n) = g(f \restriction n)$, as needed. $\square$

Further examples on the use of this version of the Recursion Theorem can be found in Chapter 4. At present, we state yet another, "parametric," version that allows us to use recursion to define functions of two variables.

3.6 Theorem *Let $a : P \to A$ and $g : P \times A \times \mathbf{N} \to A$ be functions. There exists a unique function $f : P \times \mathbf{N} \to A$ such that*
(a) $f(p, 0) = a(p)$ for all $p \in P$;
(b) $f(p, n+1) = g(p, f(p, n), n)$ for all $n \in \mathbf{N}$ and $p \in P$.

The reader may prefer the notation $f_{p,0}$ in place of $f(p, 0)$, etc.

Proof. This is just a "parametric" version of the proof of the Recursion Theorem. Define an m-step computation to be a function $t : P \times (m+1) \to A$ such that, for all $p \in P$,

$$t(p, 0) = a(p) \quad \text{and} \quad t(p, k+1) = g(p, t(p, k), k)$$

for all k such that $0 \leq k < m$. Then follow the steps in the proof of the Recursion Theorem, always carrying p along. Alternatively, one can deduce the parametric version directly from the Recursion Theorem (see Exercise 3.4). $\square$

We now have all the machinery needed to define addition of natural numbers, as well as other arithmetic operations. We do so in the next Section.

Exercises

3.1 Let f be an infinite sequence of elements of A, where A is ordered by $\prec$. Assume that $f_n \prec f_{n+1}$ for all $n \in \mathbf{N}$. Prove that $n < m$ implies $f_n \prec f_m$ for all $n, m \in \mathbf{N}$. [*Hint*: Use induction on m in the form of Exercise 2.11, with $k = n + 1$.]

3.2 Let $(A, \prec)$ be a linearly ordered set and $p, q \in A$. We say that q is a *successor* of p if $p \prec q$ and there is no $r \in A$ such that $p \prec r \prec q$. Note that each $p \in A$ can have at most one successor. Assume that $(A, \prec)$ is nonempty and has the following properties:
(a) Every $p \in A$ has a successor.
(b) Every nonempty subset of A has a $\prec$-least element.
(c) If $p \in A$ is not the $\prec$-least element of A, then p is a successor of some $q \in A$.
Prove that $(A, \prec)$ is isomorphic to $(\mathbf{N}, <)$. Show that the conclusion need not hold if one of the conditions (a)–(c) is omitted.

3.3 Give a direct proof of Theorem 3.5 in a way analogous to the proof of the Recursion Theorem.

3.4 Derive the "parametric" version of the Recursion Theorem (Theorem 3.6) from the Recursion Theorem.
[*Hint*: Define $F : \boldsymbol{N} \to A^P$ by recursion:

$$F_0 = a \in A^P;$$
$$F_{n+1} = G(F_n, n)$$

where $G : A^P \times \boldsymbol{N} \to A^P$ is defined by $G(x, n)(p) = g(p, x(p), n)$ for $x \in A^P$, $n \in \boldsymbol{N}$. Then set $f(p, n) = F_n(p)$.]

3.5 Prove the following version of the Recursion Theorem:
Let g be a function on a subset of $A \times \boldsymbol{N}$ into A, $a \in A$. Then there is a unique sequence f of elements of A such that
(a) $f_0 = a$;
(b) $f_{n+1} = g(f_n, n)$ for all $n \in \boldsymbol{N}$ such that $(n + 1) \in \operatorname{dom} f$;
(c) f is either an infinite sequence or f is a finite sequence of length $k + 1$ and $g(f_k, k)$ is undefined.
[*Hint*: Let $\overline{A} = A \cup \{\overline{a}\}$ where $\overline{a} \notin A$. Define $\overline{g} : \overline{A} \times \boldsymbol{N} \to \overline{A}$ as follows:

$$\overline{g}(x, n) = \begin{cases} g(x, n) & \text{if defined;} \\ \overline{a} & \text{otherwise.} \end{cases}$$

Use the Recursion Theorem to get the corresponding infinite sequence $\overline{f}$. If $\overline{f}_l = \overline{a}$ for some $l \in \boldsymbol{N}$, consider $\overline{f} \restriction l$ for the least such l.]

3.6 Prove: If $X \subseteq \boldsymbol{N}$, then there is a one-to-one (finite or infinite) sequence f such that $\operatorname{ran} f = X$. [*Hint*: Use Exercise 3.5.]

4. Arithmetic of Natural Numbers

As an application of the Recursion Theorem we now show how to define addition of natural numbers, and we use Induction to prove basic properties of addition. In similar fashion, one can introduce other arithmetic operations (see Exercises).

4.1 Theorem *There is a unique function* $+ : \boldsymbol{N} \times \boldsymbol{N} \to \boldsymbol{N}$ *such that*
(a) $+(m, 0) = m$ *for all* $m \in \boldsymbol{N}$*;*
(b) $+(m, n + 1) = +(m, n) + 1$ *for all* $m, n \in \boldsymbol{N}$.

Proof. In the parametric version of the Recursion Theorem let $A = P = \boldsymbol{N}$, $a(p) = p$ for all $p \in \boldsymbol{N}$ and $g(p, x, n) = x + 1$ for all $p, x, n \in \boldsymbol{N}$. □

Notice that letting $n = 0$ in (b) leads to $+(m, 0 + 1) = +(m, 0) + 1$; but $+(m, 0) = m$ by (a) and $0 + 1 = S(0) = 1$ by the definition of the number 1. Thus we have $+(m, 1) = m + 1 = S(m)$ and, as we claimed earlier, the successor of $m \in \boldsymbol{N}$ is really the sum of m and 1, justifying our notation for it. Of course, we write $m + n$ instead of $+(m, n)$ in the sequel.

The defining properties of addition can now be restated as

4.2 $m + 0 = m$.

4.3 $m + (n + 1) = (m + n) + 1$.

Properties of recursively defined functions are typically established by induction. As an example, we prove the commutative law of addition.

4.4 Theorem *Addition is commutative; i.e., for all* $m, n \in \boldsymbol{N}$,

$$m + n = n + m. \tag{4.5}$$

Proof. Let us say that n *commutes* if (4.5) holds for all $m \in \boldsymbol{N}$. We prove that every $n \in \boldsymbol{N}$ commutes, by induction on n.

To show that 0 commutes, it suffices to show that $0 + m = m$ for all m [because then $0 + m = m + 0$ by 4.2]. Clearly, $0 + 0 = 0$, and if $0 + m = m$, then $0 + (m + 1) = (0 + m) + 1 = m + 1$, so the claim follows by induction (on m).

Let us now assume that n commutes, and let us show that $n + 1$ commutes. We prove, by induction on m, that

$$m + (n + 1) = (n + 1) + m \quad \text{for all } m \in \boldsymbol{N}. \tag{4.6}$$

If $m = 0$, then (4.6) holds, as we have already shown. Thus let us assume that (4.6) holds for m, and let us prove that

$$(m + 1) + (n + 1) = (n + 1) + (m + 1). \tag{4.7}$$

We derive (4.7) as follows:

$$\begin{aligned}
(m + 1) + (n + 1) &= ((m + 1) + n) + 1 && \text{[by 4.3]} \\
&= (n + (m + 1)) + 1 && \text{[since } n \text{ commutes]} \\
&= ((n + m) + 1) + 1 && \text{[by 4.3]} \\
&= ((m + n) + 1) + 1 && \text{[since } n \text{ commutes]} \\
&= (m + (n + 1)) + 1 && \text{[by 4.3]} \\
&= ((n + 1) + m) + 1 && \text{[since (4.6) holds for } m] \\
&= (n + 1) + (m + 1) && \text{[by 4.3].}
\end{aligned}$$

□

Detailed development of the arithmetic of natural numbers is beyond the scope of this book. Readers who are so inclined may continue in it by working the exercises at the end of this section, and thus convince themselves that axiomatic set theory has the power to establish rigorously all of the usual laws of arithmetic. From there one can proceed to define such notions as divisibility and prime numbers and prove fundamental results of elementary number theory, such as the existence and uniqueness of a decomposition of each natural number into a product of prime numbers.

Having established the natural numbers and their arithmetic, one can proceed to define rigorously, first, the set $\boldsymbol{Z}$ of integers, then the set $\boldsymbol{Q}$ of rational numbers, and derive their corresponding arithmetic laws. These constructions are well known from algebra, and most readers probably encountered them there. For completeness, we describe them in some detail in Section 1 of Chapter 10, which can be studied at this point. In any case, we view the arithmetic of integers and rationals as rigorously established from now on.

We conclude this section with a remark about axiomatic arithmetic of natural numbers. The theory of arithmetic of natural numbers can be developed axiomatically. The accepted system of axioms is due to Giuseppe Peano. The undefined notions of Peano arithmetic are the constant 0, the unary operation S, and the binary operations $+$ and $\cdot$. The axioms of Peano arithmetic are the following:

(P1) If $S(n) = S(m)$, then $n = m$.
(P2) $S(n) \neq 0$.
(P3) $n + 0 = n$.
(P4) $n + S(m) = S(n + m)$.
(P5) $n \cdot 0 = 0$.
(P6) $n \cdot S(m) = (n \cdot m) + n$.
(P7) If $n \neq 0$, then $n = S(k)$ for some k.
(P8) The Induction Schema. Let $\mathbf{A}$ be an arithmetical property (i.e., a property expressible in terms of $+$, $\cdot$, S, 0). If 0 has the property $\mathbf{A}$ and if $\mathbf{A}(k)$ implies $\mathbf{A}(S(k))$ for every k, then every number has the property $\mathbf{A}$.

It is not difficult to verify that natural numbers and arithmetic operations, as we defined them, satisfy all of the Peano axioms (Exercise 4.8). Most of the needed facts have been proved already.

Exercises

4.1 Prove the associative law of addition:

$$(k + m) + n = k + (m + n) \quad \text{for all } k, m, n \in \boldsymbol{N}.$$

4.2 If $m, n, k \in \boldsymbol{N}$, then $m < n$ if and only if $m + k < n + k$.

4.3 If $m, n \in \boldsymbol{N}$ then $m \leq n$ if and only if there exists $k \in \boldsymbol{N}$ such that $n = m + k$. This k is unique, so we can denote it $n - m$, the *difference* of n and m. This is how *subtraction* of natural numbers is defined.

4.4 There is a unique function $\cdot$ (*multiplication*) from $\boldsymbol{N} \times \boldsymbol{N}$ to $\boldsymbol{N}$ such that

$$m \cdot 0 = 0 \quad \text{for all } m \in \boldsymbol{N};$$
$$m \cdot (n + 1) = m \cdot n + m \quad \text{for all } m, n \in \boldsymbol{N}.$$

4.5 Prove that multiplication is commutative, associative, and distributive over addition.

4.6 If $m, n, k \in \boldsymbol{N}$ and $k > 0$, then $m < n$ if and only if $m \cdot k < n \cdot k$.

4.7 Define exponentiation of natural numbers as follows:

$$m^0 = 1 \quad \text{for all } m \in \boldsymbol{N} \text{ (in particular, } 0^0 = 1);$$
$$m^{n+1} = m^n \cdot m \quad \text{for all } m, n \in \boldsymbol{N} \text{ (in particular, } 0^n = 0 \text{ for } n > 0).$$

Prove the usual laws of exponents.

4.8 Verify the axioms of Peano arithmetic. The needed results can be found in the text and exercises.

4.9 For each finite sequence $\langle k_i \mid 0 \leq i < n\rangle$ of natural numbers define $\sum\langle k_i \mid 0 \leq i < n\rangle$ (more usually denoted $\sum_{0\leq i<n} k_i$ or $\sum_{i=0}^{n-1} k_i$) so that

$$\sum\langle\rangle = 0;$$
$$\sum\langle k_0\rangle = k_0;$$
$$\sum\langle k_0, \ldots, k_n\rangle = \sum\langle k_0, \ldots, k_{n-1}\rangle + k_n \quad \text{for } n \geq 1.$$

5. Operations and Structures

The functions $+$, $\cdot$, etc., are usually referred to as *operations*. The aim of this section is to define the general concept of operation and to study its properties.

Each of the aforementioned operations assigns to a pair of objects (numbers, sets) a third object of the same kind (their sum, difference, union, etc.). The order may make a difference, e.g., $7 - 2$ and $2 - 7$ are two different objects. We therefore submit the following definition.

5.1 Definition A *binary operation* on S is a function mapping a subset of S^2 into S.

Nonletter symbols such as $+$, $\times$, $*$, $\triangle$, etc., are often used to denote operations. The value (result) of the operation $*$ at (x, y) is then denoted $x * y$ rather than $*(x, y)$.

There are also operations, such as square root or derivative, which are applied to one object rather than to a pair of objects. We introduce the following definitions.

5.2 Definition A *unary operation* on S is a function mapping a subset of S into S. A *ternary operation* on S is a function on a subset of S^3 into S.

5.3 Definition Let f be a binary operation on S and $A \subseteq S$. A is *closed under the operation* f if for every $x, y \in A$ such that $f(x, y)$ is defined, $f(x, y) \in A$.

One can give similar definitions in the case of unary or ternary operations.

5.4 Example

(a) Let $+$ be the operation of addition on the set of all real numbers. Then $+$ is defined for all real numbers a and b. The set of all real numbers, as well as the set of all rational numbers and the set of all integers, are closed under $+$. The set of even natural numbers is closed under $+$, but the set of odd natural numbers is not.

(b) Let $\div$ be the operation of division on the set of all real numbers. $\div$ is not defined for (a, b) where $b = 0$. The set of all rational numbers is closed under $\div$, but the set of all integers is not.

(c) Let S be a set; define binary operations $\cup_S$ and $\cap_S$ on S as follows:
   (i) If $x, y \in S$ and $x \cup y \in S$, then $x \cup_S y = x \cup y$.
   (ii) If $x, y \in S$ and $x \cap y \in S$, then $x \cap_S y = x \cap y$.
   If we take $S = \mathcal{P}(A)$ for some A, $\cup_S$ and $\cap_S$ are defined for every pair $(x, y) \in S^2$.

When developing a particular branch of mathematics, we are usually interested in sets of objects together with some relations between them and operations on them. For example, in number theory or analysis, we study the set of all (natural or real) numbers together with operations of addition and multiplication, relation of less than, etc. In geometry, we study the set of all points and lines with relations of incidence and between and operation of intersection, etc. We introduce the notion of *structure* to describe this situation abstractly.

In general, a *structure* consists of a set A and of several relations and operations on A. For example, we consider structures with two binary relations and two operations, say a unary operation and a binary operation. Let A be a set, let R_1 and R_2 be binary relations in A, and let f be a unary operation and g a binary operation on A. We make use of the five-tuple (A, R_1, R_2, f, g) to denote the structure thus defined.

5.5 Example

(a) Every ordered set is a structure with one binary relation.

(b) $(A, \cup_A, \cap_A, \subseteq_A)$ is a structure with two binary operations and one binary relation.

(c) Let $\boldsymbol{R}$ be the set of all real numbers. $(\boldsymbol{R}, +, -, \times, \div)$ is a structure with four binary operations.

We now proceed to give a general definition of a structure. In Chapter 2, we introduced unary (1-ary), binary (2-ary), and ternary (3-ary) relations. The reason we did not talk about n-ary relations for arbitrary n is simply that we could not then handle arbitrary natural numbers. This obstacle is now removed, and the present section gives precise definitions of n-tuples, n-ary relations and operations, and n-fold cartesian products, for all natural numbers n.

We start with the definition of an ordered n-tuple. Recall that ordered pair (a_0, a_1) has been defined in Section 1 of Chapter 2 as a set that uniquely determines its two coordinates, a_0 and a_1; i.e.,

$$(a_0, a_1) = (b_0, b_1) \quad \text{if and only if} \quad a_0 = b_0 \text{ and } a_1 = b_1.$$

We called a_0 the first coordinate of (a_0, a_1), and a_1 its second coordinate. In analogy, an *n-tuple* $(a_0, a_1, \dots, a_{n-1})$ should be a set that uniquely determines its n coordinates, a_0, a_1, ..., a_{n-1}; that is, we want

(*)
$$(a_0, \dots, a_{n-1}) = (b_0, \dots, b_{n-1}) \text{ if and only if } a_i = b_i \text{ for all } i = 0, \dots, n-1.$$

But we already introduced a notion that satisfies (*): it is the *sequence of length n*, $\langle a_0, a_1, \ldots, a_{n-1}\rangle$. The statement that

$$\langle a_0, \ldots, a_{n-1}\rangle = \langle b_0, \ldots, b_{n-1}\rangle \text{ if and only if } a_i = b_i \text{ for all } i = 0, \ldots, n-1$$

is just a reformulation of equality of functions, as in Lemma 3.2 in Chapter 2.

We therefore define *n-tuples* as sequences of length n. For each i, $0 \le i < n$, a_i is called the *$(i+1)$st coordinate* of $\langle a_0, a_1, \ldots, a_{n-1}\rangle$. So a_0 is the first coordinate, a_1 is the second coordinate, ... , a_{n-1} is the n-th coordinate. [Really, a more consistent terminology would be zeroth, first, ... , $(n-1)$st coordinate, but the custom is to talk about the first and second coordinates of points in the plane, etc.]

We notice that the only 0-tuple is the *empty sequence* $\langle\rangle = \emptyset$, having no coordinates. 1-tuples are sequences of the form $\langle a_0\rangle$ [i.e., sets of the form $\{(0, a_0)\}$]; it usually causes no confusion if one does not distinguish between a 1-tuple $\langle a_0\rangle$ and an element a_0.

If $\langle A_i \mid 0 \le i < n\rangle$ is a finite sequence of sets, then the n-fold cartesian product $\prod_{0 \le i < n} A_i$, as defined in Section 3 of Chapter 2, is just the set of all n-tuples $a = \langle a_0, a_1, \ldots, a_{n-1}\rangle$ such that $a_0 \in A_0$, $a_1 \in A_1$, ... , $a_{n-1} \in A_{n-1}$. If $A_i = A$ for all i, $0 \le i < n$, then $\prod_{0 \le i < n} A_i = A^n$ is the set of all n-tuples with all coordinates from A.

We note that $A^0 = \{\langle\rangle\}$ and A^1 can be identified with A. An *n-ary relation R in A* is a subset of A^n. We then write $R(a_0, a_1, \ldots, a_{n-1})$ instead of $\langle a_0, a_1, \ldots, a_{n-1}\rangle \in R$. Similarly, an *$n$-ary operation F on A* is a function on a subset of A^n into A; we write $F(a_0, a_1, \ldots, a_{n-1})$ instead of $F(\langle a_0, a_1, \ldots, a_{n-1}\rangle)$. Finally, we generalize the notation already introduced in the case of pairs and triples. If $\mathbf{P}(x_0, x_1, \ldots, x_{n-1})$ is a property with parameters x_0, x_1, ... , x_{n-1}, we write

$$\{\langle a_0, \ldots, a_{n-1}\rangle \mid a_0 \in A_0, \ldots, a_{n-1} \in A_{n-1} \text{ and } \mathbf{P}(a_0, \ldots, a_{n-1})\}$$

to denote the set

$$\{a \in \prod_{0 \le i < n} A_i \mid \text{for some } a_0, \ldots, a_{n-1},\ a = \langle a_0, \ldots, a_{n-1}\rangle \text{and } \mathbf{P}(a_0, \ldots, a_{n-1})\}.$$

We note that 1-ary relations need not be distinguished from subsets of A, and 1-ary operations can be identified with functions on a subset of A into A. 0-ary relations ($\emptyset$ and $\{\langle\rangle\}$ do not have much use, but nonempty 0-ary operations are quite useful. They are objects of the form $\{(\langle\rangle, a)\}$ where $a \in A$. We call them *constants* and in the sequel identify them with elements of A [i.e., we do not distinguish between $\{(\langle\rangle, a)\}$ and a].

The exercises at the end of this section contain further information about these and related concepts.

The problem with this approach is that the *ordered pair* from Chapter 2, $(a_0, a_1) = \{\{a_0\}, \{a_0, a_1\}\}$, is generally a different set from the just-defined *2-tuple* $\langle a_0, a_1\rangle = \{(0, a_0), (1, a_1)\}$. Consequently, we have two definitions of cartesian product, $A_0 \times A_1$ and $\prod_{0 \leq i < 2} A_i$, two definitions of binary relations and operations, etc. However, there is a canonical one-to-one correspondence between ordered pairs and 2-tuples that preserves first and second coordinates. Define $\delta((a_0, a_1)) = \langle a_0, a_1 \rangle$; then δ is a one-to-one mapping on $A_0 \times A_1$ onto $\prod_{0 \leq i < 2} A_i$ and x is a first (second, respectively) coordinate of (a_0, a_1) if and only if x is a first (second, respectively) coordinate of $\langle a_0, a_1 \rangle$. For almost all practical purposes, it makes no difference which definition one uses. Therefore, we do not distinguish between ordered pairs and 2-tuples from now on. Similar remarks hold about the relationships between triples and 3-tuples, etc. In fact, it is possible to define n-tuples for any natural number n along the lines of Chapter 2. The idea is to proceed recursively and let

$$(a_0) = a_0;$$
$$(a_0, a_1, \ldots, a_n) = ((a_0, a_1, \ldots, a_{n-1}), a_n) \quad \text{for all } n \in \boldsymbol{N},\ n \geq 1.$$

However, making this precise involves certain technical problems. We do not yet have a Recursion Theorem sufficiently general to cover this type of definition (but see Exercise 4.2 in Chapter 6). As we have no need for an alternative definition of n-tuples, we refer the interested reader to Exercise 5.17. We use notations $\langle a_0, \ldots, a_{n-1} \rangle$ and $(a_0, \ldots, a_{n-1})$ interchangeably from now on.

We continue this section by generalizing the notion of a structure. First, a *type* τ is an ordered pair $(\langle r_0, \ldots, r_{m-1} \rangle, \langle f_0, \ldots, f_{n-1} \rangle)$ of finite sequences of natural numbers, where $r_i > 0$ for all $i \leq m-1$, $i \geq 0$. A *structure of type* τ is a triple

$$\mathfrak{A} = (A, \langle R_0, \ldots, R_{m-1} \rangle, \langle F_0, \ldots, F_{n-1} \rangle)$$

where R_i is an r_i-ary relation on A for each $i \leq m-1$ and F_j is an f_j-ary operation on A for each $j \leq n-1$; in addition, we require $F_j \neq \emptyset$ if $f_j = 0$. Note that if $f_j = 0$, F_j is a 0-ary operation on A. Following our earlier remarks, F_j is a constant, i.e., just a distinguished element of A. For example, $\mathfrak{N} = (\boldsymbol{N}, \langle < \rangle, \langle 0, +, \cdot \rangle)$ is a structure of type $(\langle 2 \rangle, \langle 0, 2, 2 \rangle)$; it carries one binary relation, one constant, and two binary operations. $\mathfrak{R} = (\boldsymbol{R}, \langle \rangle, \langle 0, 1, +, -, \times, \div \rangle)$ is a structure of type $(\langle \rangle, \langle 0, 0, 2, 2, 2, 2 \rangle)$, etc. We often present the structure as a $(1+m+n)$-tuple, for example $(\boldsymbol{N}, <, 0, +, \cdot)$, when it is understood which symbols represent relations and which operations. We call A the *universe* of the structure $\mathfrak{A}$.

We now introduce a notion that is of crucial importance in the theory of structures. (In the special case of ordered sets, compare with Definition 5.17 in Chapter 2.)

5.6 Definition An *isomorphism* between structures $\mathfrak{A}$ and $\mathfrak{A}' = (A', \langle R'_0, \ldots, R'_{m-1} \rangle, \langle F'_0, \ldots, F'_{n-1} \rangle)$ of the same type τ is a one-to-one mapping h on A onto A' such that

(a) $R_i(a_0, \ldots, a_{r_i-1})$ if and only if $R'_i(h(a_0), \ldots, h(a_{r_i-1}))$ holds for all $a_0, \ldots, a_{r_i-1} \in A$ and $i \leq m-1$.

(b) $h(F_j(a_0, \ldots, a_{f_j-1})) = F'_j(h(a_0), \ldots, h(a_{f_j-1}))$ holds for all $a_0, \ldots, a_{f_j-1} \in A$ and all $j \leq n-1$ provided that either side is defined.

For example, let (A, R_1, R_2, f, g) and (A', R'_1, R'_2, f', g') be structures with two binary relations and one unary and one binary operation. Then h is an isomorphism of (A, R_1, R_2, f, g) and (A', R'_1, R'_2, f', g') if all of the following requirements hold:

(a) h is a one-to-one function on A onto A'.
(b) For all $a, b \in A$, aR_1b if and only if $h(a)R'_1h(b)$.
(c) For all $a, b \in A$, aR_2b if and only if $h(a)R'_2h(b)$.
(d) For all $a \in A$, $f(a)$ is defined if and only if $f'(h(a))$ is defined and $h(f(a)) = f'(h(a))$.
(e) For all $a, b \in A$, $g(a, b)$ is defined if and only if $g'(h(a), h(b))$ is defined and $h(g(a, b)) = g'(h(a), h(b))$.

The structures are called *isomorphic* if there is an isomorphism between them.

5.7 Example Let A be the set of all real numbers, $\leq_A$ be the usual ordering of real numbers, and $+$ be the operation of addition on A. Let A' be the set of all positive real numbers, $\leq_{A'}$ be the usual ordering of positive real numbers, and $\times$ be the operation of multiplication on A'. We show that the structures $(A, \leq_A, +)$ and $(A', \leq_{A'}, \times)$ are isomorphic. To that purpose, let h be the function

$$h(x) = e^x \quad \text{for } x \in A.$$

We prove that h is an isomorphism of $(A, \leq_A, +)$ and $(A', \leq_{A'}, \times)$. We have to prove:

(a) h is a one-to-one function on A onto A'. But clearly h is a function, $\operatorname{dom} h = A$, and $\operatorname{ran} h = A'$. If $x_1 \neq x_2$, then $e^{x_1} \neq e^{x_2}$, so h is one-to-one. (In this example, we assume knowledge of some facts from elementary calculus.)
(b) Let $x_1, x_2 \in A$; then $x_1 \leq_A x_2$ if and only if $h(x_1) \leq_{A'} h(x_2)$. But the function e^x is increasing, so $x_1 \leq x_2$ if and only if $e^{x_1} \leq e^{x_2}$ is indeed true.
(c) Let $x_1, x_2 \in A$; then $x_1 + x_2$ is defined if and only if $h(x_1) \times h(x_2)$ is defined and $h(x_1 + x_2) = h(x_1) \times h(x_2)$. First, notice that both $+$ on A and $\times$ on A' are defined for all ordered pairs. Now, $h(x_1 + x_2) = e^{x_1 + x_2} = e^{x_1} \times e^{x_2} = h(x_1) \times h(x_2)$. □

The significance of establishing an isomorphism between two structures lies in the fact that the isomorphic structures have exactly the same properties as far as the relations and operations on the structures are concerned. Therefore, if the properties of these relations and operations are our only interest, it does not make any difference which of the isomorphic structures we study.

5.8 Example Let (A, R) and (B, S) be isomorphic (R and S are binary relations). R is an ordering of A if and only if S is an ordering of B. A has a least element in R if and only if B has a least element in S.

Proof. Let h be an isomorphism of (A, R) and (B, S). Assume that R is an ordering of A. We prove that S is an ordering of B. Hence, let $b_1, b_2, b_3 \in B$ and b_1Sb_2, b_2Sb_3. Since h is onto B, there exist $a_1, a_2, a_3 \in A$ such that $b_1 = h(a_1)$,

$b_2 = h(a_2)$, and $b_2 = h(a_3)$. Because $a_1 R a_2$ holds if and only if $h(a_1) S h(a_2)$ holds, i.e., if and only if $b_1 S b_2$ holds, we conclude that $a_1 R a_2$ and similarly $a_2 R a_3$. But R is transitive in A, so $a_1 R a_3$. But then $h(a_1) S h(a_3)$, i.e., $b_1 S b_3$. We leave the proof of reflexivity and antisymmetry to the reader. Similarly, one proves that if S is an ordering of B, then R is an ordering of A. Now let A have a least element. We claim that B has a least element. Let a be the least element of A, i.e, aRx holds for all $x \in A$. Then $h(a)$ is the least element of B, as the following consideration shows. If $y \in B$, then $y = h(x)$ for some $x \in A$ (h is onto B). But, for this x, aRx holds. Correspondingly, $h(a) S h(x)$ holds; and we see that $h(a) S y$ holds for all $y \in B$. Thus $h(a)$ is the least element of B. □

An isomorphism between a structure $\mathfrak{A}$ and itself is called an *automorphism* of $\mathfrak{A}$. The identity mapping on the universe of $\mathfrak{A}$ is trivially an automorphism of $\mathfrak{A}$. The reader can easily verify that the structure $(\boldsymbol{N}, <)$ has no other automorphisms. On the other hand, the structure $(\boldsymbol{Z}, <)$, where $\boldsymbol{Z}$ is the set of all integers, has nontrivial automorphisms. In fact, they are precisely the functions f_h, $h \in \boldsymbol{Z}$, where $f_h(x) = x + h$. Some further properties of automorphisms are listed in Exercise 5.12.

Consider a fixed structure $\mathfrak{A} = (A, \langle R_0, \ldots, R_{m-1}\rangle, \langle F_0, \ldots, F_{n-1}\rangle)$. A set $B \subseteq A$ is called *closed* if the result of applying any operation to elements of B is again in B, i.e., if for all $j \le n-1$ and all $a_0, \ldots, a_{f_j-1} \in B$, $F_j(a_0, \ldots, a_{f_j-1}) \in B$ provided that it is defined. In particular, all constants of $\mathfrak{A}$ belong to B. Let $C \subseteq A$; the *closure* of C, to be denoted $\overline{C}$, is the least closed set containing all elements of C:

$$\overline{C} = \bigcap \{B \subseteq A \mid C \subseteq B \text{ and } B \text{ is closed}\}.$$

Notice that A is a closed set containing C, so the system whose intersection defines $\overline{C}$ is nonempty. It is trivial to check that $\overline{C}$ is closed; by definition, then, $\overline{C}$ is indeed the least closed set containing C.

5.9 Example

(a) Every set $B \subseteq A$ is closed if $\mathfrak{A}$ has no operations.

(b) Let $\boldsymbol{R}$ be the set of all real numbers and let $C = \{0\}$. The set of all natural numbers $\boldsymbol{N}$ is the closure of C in the structure $(\boldsymbol{R}, f)$ where f is the successor function defined by

$$f(x) = x + 1 \quad \text{for all real numbers } x.$$

(c) Let $C = \{0, 1\}$; the set of all integers $\boldsymbol{Z}$ is the closure of C in the structure $(\boldsymbol{R}, +, -)$ or in $(\boldsymbol{R}, +, -, \times)$.

(d) The set of all rationals $\boldsymbol{Q}$ is the closure of $\emptyset$ in $(\boldsymbol{R}, 0, 1, +, -, \times, \div)$.

The notion of closure is important in algebra, logic, and other areas of mathematics. The next theorem shows how to construct the closure of a set "from below."

5.10 Theorem *Let $\mathfrak{A} = (A, \langle R_0, \ldots, R_{m-1}\rangle, \langle F_0, \ldots, F_{n-1}\rangle)$ be a structure and let $C \subseteq A$. If the sequence $\langle C_i \mid i \in \boldsymbol{N}\rangle$ is defined recursively by*

$$C_0 = C;$$
$$C_{i+1} = C_i \cup F_0[C_i^{f_0}] \cup \cdots \cup F_{n-1}[C_i^{f_{n-1}}],$$

then $\overline{C} = \bigcup_{i=0}^{\infty} C_i$.

Of course, the notation $A_0 \cup \cdots \cup A_{n-1}$ is a shorthand for $\bigcup_{0 \le i < n} A_i$. Observe that $C_i \subseteq C_{i+1}$ for all i, so the sequence $\langle C_i \mid i \in \boldsymbol{N}\rangle$ is nondecreasing (see Exercise 3.1).

Proof. Let $\tilde{C} = \bigcup_{i=0}^{\infty} C_i$. We have to prove that $\overline{C} \subseteq \tilde{C}$. This in turn follows if we show that $\tilde{C}$ is closed, because $\tilde{C} \supseteq C_0 = C$. So let $j < n$ and $a_0, \ldots, a_{f_j-1} \in \tilde{C}$. From the definition of union we get that each a_r [for $0 \le r \le f_j - 1$] belongs to some C_i; let i_r be the least $i \in \boldsymbol{N}$ such that $a_r \in C_i$.

The range of the finite sequence $\langle i_r \mid 0 \le r \le f_j - 1\rangle$ of natural numbers contains a greatest element $\bar{\imath}$ (this is a fact easily proved by induction; see Exercise 5.13). Since $\langle C_i \mid i \in \boldsymbol{N}\rangle$ is nondecreasing, we have $a_r \in C_{i_r} \subseteq C_{\bar{\imath}}$ for all $r \le f_j - 1$, $r \ge 0$. We conclude that if $F_j(a_0, \ldots, a_{f_j-1})$ is defined, then it belongs to $F_j[C_{\bar{\imath}}^{f_j}] \subseteq C_{\bar{\imath}+1} \subseteq \tilde{C}$, so $\tilde{C}$ is closed.

It remains to prove the opposite inclusion $\tilde{C} \subseteq \overline{C}$. Clearly, $C_0 = C \subseteq \overline{C}$. If $C_i \subseteq \overline{C}$, then $F_j[C_i^{f_j}] \subseteq \overline{C}$ for each $j \le n-1$ because $\overline{C}$ is closed, and therefore also $C_{i+1} \subseteq \overline{C}$. We conclude using the Induction Principle that $C_i \subseteq \overline{C}$ for all $i \in \boldsymbol{N}$ and, finally, $\tilde{C} = \bigcup_{i=0}^{\infty} C_i \subseteq \overline{C}$, as required. □

We close this section with a theorem that is often used to prove that all elements of a closure have some property **P**. It can be viewed as a generalization of the Induction Principle, which is its special case for the structure $(\boldsymbol{N}, S)$ (S is the successor operation) and $C = \{0\}$.

5.11 Theorem *Let $\mathbf{P}(x)$ be a property. Assume that*

(a) $\mathbf{P}(a)$ holds for all $a \in C$.

(b) For each $j \le n-1$, if $\mathbf{P}(a_0), \ldots, \mathbf{P}(a_{f_j-1})$ hold and $F_j(a_0, \ldots, a_{f_j-1})$ is defined, then $\mathbf{P}(F_j(a_0, \ldots, a_{f_j-1}))$ holds.

Then $\mathbf{P}(x)$ holds for all $x \in \overline{C}$.

Proof. The assumptions (a) and (b) postulate that the set $B = \{x \in A \mid \mathbf{P}(x)\}$ is closed and $C \subseteq B$. □

Exercises

5.1 Which of the following sets are closed under operations of addition, subtraction, multiplication, and division of real numbers?

(a) The set of all positive integers.

(b) The set of all integers.

(c) The set of all rational numbers.
(d) The set of all negative rational numbers.
(e) The empty set.

5.2 Let $*$ be a binary operation on A.
(a) $*$ is called *commutative* if, for all $a, b \in A$, whenever $a * b$ is defined, then $b * a$ is also defined and $a * b = b * a$.
(b) $*$ is called *associative* if, for all $a, b, c \in A$, $(a*b)*c = a*(b*c)$ holds whenever the expression on one side of the equality sign is defined (the other expression then must also be defined).

Which of the operations mentioned in this section are commutative or associative?

5.3 Let $*$ and $\triangle$ be operations in A. We say that $*$ is *distributive over* $\triangle$ if $a * (b \triangle c) = (a * b) \triangle (a * c)$ for all $a, b, c \in A$ for which the expression on one (and then also on the other) side is defined. For example, multiplication of real numbers is distributive over addition, but addition is not distributive over multiplication.

5.4 Let $A \neq \emptyset$, $B = \mathcal{P}(A)$. Show that $(B, \cup_B, \cap_B)$ and $(B, \cap_B, \cup_B)$ are isomorphic structures. [*Hint*: Set $h(x) = B - x$; notice that $\cup_B$ in the first structure corresponds to $\cap_B$ in the second structure and vice versa.]

5.5 Refer to Example 5.7 for notation.
(a) There is a real number $a \in A$ such that $a + a = a$ (namely, $a = 0$). Prove from this that there is $a' \in A'$ such that $a' \times a' = a'$. Find this a'.
(b) For every $a \in A$ there is $b \in A$ such that $a + b = 0$. Show that for every $a' \in A'$ there is $b' \in A'$ such that $a' \times b' = 1$. Find this b'.

5.6 Let $\boldsymbol{Z}^+$ and $\boldsymbol{Z}^-$ be, respectively, the sets of all positive and negative integers. Show that $(\boldsymbol{Z}^+, <, +)$ is isomorphic to $(\boldsymbol{Z}^-, >, +)$ (where $<$ is the usual ordering of integers).

5.7 Let R be a set all elements of which are n-tuples. Prove that R is an n-ary relation in A for some A.

5.8 For every n-ary operation F on A there is a unique $(n + 1)$-ary relation R in A such that $F(a_0, \ldots, a_{n-1}) = a_n$ if and only if $R(a_0, \ldots, a_{n-1}, a_n)$ holds.

5.9 Let $B = \prod_{0 \leq i < n} A_i$; the *projection of* B *onto its* $(i+1)$*-st coordinate set* is the function $\pi_i : B \to A_i$ defined by $\pi_i(a) = a_i$. Prove that π_i is onto A_i and that $a = \langle \pi_0(a), \ldots, \pi_{n-1}(a) \rangle$. More generally,
(a) For any $f : C \to B$, let $f_i = \pi_i \circ f$, $0 \leq i < n$. Then $f_i : C \to A_i$ and $f(c) = \langle f_0(c), \ldots, f_{n-1}(c) \rangle$.
(b) Given $f_i : C \to A_i$, $0 \leq i < n$, define $f : C \to B$ by $f(c) = \langle f_0(c), \ldots, f_{n-1}(c) \rangle$. Then $f_i = \pi_i \circ f$.

5.10 $\prod_{0 \leq i < n} A_i \neq \emptyset$ if and only if $A_i \neq \emptyset$ for all $i \geq 0$, $i < n$. [*Hint*: Induction.]

5.11 Let $B = \mathrm{Seq}(A)$; the function length $: B \to \boldsymbol{N}$, and operations head, tail, $*$ (concatenation), and conv (converse) are defined on B as follows:

$\mathrm{length}(\langle a_0, \ldots, a_{m-1} \rangle) = m$;
$\mathrm{head}(\langle a_0, \ldots, a_{m-1} \rangle) = a_0$ (if $m \geq 1$);
$\mathrm{tail}(\langle a_0, \ldots, a_{m-1} \rangle) = \langle a_1, \ldots, a_{m-1} \rangle$ (if $m \geq 1$);

$\langle a_0, \dots, a_{m-1}\rangle * \langle b_0, \dots, b_{n-1}\rangle = \langle c_0, \dots, c_{m+n-1}\rangle$ where $c_i = a_i$ for $0 \le i < m$ and $c_i = b_{i-m}$ for $m \le i \le m+n-1$;
$\mathrm{conv}(\langle a_0, \dots, a_{m-1}\rangle) = \langle b_0, \dots, b_{m-1}\rangle$ where $b_i = a_{m-i-1}$ for $0 \le i \le m-1$.

Prove that for those $a, b, c \in B$ where the appropriate operations are defined

$\mathrm{length}(a * b) = \mathrm{length}(a) + \mathrm{length}(b)$;
$\mathrm{length}(\mathrm{tail}(a)) = \mathrm{length}(a) - 1$;
$a = \mathrm{head}(a) * \mathrm{tail}(a)$;
$\mathrm{head}(a * b) = \mathrm{head}(a)$;
$\mathrm{tail}(a * b) = \mathrm{tail}(a) * b$;
$a * (b * c) = (a * b) * c$;
$\mathrm{conv}(\mathrm{conv}(a)) = a$;
$\mathrm{conv}(a) = \mathrm{conv}(\mathrm{tail}(a)) * \mathrm{head}(a)$.

5.12 Let $\mathfrak{A}$, $\mathfrak{B}$, and $\mathfrak{C}$ be structures of type τ. Prove that
(a) $\mathfrak{A}$ is isomorphic to $\mathfrak{A}$.
(b) If $\mathfrak{A}$ is isomorphic to $\mathfrak{B}$ then $\mathfrak{B}$ is isomorphic to $\mathfrak{A}$.
(c) If $\mathfrak{A}$ is isomorphic to $\mathfrak{B}$ and $\mathfrak{B}$ is isomorphic to $\mathfrak{C}$, then $\mathfrak{A}$ is isomorphic to $\mathfrak{C}$.

Prove further that
(d) The identity function on the universe of $\mathfrak{A}$ is an automorphism of $\mathfrak{A}$.
(e) If f and g are automorphisms of $\mathfrak{A}$, then $f \circ g$ is also an automorphism of $\mathfrak{A}$.
(f) If f is an automorphism of $\mathfrak{A}$, then f^{-1} is also an automorphism of $\mathfrak{A}$.

5.13 Let $\langle k_0, \dots, k_{n-1}\rangle$ be a finite sequence of natural numbers of length $n \ge 1$. Then its range $\{k_0, \dots, k_{n-1}\}$ has a greatest element (in the standard ordering of natural numbers by size). [*Hint*: Use induction on the length of the sequence.]

5.14 Construct the sets C_0, C_1, C_2, and C_3 in Theorem 5.10 for
(a) $\mathfrak{A} = (\boldsymbol{R}, S)$ and $C = \{0\}$.
(b) $\mathfrak{A} = (\boldsymbol{R}, +, -)$ and $C = \{0, 1\}$.

5.15 Let $R \subseteq A^2$ be a binary relation. Define a binary operation F_R on A^2 by

$$F_R((a_1, a_2), (b_1, b_2)) = (a_1, b_2) \text{ if } a_2 = b_1 \text{ (and is undefined otherwise).}$$

Show that the closure of R in (A^2, F_R) is a transitive relation. Show that if R is reflexive and symmetric, then its closure is an equivalence relation.

5.16 Let $B = \mathrm{Seq}(A)$ where $A = \{a, b, c, \dots, x, y, z\}$ is the set of all letters in the English alphabet. Refer to Exercise 5.11. Identify sequences of length 1 with elements of A, so that $A \subseteq B$. Now let F be a binary operation defined as follows:

$$\text{if } \overline{x}, \overline{y} \in B \text{ and } \overline{y} \in A \text{ (i.e., it has length 1), then } F(\overline{x}, \overline{y}) = \overline{y} * \overline{x} * \overline{y}$$
$$(F \text{ is undefined otherwise}).$$

Let C be the closure of A in the structure (B, F). Prove that $c = \mathrm{conv}(c)$

for all $c \in C$. (The elements of C are called *palindromes*.) [*Hint*: Use Theorem 5.11.]

5.17 Define n-tuples so that

(a) $(a_0) = a_0$.

(b) $(a_0, a_1, \ldots, a_n) = ((a_0, \ldots, a_{n-1}), a_n)$ for all $n \geq 1$.

[*Hint*: Say that a *is an n-tuple* if there exist finite sequences f and $\langle a_0, \ldots, a_{n-1} \rangle$ of length n such that

(i) $f_0 = a_0$.

(ii) $f_{i+1} = (f_i, a_{i+1})$ for all $i < n-1$, $i \geq 0$.

(iii) $a = f_{n-1}$.

Show that for each n-tuple a there is a unique pair of finite sequences f, $\langle a_0, \ldots, a_{n-1} \rangle$ of length n such that (i), (ii), and (iii) hold; we then write $a = (a_0, \ldots, a_{n-1})$. Show that for every finite sequence $\langle a_0, \ldots, a_{n-1} \rangle$ of length n there is a unique n-tuple $a = (a_0, \ldots, a_{n-1})$. Then prove (a) and (b).]

Chapter 4

Finite, Countable, and Uncountable Sets

1. Cardinality of Sets

From the point of view of pure set theory, the most basic question about a set is: How many elements does it have? It is a fundamental observation that we can define the statement "sets A and B have the same number of elements" without knowing anything about numbers.

To see how that is done, consider the problem of determining whether the set of all patrons of some theater performance has the same number of elements as the set of all seats. To find the answer, the ushers need not count the patrons or the seats. It is enough if they check that each patron sits in one, and only one, seat, and each seat is occupied by one, and only one, theatergoer.

1.1 Definition Sets A and B are *equipotent* (have *the same cardinality*) if there is a one-to-one function f with domain A and range B. We denote this by $|A| = |B|$.

1.2 Example

(a) $\{\emptyset, \{\emptyset\}\}$ and $\{\{\{\emptyset\}\}, \{\{\{\emptyset\}\}\}\}$ are equipotent; let $f(\emptyset) = \{\{\emptyset\}\}$, $f(\{\emptyset\}) = \{\{\{\emptyset\}\}\}$.

(b) $\{\emptyset\}$ and $\{\emptyset, \{\emptyset\}\}$ are not equipotent.

(c) The set of all positive real numbers is equipotent with the set of all negative real numbers; set $f(x) = -x$ for all positive real numbers x.

1.3 Theorem

(a) A is equipotent to A.

(b) If A is equipotent to B, then B is equipotent to A.

(c) If A is equipotent to B and B is equipotent to C, then A is equipotent to C.

Proof.
(a) Id_A is a one-to-one mapping of A onto A.
(b) If f is a one-to-one mapping of A onto B, f^{-1} is a one-to-one mapping of B onto A.
(c) If f is a one-to-one mapping of A onto B and g is a one-to-one mapping of B onto C, then $g \circ f$ is a one-to-one mapping of A onto C.

□

Similarly, the following definition is very intuitive.

1.4 Definition *The cardinality of A is less than or equal to the cardinality of B* (notation: $|A| \leq |B|$) if there is a one-to-one mapping of A into B.

Notice that $|A| \leq |B|$ means that $|A| = |C|$ for some subset C of B. We also write $|A| < |B|$ to mean that $|A| \leq |B|$ and not $|A| = |B|$, i.e., that there is a one-to-one mapping of A onto a subset of B, but there is no one-to-one mapping of A onto B. Notice that this is not the same thing as saying that there exists a one-to-one mapping of A onto a proper subset of B; for example, there exists a one-to-one mapping of the set $\boldsymbol{N}$ onto its proper subset (Exercise 2.3 in Chapter 3), while of course $|\boldsymbol{N}| = |\boldsymbol{N}|$.

Theorem 1.3 shows that the property $|A| = |B|$ behaves like an equivalence relation: it is reflexive, symmetric, and transitive. We show next that the property $|A| \leq |B|$ behaves like an ordering on the "equivalence classes" under equipotence.

1.5 Lemma
(a) If $|A| \leq |B|$ and $|A| = |C|$, then $|C| \leq |B|$.
(b) If $|A| \leq |B|$ and $|B| = |C|$, then $|A| \leq |C|$.
(c) $|A| \leq |A|$.
(d) If $|A| \leq |B|$ and $|B| \leq |C|$, then $|A| \leq |C|$.

Proof. Exercise 1.1. □

We see that $\leq$ is reflexive and transitive. It remains to establish antisymmetry. Unlike the other properties, this is a major theorem.

1.6 Cantor-Bernstein Theorem *If $|X| \leq |Y|$ and $|Y| \leq |X|$, then $|X| = |Y|$.*

Proof. If $|X| \leq |Y|$, then there is a one-to-one function f that maps X into Y; if $|Y| \leq |X|$, then there is a one-to-one function g that maps Y into X. To show that $|X| = |Y|$ we have to exhibit a one-to-one function which maps X onto Y.

Let us apply first f and then g; the function $g \circ f$ maps X into X and is one-to-one. Clearly, $g[f[X]] \subseteq g[Y] \subseteq X$; moreover, since f and g are one-to-one, we have $|X| = |g[f[X]]|$ and $|Y| = |g[Y]|$. Thus the theorem follows from the next Lemma 1.7. (Let $A = X$, $B = g[Y]$, $A_1 = g[f[X]]$.) □

1.7 Lemma *If $A_1 \subseteq B \subseteq A$ and $|A_1| = |A|$, then $|B| = |A|$.*

The proof can be followed with the help of this diagram:

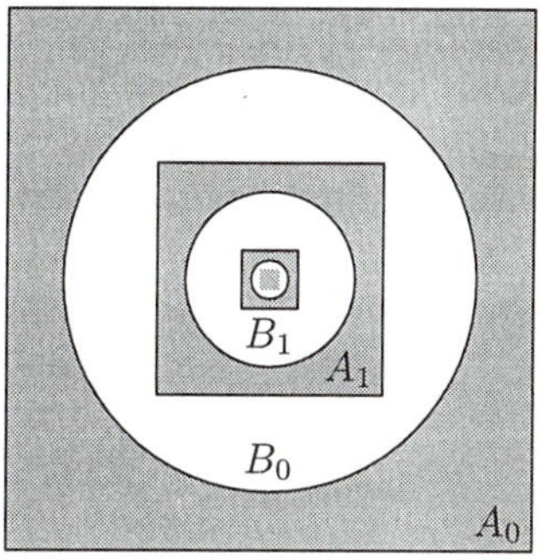

Proof. Let f be a one-to-one mapping of A onto A_1. By recursion, we define two sequences of sets:

$$A_0, A_1, \dots, A_n, \dots$$

and

$$B_0, B_1, \dots, B_n, \dots$$

(In the diagram, then A_n's are the squares and the B_n's are the disks.)

Let

$$A_0 = A, \quad B_0 = B,$$

and for each n,

$$(*) \qquad A_{n+1} = f[A_n], \quad B_{n+1} = f[B_n].$$

Since $A_0 \supseteq B_0 \supseteq A_1$, it follows from (*),by induction, that for each n, $A_n \supseteq A_{n+1}$. We let, for each n,

$$C_n = A_n - B_n,$$

and

$$C = \bigcup_{n=0}^{\infty} C_n, \quad D = A - C.$$

(C is the shaded part of the diagram.) By (*), we have $f[C_n] = C_{n+1}$, so

$$f[C] = \bigcup_{n=1}^{\infty} C_n.$$

Now we are ready to define a one-to-one mapping g of A onto B:

$$g(x) = \begin{cases} f(x) & \text{if } x \in C; \\ x & \text{if } x \in D. \end{cases}$$

Both $g \restriction C$ and $g \restriction D$ are one-to-one functions, and their ranges are disjoint. Thus g is a one-to-one function and maps A onto $f[C] \cup D = B$. □

We now see that $\leq$ has all the attributes of an ordering. A natural question to ask is whether it is linear, i.e., whether

(e) $|A| \leq |B|$ or $|B| \leq |A|$ holds for all A and B.

It is known that the proof of (e) requires the Axiom of Choice. We return to this subject in Chapter 8; in the meantime, we refrain from using (e) in any proofs.

By now, we have established the basic properties of cardinalities, without actually defining what cardinalities are. In principle, it is possible to continue the study of the properties $|A| = |B|$ and $|A| \leq |B|$ in this vein, without ever defining $|A|$; one can view $|A| = |B|$ simply as shorthand for the property "A is equipotent to B," etc. However, it is both conceptually and notationally useful to define $|A|$, "the number of elements of the set A," as an actual object of set theory, i.e., a set. We therefore make the following assumption.

1.8 Assumption There are sets called *cardinal numbers* (or *cardinals*) with the property that for every set X there is a unique cardinal $|X|$ (the *cardinal number of* X, the *cardinality of* X) and sets X and Y are equipotent if and only if $|X|$ is equal to $|Y|$.

In effect, we are assuming existence of a unique "representative" for each class of mutually equipotent sets. The Assumption is harmless in the sense that we only use it for convenience, and could formulate and prove all our theorems without it. The Assumption can actually be proved with the help of the Axiom of Choice, and we do that in Chapter 8. Moreover, for certain classes of sets, cardinal numbers can be defined, and the Assumption proved, even without the Axiom of Choice. The most important case is that of finite sets.

We study finite sets and their cardinal numbers in detail in the next section.

Exercises

1.1 Prove Lemma 1.5.

1.2 Prove:

(a) If $|A| < |B|$ and $|B| \leq |C|$, then $|A| < |C|$.

(b) If $|A| \leq |B|$ and $|B| < |C|$, then $|A| < |C|$.

1.3 If $A \subseteq B$, then $|A| \leq |B|$.

1.4 Prove:

(a) $|A \times B| = |B \times A|$.

(b) $|(A \times B) \times C| = |A \times (B \times C)|$.

(c) $|A| \leq |A \times B|$ if $B \neq \emptyset$.

1.5 Show that $|S| \leq |\mathcal{P}(S)|$. [*Hint*: $|S| = |\{\{a\} \mid a \in S\}|$.]

1.6 Show that $|A| \leq |A^S|$ for any A and any $S \neq \emptyset$. [*Hint*: Consider constant functions.]

1.7 If $S \subseteq T$, then $|A^S| \leq |A^T|$; in particular, $|A^n| \leq |A^m|$ if $n \leq m$. [*Hint*: Consider functions that have a fixed constant value on $T - S$.]

1.8 $|T| \le |S^T|$ if $|S| \ge 2$. [*Hint*: Pick $u, v \in S$, $u \ne v$, and, for each $t \in T$, consider $f_t : T \to S$ such that $f_t(t) = u$, $f_t(x) = v$ otherwise.]

1.9 If $|A| \le |B|$ and if A is nonempty then there exists a mapping f of B onto A.

It is somewhat peculiar that the proof of a fundamental, general result, like the Cantor-Bernstein Theorem, should require the use of such specific sets as natural numbers. In fact, it does not. The following sequence of exercises leads to an alternative proof, and in the process, acquaints the reader with the fixed-point property of monotone functions, an important result in itself.

Let F be a function on $\mathcal{P}(A)$ into $\mathcal{P}(A)$. A set $X \subseteq A$ is called a *fixed point* of F if $F(X) = X$. The function F is called *monotone* if $X \subseteq Y \subseteq A$ implies $F(X) \subseteq F(Y)$.

1.10 Let $F : \mathcal{P}(A) \to \mathcal{P}(A)$ be monotone. Then F has a fixed point. [*Hint*: Let $T = \{X \subseteq A \mid F(X) \subseteq X\}$. Notice that $T \ne \emptyset$. Let $\overline{X} = \bigcap T$ and prove that $\overline{X} \in T$, $F(\overline{X}) \in T$. Therefore, $F(\overline{X}) \subset \overline{X}$ is impossible.]

1.11 Use Exercise 1.10 to give an alternative proof of the Cantor-Bernstein Theorem. [*Hint*: Prove Lemma 1.7 as follows: let $F : \mathcal{P}(A) \to \mathcal{P}(A)$ be defined by $F(X) = (A - B) \cup f[X]$. Show that F is monotone. Let C be a fixed point of F, i.e., $C = (A - B) \cup f[C]$, and let $D = A - C$. Define g as in the original proof and show that it is one-to-one and onto B.]

1.12 Prove that $\overline{X}$ in Exercise 1.10 is the least fixed point of F, i.e., if $F(X) = X$ for some $X \subseteq A$, then $\overline{X} \subseteq X$.

The remaining exercises show that the two proofs are not that different, after all.

A function $F : \mathcal{P}(A) \to \mathcal{P}(A)$ is *continuous* if $F(\bigcup_{i \in \mathbf{N}} X_i) = \bigcup_{i \in \mathbf{N}} F(X_i)$ holds for any nondecreasing sequence of subsets of A. ($\langle X_i \mid i \in \mathbf{N} \rangle$ is *nondecreasing* if $X_i \subseteq X_j$ holds whenever $i \le j$.)

1.13 Prove that F used in Exercise 1.11 is continuous. [*Hint*: See Exercise 3.12 in Chapter 2.]

1.14 Prove that if $\overline{X}$ is the least fixed point of a monotone continuous function $F : \mathcal{P}(A) \to \mathcal{P}(A)$, then $\overline{X} = \bigcup_{i \in \mathbf{N}} X_i$ where we define recursively $X_0 = \emptyset$, $X_{i+1} = F(X_i)$.

The reader should compare the proof of Lemma 1.7 with the construction of the least fixed point for $F(X) = (A - B) \cup f[X]$ in Exercise 1.14.

2. Finite Sets

Finite sets can be defined as those sets whose size is a natural number.

2.1 Definition A set S is *finite* if it is equipotent to some natural number $n \in \mathbf{N}$. We then define $|S| = n$ and say that S *has n elements*. A set is *infinite* if it is not finite.

By our definition, cardinal numbers of finite sets are the natural numbers. Obviously, natural numbers are themselves finite sets, and $|n| = n$ for all $n \in \boldsymbol{N}$. However, we have to verify that the cardinal number of a finite set is unique. This follows from Lemma 2.2

2.2 Lemma *If $n \in \boldsymbol{N}$, then there is no one-to-one mapping of n onto a proper subset $X \subset n$.*

Proof. By induction on n. For $n = 0$, the assertion is trivially true. Assuming that it is true for n, let us prove it for $n+1$. If the assertion is false for $n+1$, then there is a one-to-one mapping f of $n+1$ onto some $X \subset n+1$. There are two possible cases: Either $n \in X$ or $n \notin X$. If $n \notin X$, then $X \subseteq n$, and $f \restriction n$ maps n onto a proper subset $X - \{f(n)\}$ of n, a contradiction. If $n \in X$, then $n = f(k)$ for some $k \leq n$. We consider the function g on n defined as follows:

$$g(i) = \begin{cases} f(i) & \text{for all } i \neq k,\ i < n; \\ f(n) & \text{if } i = k < n. \end{cases}$$

The function g is one-to-one and maps n onto $X - \{n\}$, a proper subset of n, a contradiction. □

2.3 Corollary
(a) If $n \neq m$, then there is no one-to-one mapping of n onto m.
(b) If $|S| = n$ and $|S| = m$, then $n = m$.
(c) $\boldsymbol{N}$ is infinite.

Proof.
(a) If $n \neq m$, then either $n \subset m$ or $m \subset n$, by Exercise 2.7 in Chapter 3, so there is no one-to-one mapping of n onto m.
(b) Immediate from (a).
(c) By Exercise 2.3 in Chapter 3, the successor function is a one-to-one mapping of $\boldsymbol{N}$ onto its proper subset $\boldsymbol{N} - \{0\}$.

□

Another noteworthy observation is that if $m, n \in \boldsymbol{N}$ and $m < n$ (in the usual ordering of natural numbers by size defined in Chapter 3), then $m \subset n$, so $m = |m| < |n| = n$, where $<$ is the ordering of cardinal numbers defined in the preceding section. Hence there is no need to distinguish between the two orderings, and we denote both by $<$.

The rest of the section studies properties of finite sets and their cardinals in more detail.

2.4 Theorem *If X is a finite set and $Y \subseteq X$, then Y is finite. Moreover, $|Y| \leq |X|$.*

Proof. We may assume that $X = \{x_0, \dots, x_{n-1}\}$, where $\langle x_0, \dots, x_{n-1}\rangle$ is a one-to-one sequence, and that Y is not empty. To show that Y is finite, we construct a one-to-one finite sequence whose range is Y. We use the Recursion Theorem in the version from Exercise 3.5 in Chapter 3. Let

$$k_0 = \text{the least } k \text{ such that } x_k \in Y;$$
$$k_{i+1} = \text{the least } k \text{ such that } k > k_i,\ k < n, \text{ and } x_k \in Y \text{ (if such } k \text{ exists).}$$

We leave it to the reader to verify that this definition fits the pattern of Exercise 3.5 in Chapter 3. [*Hint*: $A = n = \{0, 1, \dots, n-1\}$; $a = \min\{k \in n \mid x_k \in Y\}$; $g(t, i) = \min\{k \in n \mid k > t \text{ and } x_k \in Y\}$ if such k exists; undefined otherwise.] This defines a sequence $\langle k_0, \dots, k_{m-1}\rangle$. When we let $y_i = x_{k_i}$ for all $i < m$, then $Y = \{y_i \mid i < m\}$. The reader should verify that $m \le n$ (by induction, $k_i \ge i$ whenever defined, so especially $m - 1 \le k_{m-1} \le n - 1$). □

2.5 Theorem *If X is a finite set and f is a function, then $f[X]$ is finite. Moreover, $|f[X]| \le |X|$.*

Proof. Let $X = \{x_0, \dots, x_{n-1}\}$. Again, we use recursion to construct a finite one-to-one sequence whose range is $f[X]$. Actually, here we use the version with $f(n+1) = g(f \restriction n)$. We ask the reader to supply the details; the construction goes as follows:

$$k_0 = 0,$$
$$k_{i+1} = \text{the least } k > k_i \text{ such that } k < n \text{ and } f(x_k) \ne f(x_{k_j}) \text{ for all } j \le i$$

(if it exists), and $y_i = f(x_{k_i})$. Then $f[X] = \{y_0, \dots, y_{m-1}\}$ for some $m \le n$. □

As a consequence, if $\langle a_i \mid i < n\rangle$ is any finite sequence (with or without repetition), then the set $\{a_i \mid i < n\}$ is finite.

All the constructions made possible by the Axioms of Comprehension when applied to finite sets yield finite sets. We now show that if X is finite, then $\mathcal{P}(X)$ is finite, and if X is a finite collection of finite sets, then $\bigcup X$ is finite. Hence addition of the Axiom of Infinity is necessary in order to obtain infinite sets.

2.6 Lemma *If X and Y are finite, then $X \cup Y$ is finite. Moreover, $|X \cup Y| \le |X| + |Y|$, and if X and Y are disjoint, then $|X \cup Y| = |X| + |Y|$.*

Proof. If $X = \{x_0, \dots, x_{n-1}\}$ and if $Y = \{y_0, \dots, y_{m-1}\}$, where $\langle x_0, \dots, x_{n-1}\rangle$ and $\langle y_0, \dots, y_{m-1}\rangle$ are one-to-one finite sequences, consider the finite sequence $z = \langle x_0, \dots, x_{n-1}, y_0, \dots, y_{m-1}\rangle$ of length $n + m$. (Precisely, define $z = \langle z_i \mid 0 \le i < n + m\rangle$ by

$$z_i = x_i \quad \text{for } 0 \le i < n, \qquad z_i = y_{i-n} \quad \text{for } n \le i < n + m.)$$

Clearly, z maps $n+m$ onto $X \cup Y$, so $X \cup Y$ is finite and $|X \cup Y| \leq n+m$ by Theorem 2.5. If X and Y are disjoint, z is one-to-one and $|X \cup Y| = n+m$.

□

2.7 Theorem *If S is finite and if every $X \in S$ is finite, then $\bigcup S$ is finite.*

Proof. We proceed by induction on the number of elements of S. The statement is true if $|S| = 0$. Thus assume that it is true for all S with $|S| = n$, and let $S = \{X_0, \ldots, X_{n-1}, X_n\}$ be a set with $n+1$ elements, each $X_i \in S$ being a finite set. By the induction hypothesis, $\bigcup_{i=0}^{n-1} X_i$ is finite, and we have

$$\bigcup S = \left(\bigcup_{i=0}^{n-1} X_i \right) \cup X_n,$$

which is, therefore, finite by Lemma 2.6. □

2.8 Theorem *If X is finite, then $\mathcal{P}(X)$ is finite.*

Proof. By induction on $|X|$. If $|X| = 0$, i.e., $X = \emptyset$, then $\mathcal{P}(X) = \{\emptyset\}$ is finite. Assume that $\mathcal{P}(X)$ is finite whenever $|X| = n$, and let Y be a set with $n+1$ elements: $Y = \{y_0, \ldots, y_n\}$. Let $X = \{y_0, \ldots, y_{n-1}\}$. We note that $\mathcal{P}(Y) = \mathcal{P}(X) \cup U$, where $U = \{u \mid u \subseteq Y \text{ and } y_n \in u\}$. We further note that $|U| = |\mathcal{P}(X)|$ because there is a one-to-one mapping of U onto $\mathcal{P}(X)$: $f(u) = u - \{y_n\}$ for all $u \in U$. Hence $\mathcal{P}(Y)$ is a union of two finite sets and, consequently, finite. □

The final theorem of the section shows that infinite sets really have more elements than finite sets.

2.9 Theorem *If X is infinite, then $|X| > n$ for all $n \in \boldsymbol{N}$.*

Proof. It suffices to show that $|X| \geq n$ for all $n \in \boldsymbol{N}$. This can be done by induction. Certainly $0 < |X|$. Assume that $|X| \geq n$: then there is a one-to-one function $f : n \to X$. Since X is infinite, there exists $x \in (X - \operatorname{ran} f)$. Define $g = f \cup \{(n, x)\}$; g is a one-to-one function on $n+1$ into X. We conclude that $|X| \geq n+1$. □

To conclude this section, we briefly discuss another approach to finiteness. The following definition of finite sets does not use natural numbers: A set X is *finite* if and only if there exists a relation $\prec$ such that
(a) $\prec$ is a linear ordering of X.
(b) Every nonempty subset of X has a least and a greatest element in $\prec$.
Note that this notion of finiteness agrees with the one we defined using finite sequences: If $X = \{x_0, \ldots, x_{n-1}\}$, then $x_0 \prec \cdots \prec x_{n-1}$ describes a linear ordering of X with the foregoing properties. On the other hand, if $(X, \prec)$ satisfies (a) and (b), we construct, by recursion, a sequence $\langle f_0, f_1, \ldots \rangle$, as

in Theorem 3.4 in Chapter 3. As in that theorem, the sequence exhausts all elements of X, but the construction must come to a halt after a finite number of steps. Otherwise, the infinite set $\{f_0, f_1, f_2, \dots\}$ has no greatest element in $(X, \prec)$.

We mention another definition of finiteness that does not involve natural numbers. We say that X is *finite* if every nonempty family of subsets of X has a $\subseteq$-maximal element; i.e., if $\emptyset \neq U \subseteq \mathcal{P}(X)$, then there exists $z \in U$ such that for no $y \in U$, $z \subset y$. Exercise 2.6 shows equivalence of this definition with Definition 2.1.

Finally, let us consider yet another possible approach to finiteness. It follows from Lemma 2.2 that if X is a finite set, then there is no one-to-one mapping of X onto its proper subset. On the other hand, infinite sets, such as the set of all natural numbers $\boldsymbol{N}$, have one-to-one mappings onto a proper subset [e.g., $f(n) = n + 1$]. One might attempt to define finite sets as those sets which are not equipotent to any of their proper subsets. However, it is impossible to prove equivalence of this definition with Definition 2.1 without using the Axiom of Choice (see Exercise 1.9 in Chapter 8).

Exercises

2.1 If $S = \{X_0, \dots, X_{n-1}\}$ and the elements of S are mutually disjoint, then $|\bigcup S| = \sum_{i=0}^{n-1} |X_i|$.

2.2 If X and Y are finite, then $X \times Y$ is finite, and $|X \times Y| = |X| \cdot |Y|$.

2.3 If X is finite, then $|\mathcal{P}(X)| = 2^{|X|}$.

2.4 If X and Y are finite, then X^Y has $|X|^{|Y|}$ elements.

2.5 If $|X| = n \geq k = |Y|$, then the number of one-to-one functions $f : Y \to X$ is $n \cdot (n-1) \cdot \dots \cdot (n-k+1)$.

2.6 X is finite if and only if every nonempty system of subsets of X has a $\subseteq$-maximal element. [*Hint*: If X is finite, $|X| = n$ for some n. If $U \subseteq \mathcal{P}(X)$, let m be the greatest number in $\{|Y| \mid Y \in U\}$. If $Y \in U$ and $|Y| = m$, then Y is maximal. On the other hand, if X is infinite, let $U = \{Y \subseteq X \mid Y \text{ is finite}\}$.]

2.7 Use Lemma 2.6 and Exercises 2.2 and 2.4 to give easy proofs of commutativity and associativity for addition and multiplication of natural numbers, distributivity of multiplication over addition, and the usual arithmetic properties of exponentiation. [*Hint*: To prove, e.g., the commutativity of multiplication, pick X and Y such that $|X| = m$, $|Y| = n$. By Exercise 2.2, $m \cdot n = |X \times Y|$, $n \cdot m = |Y \times X|$. But $X \times Y$ and $Y \times X$ are equipotent.]

2.8 If A, B are finite and $X \subseteq A \times B$, then $|X| = \sum_{a \in A} k_a$ where $k_a = |X \cap (\{a\} \times B)|$.

3. Countable Sets

The Axiom of Infinity provides us with an example of an infinite set — the set $\boldsymbol{N}$ of all natural numbers. In this section, we investigate the cardinality of $\boldsymbol{N}$; that is, we are interested in sets that are equipotent to the set $\boldsymbol{N}$.

3.1 Definition A set S is *countable* if $|S| = |\boldsymbol{N}|$. A set S is *at most countable* if $|S| \leq |\boldsymbol{N}|$.

Thus a set S is countable if there is a one-to-one mapping of $\boldsymbol{N}$ onto S, that is, if S is the range of an infinite one-to-one sequence.

3.2 Theorem *An infinite subset of a countable set is countable.*

Proof. Let A be a countable set, and let $B \subseteq A$ be infinite. There is an infinite one-to-one sequence $\langle a_n \rangle_{n=0}^{\infty}$, whose range is A. We let $b_0 = a_{k_0}$, where k_0 is the least k such that $a_k \in B$. Having constructed b_n, we let $b_{n+1} = a_{k_{n+1}}$, where k_{n+1} is the least k such that $a_k \in B$ and $a_k \neq b_i$ for every $i \leq n$. Such k exists since B is infinite. The existence of the sequence $\langle b_n \rangle_{n=0}^{\infty}$ follows easily from the Recursion Theorem stated as Theorem 3.5 in Chapter 3. It is easily seen that $B = \{b_n \mid n \in \boldsymbol{N}\}$ and that $\langle b_n \rangle_{n=0}^{\infty}$ is one-to-one. Thus B is countable. □

If a set S is at most countable then it is equipotent to a subset of a countable set; by Theorem 3.2, this is either finite or countable. Thus we have Corollary 3.3.

3.3 Corollary *A set is at most countable if and only if it is either finite or countable.*

The range of an infinite one-to-one sequence is countable. If $\langle a_n \rangle_{n=0}^{\infty}$ is an infinite sequence which is not one-to-one, then the set $\{a_n\}_{n=0}^{\infty}$ may be finite (e.g., this happens if it is a constant sequence). However, if the range is infinite, then it is countable.

3.4 Theorem *The range of an infinite sequence $\langle a_n \rangle_{n=0}^{\infty}$ is at most countable, i.e., either finite or countable. (In other words, the image of a countable set under any mapping is at most countable.)*

Proof. By recursion, we construct a sequence $\langle b_n \rangle$ (with either finite or infinite domain) which is one-to-one and has the same range as $\langle a_n \rangle_{n=0}^{\infty}$. We let $b_0 = a_0$, and, having constructed b_n, we let $b_{n+1} = a_{k_{n+1}}$, where k_{n+1} is the least k such that $a_k \neq b_i$ for all $i \leq n$. (If no such k exists, then we consider the finite sequence $\langle b_i \mid i \leq n \rangle$.) The sequence $\langle b_i \rangle$ thus constructed is one-to-one and its range is $\{a_n\}_{n=0}^{\infty}$. □

One should realize that not all properties of size carry over from finite sets to the infinite case. For instance, a countable set S can be decomposed into two disjoint parts, A and B, such that $|A| = |B| = |S|$; that is inconceivable if S is finite (unless $S = \emptyset$).

Namely, consider the set $E = \{2k \mid k \in \boldsymbol{N}\}$ of all even numbers, and the set $O = \{2k+1 \mid k \in \boldsymbol{N}\}$ of all odd numbers. Both E and O are infinite, hence countable; thus we have $|\boldsymbol{N}| = |E| = |O|$ while $\boldsymbol{N} = E \cup O$ and $E \cap O = \emptyset$.

We can do even better. Let p_n denote the nth prime number (i.e., $p_0 = 2$, $p_1 = 3$, etc.). Let

$$S_0 = \{2^k \mid k \in \boldsymbol{N}\},\ S_1 = \{3^k \mid k \in \boldsymbol{N}\},\ \ldots,\ S_n = \{p_n^k \mid k \in \boldsymbol{N}\},\ \ldots.$$

The sets S_n $(n \in \boldsymbol{N})$ are mutually disjoint countable subsets of $\boldsymbol{N}$. Thus we have $\boldsymbol{N} \supseteq \bigcup_{n=0}^{\infty} S_n$, where $|S_n| = |\boldsymbol{N}|$ and the S_n's are mutually disjoint.

The following two theorems show that simple operations applied to countable sets yield countable sets.

3.5 Theorem *The union of two countable sets is a countable set.*

Proof. Let $A = \{a_n \mid n \in \boldsymbol{N}\}$ and $B = \{b_n \mid n \in \boldsymbol{N}\}$ be countable. We construct a sequence $\langle c_n \rangle_{n=0}^{\infty}$ as follows:

$$c_{2k} = a_k \text{ and } c_{2k+1} = b_k \quad \text{for all } k \in \boldsymbol{N}.$$

Then $A \cup B = \{c_n \mid n \in \boldsymbol{N}\}$ and since it is infinite, it is countable. □

3.6 Corollary *The union of a finite system of countable sets is countable.*

Proof. Corollary 3.6 can be proved by induction (on the size of the system). □

One might be tempted to conclude that the union of a countable system of countable sets in countable. However, this can be proved only if one uses the Axiom of Choice (see Theorem 1.7 in Chapter 8). Without the Axiom of Choice, one cannot even prove the following "evident" theorem: If $S = \{A_n \mid n \in \boldsymbol{N}\}$ and $|A_n| = 2$ for each n, then $\bigcup_{n=0}^{\infty} A_n$ is countable!

The difficulty is in *choosing* for each $n \in \boldsymbol{N}$ *a unique* sequence enumerating A_n. If such a choice can be made, the result holds, as is shown by Theorem 3.9. But first we need another important result.

3.7 Theorem *If A and B are countable, then $A \times B$ is countable.*

Proof. It suffices to show that $|\boldsymbol{N} \times \boldsymbol{N}| = |\boldsymbol{N}|$, i.e., to construct either a one-to-one mapping of $\boldsymbol{N} \times \boldsymbol{N}$ onto $\boldsymbol{N}$ or a one-to-one sequence with range $\boldsymbol{N} \times \boldsymbol{N}$. [See Exercise 3.1(b).]

(a) Consider the function

$$f(k, n) = 2^k \cdot (2n+1) - 1.$$

We leave it to the reader to verify that f is one-to-one and that the range of f is $\boldsymbol{N}$. □

(b) Here is another proof: Construct a sequence of elements of $\boldsymbol{N} \times \boldsymbol{N}$ in the manner prescribed by the diagram.

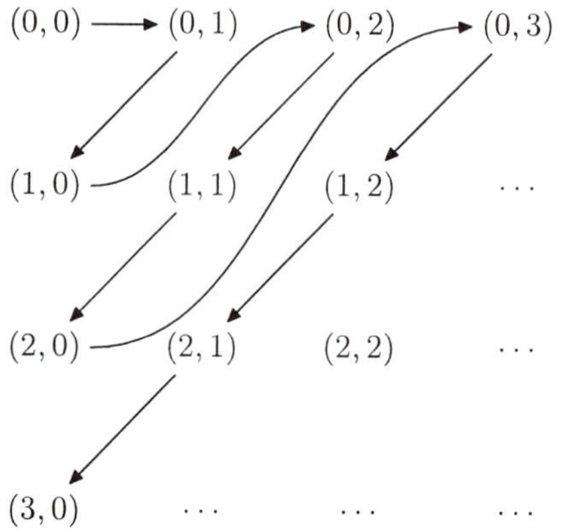

We have yet another proof:

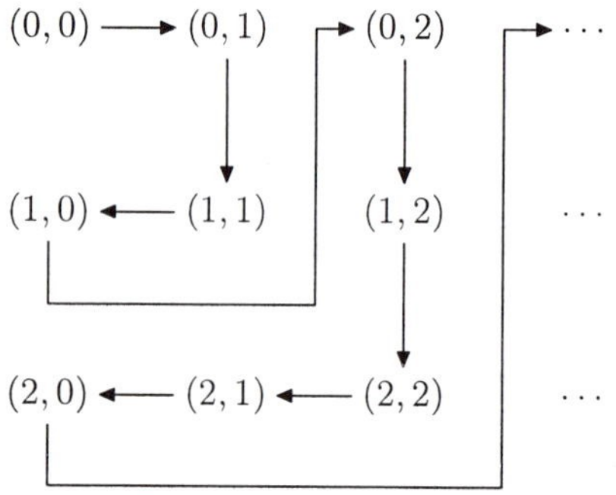

□

3.8 Corollary *The cartesian product of a finite number of countable sets is countable. Consequently, $\boldsymbol{N}^m$ is countable, for every $m > 0$.*

Proof. Corollary 3.8 can be proved by induction. □

3.9 Theorem *Let $\langle A_n \mid n \in \boldsymbol{N} \rangle$ be a countable system of at most countable sets, and let $\langle a_n \mid n \in \boldsymbol{N} \rangle$ be a system of enumerations of A_n; i.e., for each $n \in \boldsymbol{N}$, $a_n = \langle a_n(k) \mid k \in \boldsymbol{N} \rangle$ is an infinite sequence, and $A_n = \{a_n(k) \mid k \in \boldsymbol{N}\}$. Then $\bigcup_{n=0}^{\infty} A_n$ is at most countable.*

Proof. Define $f : \boldsymbol{N} \times \boldsymbol{N} \to \bigcup_{n=0}^{\infty} A_n$ by $f(n,k) = a_n(k)$. f maps $\boldsymbol{N} \times \boldsymbol{N}$ onto $\bigcup_{n=0}^{\infty} A_n$, so the latter is at most countable by Theorems 3.4 and 3.7. □

As a corollary of this result we can now prove

3.10 Theorem *If A is countable, then the set* $\operatorname{Seq}(A)$ *of all finite sequences of elements of A is countable.*

Proof. It is enough to prove the theorem for $A = \boldsymbol{N}$. As $\operatorname{Seq}(\boldsymbol{N}) = \bigcup_{n=0}^{\infty} \boldsymbol{N}^n$, the theorem follows from Theorem 3.9, if we can produce a sequence $\langle a_n \mid n \geq 1 \rangle$ of enumerations of $\boldsymbol{N}^n$. We do that by recursion.

Let g be a one-to-one mapping of $\boldsymbol{N}$ onto $\boldsymbol{N} \times \boldsymbol{N}$. Define recursively

$$\begin{aligned} a_1(i) &= \langle i \rangle \text{ for all } i \in \boldsymbol{N}; \\ a_{n+1}(i) &= \langle b_0, \ldots, b_{n-1}, i_2 \rangle \text{ where } g(i) = (i_1, i_2) \text{ and} \\ &\qquad \langle b_0, \ldots, b_{n-1} \rangle = a_n(i_1), \text{ for all } i \in \boldsymbol{N}. \end{aligned}$$

The idea is to let $a_{n+1}(i)$ be the $(n+1)$-tuple resulting from the concatenation of the (i_1)th n-tuple (in the previously constructed enumeration of n-tuples, a_n) with i_2. An easy proof by induction shows that a_n is onto $\boldsymbol{N}^n$, for all $n \geq 1$, and therefore $\bigcup_{n=1}^{\infty} \boldsymbol{N}^n$ is countable. Since $\boldsymbol{N}^0 = \{\langle\rangle\}$, $\bigcup_{n=0}^{\infty} \boldsymbol{N}^n$ is also countable. □

3.11 Corollary *The set of all finite subsets of a countable set is countable.*

Proof. The function F defined by $F(\langle a_0, \ldots, a_{n-1} \rangle) = \{a_0, \ldots, a_{n-1}\}$ maps the countable set $\operatorname{Seq}(A)$ onto the set of all finite subsets of A. The conclusion follows from Theorem 3.4. □

Other useful results about countable sets are the following.

3.12 Theorem *The set of all integers $\boldsymbol{Z}$ and the set of all rational numbers $\boldsymbol{Q}$ are countable.*

Proof. $\boldsymbol{Z}$ is countable because it is the union of two countable sets:

$$\boldsymbol{Z} = \{0, 1, 2, 3, \ldots\} \cup \{-1, -2, -3, \ldots\}.$$

$\boldsymbol{Q}$ is countable because the function $f : \boldsymbol{Z} \times (\boldsymbol{Z} - \{0\}) \to \boldsymbol{Q}$ defined by $f(p, q) = p/q$ maps a countable set onto $\boldsymbol{Q}$. □

3.13 Theorem *An equivalence relation on a countable set has at most countably many equivalence classes.*

Proof. Let E be an equivalence relation on a countable set A. The function F defined by $F(a) = [a]_E$ maps the countable set A onto the set A/E. By Theorem 3.4, A/E is at most countable. □

3.14 Theorem *Let $\mathfrak{A}$ be a structure with the universe A, and let $C \subseteq A$ be at most countable. Then $\overline{C}$, the closure of C, is also at most countable.*

Proof. Theorem 5.10 in Chapter 3 shows that $\overline{C} = \bigcup_{n=0}^{\infty} C_i$, where $C_0 = C$ and $C_{i+1} = C_i \cup F_0[C_i^{f_0}] \cup \cdots \cup F_{n-1}[C_i^{f_{n-1}}]$. It therefore suffices to produce a system of enumerations of $\langle C_i \mid i \in \boldsymbol{N}\rangle$.

Let $\langle c(k) \mid k \in \boldsymbol{N}\rangle$ be an enumeration of C, and let g be a mapping of $\boldsymbol{N}$ onto the countable set $(n+1) \times \boldsymbol{N} \times \boldsymbol{N}^{f_0} \times \cdots \times \boldsymbol{N}^{f_{n-1}}$. We define a system of enumerations $\langle a_i \mid i \in \boldsymbol{N}\rangle$ recursively as follows:

$$a_0(k) = c(k) \quad \text{for all } k \in \boldsymbol{N};$$

$$a_{i+1}(k) = \begin{cases} F_p(a_i(r_p^0), \dots, a_i(r_p^{f_p-1})) & \text{if } 0 \le p \le n-1; \\ a_i(q) & \text{if } p = n, \end{cases}$$

$$\text{where } g(k) = \langle p, q, \langle r_0^0, \dots, r_0^{f_0-1}\rangle, \dots, \langle r_{n-1}^0, \dots, r_{n-1}^{f_{n-1}-1}\rangle\rangle.$$

The definition of a_{i+1} is designed so as to make it transparent that if a_i enumerates C_i, a_{i+1} enumerates C_{i+1} (with many repetitions). By induction, a_i enumerates C_i for each $i \in \boldsymbol{N}$, as required. □

We conclude this section with the definition of the cardinal number of countable sets.

3.15 Definition $|A| = \boldsymbol{N}$ for all countable sets A.

We use the symbol $\aleph_0$ (aleph-naught) to denote the cardinal number of countable sets (i.e., the set of natural numbers, when it is used as a cardinal number).

Here is a reformulation of some of the results of this section in terms of the new notation.

3.16 (a) $\aleph_0 > n$ for all $n \in \boldsymbol{N}$;
if $\aleph_0 \ge \kappa$ for some cardinal number κ, then $\kappa = \aleph_0$ or $\kappa = n$ for some $n \in \boldsymbol{N}$ (this is Corollary 3.3).

(b) If $|A| = \aleph_0$, $|B| = \aleph_0$, then $|A \cup B| = \aleph_0$, $|A \times B| = \aleph_0$ (Theorems 3.5 and 3.7).

(c) If $|A| = \aleph_0$, then $|\operatorname{Seq}(A)| = \aleph_0$ (Theorem 3.10).

Exercises

3.1 Let $|A_1| = |A_2|$, $|B_1| = |B_2|$. Prove:
 (a) If $A_1 \cap A_2 = \emptyset$, $B_1 \cap B_2 = \emptyset$, then $|A_1 \cup A_2| = |B_1 \cup B_2|$.
 (b) $|A_1 \times A_2| = |B_1 \times B_2|$.
 (c) $|\operatorname{Seq}(A_1)| = |\operatorname{Seq}(A_2)|$.

3.2 The union of a finite set and a countable set is countable.

3.3 If $A \neq \emptyset$ is finite and B is countable, then $A \times B$ is countable.

3.4 If $A \neq \emptyset$ is finite, then $\operatorname{Seq}(A)$ is countable.

3.5 Let A be countable. The set $[A]^n = \{S \subseteq A \mid |S| = n\}$ is countable for all $n \in \boldsymbol{N}$, $n \neq 0$.

3.6 A sequence $\langle s_n \rangle_{n=0}^{\infty}$ of natural numbers is *eventually constant* if there is $n_0 \in \boldsymbol{N}$, $s \in \boldsymbol{N}$ such that $s_n = s$ for all $n \geq n_0$. Show that the set of eventually constant sequences of natural numbers is countable.

3.7 A sequence $\langle s_n \rangle_{n=0}^{\infty}$ of natural numbers is (eventually) *periodic* if there are $n_0, p \in \boldsymbol{N}$, $p \geq 1$, such that for all $n \geq n_0$, $s_{n+p} = s_n$. Show that the set of all periodic sequences of natural numbers is countable.

3.8 A sequence $\langle s_n \rangle_{n=0}^{\infty}$ of natural numbers is called an *arithmetic progression* if there is $d \in \boldsymbol{N}$ such that $s_{n+1} = s_n + d$ for all $n \in \boldsymbol{N}$. Prove that the set of all arithmetic progressions is countable.

3.9 For every $s = \langle s_0, \ldots, s_{n-1} \rangle \in \operatorname{Seq}(\boldsymbol{N} - \{0\})$, let $f(s) = p_0^{s_0} \cdots p_{n-1}^{s_{n-1}}$ where p_i is the ith prime number. Show that f is one-to-one and use this fact to give another proof of $|\operatorname{Seq}(\boldsymbol{N})| = \aleph_0$.

3.10 Let $(S, <)$ be a linearly ordered set and let $\langle A_n \mid n \in \boldsymbol{N} \rangle$ be an infinite sequence of finite subsets of S. Then $\bigcup_{n=0}^{\infty} A_n$ is at most countable. [*Hint*: For every $n \in \boldsymbol{N}$ consider the unique enumeration $\langle a_n(k) \mid k < |A_n| \rangle$ of A_n in the increasing order.]

3.11 Any partition of an at most countable set has a set of representatives.

4. Linear Orderings

In the preceding section we proved countability of various familiar sets, such as the set of all integers $\boldsymbol{Z}$ and the set of all rational numbers $\boldsymbol{Q}$. The important point we want to make is that we cannot distinguish among the sets $\boldsymbol{N}$, $\boldsymbol{Z}$, and $\boldsymbol{Q}$ solely on the basis of their cardinality; nevertheless, the three sets "look" quite different (visualize them as subsets of the real line!). In order to capture this difference, we have to consider the way they are ordered. Then it becomes apparent that the ordering of $\boldsymbol{N}$ by size is quite different from the usual ordering of $\boldsymbol{Z}$ (for example, $\boldsymbol{N}$ has a least element and $\boldsymbol{Z}$ does not), and both are quite different from the usual ordering of $\boldsymbol{Q}$ (for example, between any two distinct rational numbers lie infinitely many rationals, while between any two distinct integers lie only finitely many integers). Linear orderings are an important tool for deeper study of properties of sets; therefore, we devote this section to the theory of linearly ordered sets, using countable sets for illustration.

4.1 Definition Linearly ordered sets $(A, <)$ and $(B, \prec)$ are *similar* (*have the same order type*) if they are isomorphic, i.e., if there is a one-to-one mapping f on A onto B such that for all $a_1, a_2 \in A$, $a_1 < a_2$ holds if and only if $f(a_1) \prec f(a_2)$ holds. (See Definition 5.17 in Chapter 2 and the subsequent Lemma.)

Similar ordered sets "look alike"; their orderings have the same properties [see Example 5.8 in Chapter 3]. It follows that $(\boldsymbol{N}, <)$ and $(\boldsymbol{Z}, <)$ are not similar; likewise, $(\boldsymbol{Z}, <)$ and $(\boldsymbol{Q}, <)$ are not similar, and neither are $(\boldsymbol{N}, <)$ and $(\boldsymbol{Q}, <)$. (Here, $<$ is the usual ordering of numbers.)

It is trivial to show that the property of similarity behaves like an equivalence relation:

(a) $(A, <)$ is similar to $(A, <)$.
(b) If $(A, <)$ is similar to $(B, \prec)$, then $(B, \prec)$ is similar to $(A, <)$.
(c) If $(A_1, <_1)$ is similar to $(A_2, <_2)$, and $(A_2, <_2)$ is similar to $(A_3, <_3)$, then $(A_1, <_1)$ is similar to $(A_3, <_3)$.

Just as in the case of cardinal numbers, it is possible to assume that with each linearly ordered set there is associated an object called its *order type* so that similar ordered sets have the same order type. To avoid technical problems connected with a formal definition of order types, we use them only as a figure of speech, which can be avoided by talking about similar sets instead. We define order types of well-ordered sets (the most important special case) rigorously in Chapter 6.

We begin our study of linear orderings by establishing that a finite set can be linearly ordered in only one way, up to isomorphism.

4.2 Lemma *Every linear ordering on a finite set is a well-ordering.*

Proof. We prove that every nonempty finite subset B of a linearly ordered set $(A, <)$ has a least element by induction on the number of elements of B. If B has 1 element, the claim is clearly true. Assume that it is true for all n-element sets and let B have $n+1$ elements. Then $B = \{b\} \cup B'$ where B' has n elements and $b \notin B'$. By the inductive assumption, B' has a least element b'. If $b' < b$, then b' is the least element of B; otherwise, b is the least element of B. In either case, B has a least element. □

4.3 Theorem *If $(A_1, <_1)$ and $(A_2, <_2)$ are linearly ordered sets and $|A_1| = |A_2|$ is finite, then $(A_1, <_1)$ and $(A_2, <_2)$ are similar.*

Proof. We proceed by induction on $n = |A_1| = |A_2|$. If $n = 0$, then $A_1 = A_2 = \emptyset$ and $(A_1, <_1)$, $(A_2, <_2)$ are clearly isomorphic. Assume that the claim is true for all linear orderings of n-element sets. Let $|A_1| = |A_2| = n+1$. We proved that $<_1$ and $<_2$ are well-orderings, so let a_1 (a_2, respectively) be the least element of $(A_1, <_1)$ [$(A_2, <_2)$, respectively]. Now $|A_1 - \{a_1\}| = |A_2 - \{a_2\}| = n$, so by the inductive assumption, there is an isomorphism g between $(A_1 - \{a_1\}, <_1 \cap (A_1 - \{a_1\})^2)$ and $(A_2 - \{a_2\}, <_2 \cap (A_2 - \{a_2\})^2)$. Define $f : A_1 \to A_2$ by

$$f(a_1) = a_2;$$
$$f(a) = g(a) \quad \text{for } a \in A_1 - \{a_1\}.$$

It is easy to check that f is an isomorphism between $(A_1, <_1)$ and $(A_2, <_2)$. □

We conclude that for finite sets, order types correspond to cardinal numbers. As the examples at the beginning of this section show, linear orderings of infinite sets are much more interesting. We next look at some ways of producing linear orderings that will be useful later.

4.4 Lemma *If $(A, <)$ is a linear ordering, then $(A, <^{-1})$ is also a linear ordering.*

Proof. Left to the reader (see Exercise 5.3 in Chapter 2). □

For example, the inverse of the ordering $(\boldsymbol{N}, <)$ is the ordering $(\boldsymbol{N}, <^{-1})$: $\cdots 4 <^{-1} 3 <^{-1} 2 <^{-1} 1 <^{-1} 0$; notice that it is similar to the ordering of negative integers by size:

$$\cdots - 4 < -3 < -2 < -1$$

and that it is not a well-ordering.

4.5 Lemma *Let $(A_1, <_1)$ and $(A_2, <_2)$ be linearly ordered sets and $A_1 \cap A_2 = \emptyset$. The relation $<$ on $A = A_1 \cup A_2$ defined by*

$$\begin{aligned} a < b \textit{ if and only if } & a, b \in A_1 \textit{ and } a <_1 b \\ & \textit{or } a, b \in A_2 \textit{ and } a <_2 b \\ & \textit{or } a \in A_1,\ b \in A_2 \end{aligned}$$

is a linear ordering.

Proof. This is Exercise 5.6 in Chapter 2, so again we leave it to the reader. □

The set A is ordered by putting all elements of A_1 before all elements of A_2. We say that the linearly ordered set $(A, <)$ is the *sum* of the linearly ordered sets $(A_1, <_1)$ and $(A_2, <_2)$.

Notice that the order type of the sum does not depend on the particular orderings $(A_1, <_1)$ and $(A_2, <_2)$, only on their types (see Exercise 4.1). As an example, the linearly ordered set $(\boldsymbol{Z}, <)$ of all integers is similar to the sum of the linearly ordered sets $(\boldsymbol{N}, <^{-1})$ and $(\boldsymbol{N}, <)$ ($<$ denotes the usual ordering of numbers by size).

Next, let us consider ways to order cartesian product.

4.6 Lemma *Let $(A_1, <_1)$ and $(A_2, <_2)$ be linearly ordered sets. The relation $<$ on $A = A_1 \times A_2$ defined by*

$$(a_1, a_2) < (b_1, b_2) \quad \textit{if and only if} \quad a_1 <_1 b_1 \textit{ or } (a_1 = b_1 \textit{ and } a_2 <_2 b_2)$$

is a linear ordering.

Proof. *Transitivity*: If $(a_1, a_2) < (b_1, b_2)$ and $(b_1, b_2) < (c_1, c_2)$, we have either $a_1 <_1 b_1$ or ($a_1 = b_1$ and $a_2 <_2 b_2$).

In the first case $a_1 <_1 b_1$ and $b_1 \leq_1 c_1$ gives $a_1 <_1 c_1$. In the second case, either $b_1 <_1 c_1$ and $a_1 <_1 c_1$ again, or $b_1 = c_1$ and $b_2 <_2 c_2$, so that $a_1 = c_1$ and $a_2 <_2 c_2$. In either case we conclude that $(a_1, a_2) < (c_1, c_2)$.

Asymmetry: This follows immediately from asymmetry of $<_1$ and $<_2$.

Linearity: Given (a_1, a_2) and (b_1, b_2), one of the following cases has to occur:

(a) $a_1 <_1 b_1$ [so $(a_1, a_2) < (b_1, b_2)$].
(b) $b_1 <_1 a_1$ [so $(b_1, b_2) < (a_1, a_2)$].
(c) $a_1 = b_1$ and $a_2 <_2 b_2$ [so $(a_1, a_2) < (b_1, b_2)$].
(d) $a_1 = b_1$ and $b_2 <_2 a_2$ [so $(b_1, b_2) < (a_1, a_2)$].
(e) $a_1 = b_1$ and $a_2 = b_2$ [so $(a_1, a_2) = (b_1, b_2)$].
In each case (a_1, a_2) and (b_1, b_2) are comparable in $<$. □

We call $<$ the *lexicographic ordering* (*lexicographic product*) of $A_1 \times A_2$; the terminology comes from the observation that if $A_1 = A_2 = \{a, b, \ldots, z\}$ is the set of all letters and $<_1 = <_2$ is the alphabetic ordering $a <_1 b <_1 c <_1 \cdots <_1 z$, then $<$ orders elements of $A_1 \times A_2$ ("two-letter words") as they would be in a dictionary.

One can easily generalize the notion of lexicographic ordering to a product of any finite or infinite sequence of linearly ordered sets.

4.7 Theorem *Let $\langle (A_i, <_i) \mid i \in I \rangle$ be an indexed system of linearly ordered sets, where $I \subseteq \boldsymbol{N}$. The relation $\prec$ on $\prod_{i \in I} A_i$ defined by*

$$f \prec g \quad \textit{if and only if } \operatorname{diff}(f, g) = \{i \in I \mid f_i \neq g_i\} \neq \emptyset \textit{ and } f_{i_0} <_{i_0} g_{i_0}$$
$$\textit{where } i_0 \textit{ is the least element of } \operatorname{diff}(f, g)$$
$$\textit{(in the usual ordering } < \textit{ of natural numbers)}$$

is a linear ordering of $\prod_{i \in I} A_i$ (it is called its lexicographic ordering*).*

Proof. *Transitivity*: Assume that $f \prec g$ and $g \prec h$ and let i_0 [j_0, respectively] be the least element of $\operatorname{diff}(f, g)$ [$\operatorname{diff}(g, h)$, respectively]. If $i_0 < j_0$, we have $f_{i_0} < g_{i_0}$ and $g_{i_0} = h_{i_0}$, so $f_{i_0} < h_{i_0}$ and i_0 is the least element of $\operatorname{diff}(f, h)$. We conclude that $f \prec h$. The cases $i_0 = j_0$ and $i_0 > j_0$ are similar.

Asymmetry: $f \prec g$ and $g \prec f$ is impossible because it would mean that $f_{i_0} < g_{i_0}$ and $g_{i_0} < f_{i_0}$ for $i_0 =$ the least element of $\operatorname{diff}(f, g) = \operatorname{diff}(g, f)$.

Linearity: If $\operatorname{diff}(f, g) = \emptyset$, we have $f = g$. Otherwise, if i_0 is the least element of $\operatorname{diff}(f, g)$, either $f_{i_0} < g_{i_0}$ or $f_{i_0} > g_{i_0}$ holds and, consequently, either $f \prec g$ or $f \succ g$. □

In particular, if $(A_i, <_i) = (A, <)$ for all $i \in I = \boldsymbol{N}$, $\prec$ is the lexicographic ordering of the set $A^{\boldsymbol{N}}$ of all infinite sequences of elements of A.

One can also choose to compare second coordinates before comparing the first coordinates and so define the *antilexicographic ordering* $\prec$ of $A_1 \times A_2$:

$$(a_1, a_2) \prec (b_1, b_2) \quad \text{if and only if} \quad a_2 <_2 b_2 \text{ or } (a_2 = b_2 \text{ and } a_1 <_1 b_1).$$

The proof that $\prec$ is a linear ordering is entirely analogous to the lexicographic case. The two orderings are generally quite different; compare, for example, the lexicographic and antilexicographic products of $A_1 = \boldsymbol{N} = \{0, 1, 2, \ldots\}$ and $A_2 = \{0, 1\}$ (both ordered by size):

ordered lexicographically:

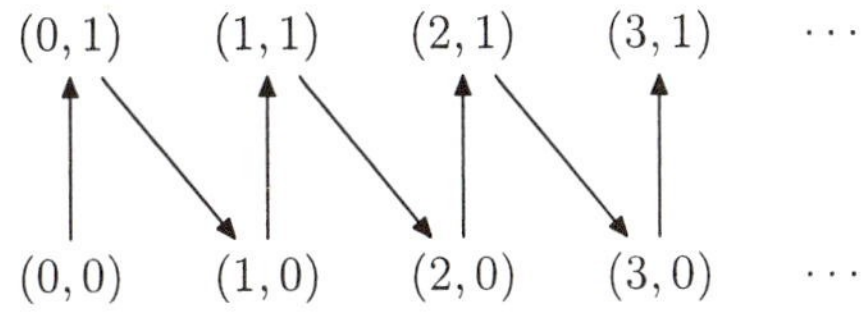

ordered antilexicographically:

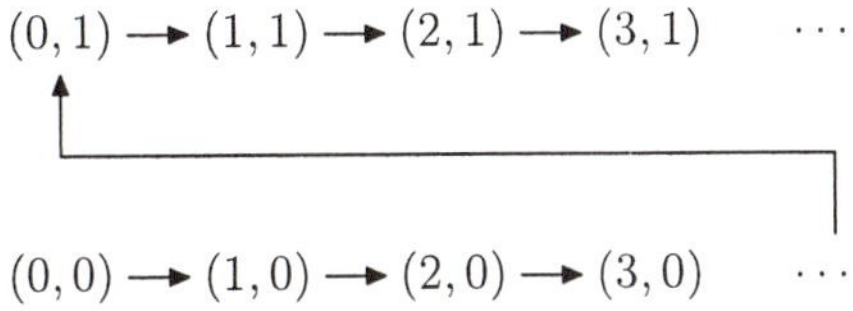

The first ordering is similar to $(\boldsymbol{N}, <)$ and the second is not [it is the sum of two copies of $(\boldsymbol{N}, <)$].

The previous results show that there is a rich variety of types of linear orderings on countable sets. It is thus rather surprising to learn that there is a *universal* linear ordering of countable sets, i.e., such that every countable linearly ordered set is similar to one of its subsets. The rest of this section is devoted to the proof of this important result.

4.8 Definition An ordered set $(X, <)$ is *dense* if it has at least two elements and if for all $a, b \in X$, $a < b$ implies that there exists $x \in X$ such that $a < x < b$.

Let us call the least and the greatest elements of a linearly ordered set (if they exist) the *endpoints* of the set.

The most important example of a countable dense linearly ordered set is the set $\boldsymbol{Q}$ of all rational numbers, ordered by size. The ordering is dense because, if r, s are rational numbers and $r < s$, then $x = (r+s)/2$ is also a rational number, and $r < x < s$. Moreover, $(\boldsymbol{Q}, <)$ has no endpoints (if $r \in \boldsymbol{Q}$ then $r+1, r-1 \in \boldsymbol{Q}$ and $r - 1 < r < r + 1$). Other examples of countable dense linearly ordered sets are in Exercises 4.6 and 4.7. However, we prove that all countable linearly ordered sets without endpoints have the same order type.

4.9 Theorem *Let $(P, \prec)$ and $(Q, <)$ be countable dense linearly ordered sets without endpoints. Then $(P, \prec)$ and $(Q, <)$ are similar.*

Proof. Let $\langle p_n \mid n \in \boldsymbol{N} \rangle$ be a one-to-one sequence such that $P = \{p_n \mid n \in \boldsymbol{N}\}$, and let $\langle q_n \mid n \in \boldsymbol{N} \rangle$ be a one-to-one sequence such that $Q = \{q_n \mid n \in \boldsymbol{N}\}$. A function h on a subset of P into Q is called a *partial isomorphism* from P to Q if $p \prec p'$ if and only if $h(p) < h(p')$ holds for all $p, p' \in \operatorname{dom} h$.

We need the following *claim*: If h is a partial isomorphism from P to Q such that $\operatorname{dom} h$ is finite, and if $p \in P$ and $q \in Q$, then there is a partial isomorphism $h_{p,q} \supseteq h$ such that $p \in \operatorname{dom} h_{p,q}$, and $q \in \operatorname{ran} h_{p,q}$.

Proof of the claim. Let $h = \{(p_{i_1}, q_{i_1}), \dots, (p_{i_k}, q_{i_k})\}$ where $p_{i_1} \prec p_{i_2} \prec \dots \prec p_{i_k}$, and thus also $q_{i_1} < q_{i_2} < \dots < q_{i_k}$. If $p \notin \operatorname{dom} h$, we have either $p \prec p_{i_1}$ or $p_{i_e} \prec p \prec p_{i_{e+1}}$ for some $1 \le e \le k$, or $p_{i_k} \prec p$. Take the least natural number n such that q_n is in the same relationship to $q_{i_1}, \dots, q_{i_k}$ as p is to $p_{i_1}, \dots, p_{i_k}$; more precisely, q_n is such that:

if $p \prec p_{i_1}$, then $q_n < q_{i_1}$;

if $p_{i_e} \prec p \prec p_{i_{e+1}}$, then $q_{i_e} < q_n < q_{i_{e+1}}$; and

if $p_{i_k} \prec p$, then $q_{i_k} < q_n$.

The possibility of doing this is guaranteed by the fact that $(Q, <)$ is a dense linear ordering without endpoints. It is clear that $h' = h \cup \{(p, q_n)\}$ is a partial isomorphism. If $q \in \operatorname{ran} h'$, then we are done. If $q \notin \operatorname{ran} h'$, then by the same argument as before (with the roles of P and Q reversed), there is $p_m \in P$ such that $h' \cup \{(p_m, q)\}$ is a partial isomorphism, and we take the least such m, and let $h_{p,q} = h' \cup \{(p_m, q)\}$. The claim is now proved. □

We next construct a sequence of compatible partial isomorphisms by recursion:

$$h_0 = \emptyset,$$
$$h_{n+1} = (h_n)_{p_n, q_n}$$

where $(h_n)_{p_n,q_n}$ is the extension of h_n (as provided by the claim) such that $p_n \in \operatorname{dom}(h_n)_{p_n,q_n}$, $q_n \in \operatorname{ran}(h_n)_{p_n,q_n}$. Let $h = \bigcup_{n \in N} h_n$. It is trivial to verify that $h : P \to Q$ is an isomorphism between $(P, \prec)$ and $(Q, <)$. □

4.10 Theorem *Every countable linearly ordered set can be mapped isomorphically into any countable dense linearly ordered set without endpoints.*

Proof. Theorem 4.10 requires a one-sided version of the previous proof. Let $(P, \prec)$ be a countable linearly ordered set and let $(Q, <)$ be a countable dense linearly ordered set without endpoints. For every partial isomorphism h from the ordered set $(P, \prec)$ into Q and for every $p \in P$, we define a partial isomorphism $h_p \supseteq h$ such that $p \in \operatorname{dom} h_p$. Then we use recursion again. □

Exercises

4.1 Assume that $(A_1, <_1)$ is similar to $(B_1, \prec_1)$ and $(A_2, <_2)$ is similar to $(B_2, \prec_2)$.

(a) The sum of $(A_1, <_1)$ and $(A_2, <_2)$ is similar to the sum of $(B_1, \prec_1)$ and $(B_2, \prec_2)$, assuming that $A_1 \cap A_2 = \emptyset = B_1 \cap B_2$.

(b) The lexicographic product of $(A_1, <_1)$ and $(A_2, <_2)$ is similar to the lexicographic product of $(B_1, \prec_1)$ and $(B_2, \prec_2)$.

4.2 Give an example of linear orderings $(A_1, <_1)$ and $(A_2, <_2)$ such that the sum of $(A_1, <_1)$ and $(A_2, <_2)$ does not have the same order type as the sum of $(A_2, <_2)$ and $(A_1, <_1)$ ("addition of order types is not commutative"). Do the same thing for lexicographic product.

4.3 Prove that the sum and the lexicographic product of two well-orderings are well-orderings.

4.4 If $\langle A_i \mid i \in \boldsymbol{N} \rangle$ is an infinite sequence of linearly ordered sets of natural numbers and $|A_i| \geq 2$ for all $i \in \boldsymbol{N}$, then the lexicographic ordering of $\prod_{i \in \boldsymbol{N}} A_i$ is *not* a well-ordering.

4.5 Let $\langle (A_i, <_i) \mid i \in I \rangle$ be an indexed system of mutually disjoint linearly ordered sets, $I \subseteq \boldsymbol{N}$. The relation $\prec$ on $\bigcup_{i \in I} A_i$ defined by: $a \prec b$ if and only if either $a, b \in A_i$ and $a <_i b$ for some $i \in I$ or $a \in A_i$, $b \in A_j$ and $i < j$ (in the usual ordering of natural numbers) is a linear ordering. If all $<_i$ are well-orderings, so is $\prec$.

4.6 Let $(\boldsymbol{Z}, <)$ be the set of all integers with the usual linear ordering. Let $\prec$ be the lexicographic ordering of $\boldsymbol{Z}^{\boldsymbol{N}}$ as defined in Theorem 4.7. Finally, let $FS \subseteq \boldsymbol{Z}^{\boldsymbol{N}}$ be the set of all eventually constant elements of $\boldsymbol{Z}^{\boldsymbol{N}}$; i.e., $\langle a_i \mid i \in \boldsymbol{N} \rangle \in FS$ if and only if there exist $n_0 \in \boldsymbol{N}$, $a \in \boldsymbol{Z}$ such that $a_i = a$ for all $i \geq n_0$ (compare with Exercise 3.6). Prove that FS is countable and $(FS, \prec \cap FS^2)$ is a dense linearly ordered set without endpoints.

4.7 Let $\prec$ be the lexicographic ordering of $\boldsymbol{N}^{\boldsymbol{N}}$ (where $\boldsymbol{N}$ is assumed to be ordered in the usual way) and let $P \subseteq \boldsymbol{N}^{\boldsymbol{N}}$ be the set of all eventually periodic, but not eventually constant, sequences of natural numbers (see Exercises 3.6 and 3.7 for definitions of these concepts). Show that $(P, \prec \cap P^2)$ is a countable dense linearly ordered set without endpoints.

4.8 Let $(A, <)$ be linearly ordered. Define $\prec$ on Seq(A) by: $\langle a_0, \dots, a_{m-1} \rangle \prec \langle b_0, \dots, b_{n-1} \rangle$ if and only if there is $k < n$ such that $a_i = b_i$ for all $i < k$ and either $a_k < b_k$ or a_k is undefined (i.e., $k = m < n$). Prove that $\prec$ is a linear ordering. If $(A, <)$ is well-ordered, (Seq$(A), \prec$) is also well-ordered. (In this ordering, if a finite sequence extends a shorter sequence, the shorter one comes before the longer one.)

4.9 Let $(A, <)$ be linearly ordered. Define $\prec$ on Seq(A) by: $\langle a_0, \dots, a_{m-1} \rangle \prec \langle b_0, \dots, b_{n-1} \rangle$ if and only if there is $k < m$ such that $a_i = b_i$ for all $i < k$ and either $a_k < b_k$ or b_k is undefined (i.e., $k = n < m$). Prove that $\prec$ is a linear ordering. If $|A| \geq 2$, it is not a well-ordering. (In this ordering, if a finite sequence extends a shorter one, the longer one comes before the shorter one.) If $A = \boldsymbol{N}$ and $<$ is the usual ordering of natural numbers, $\prec$ is called the Brouwer-Kleene ordering of Seq$(\boldsymbol{N})$; it is a dense linear ordering with no least element and a greatest element $\langle \rangle$.

4.10 Let $(A, <)$ be a linearly ordered set without endpoints, $A \neq \emptyset$. A *closed interval* $[a, b]$ is defined for $a, b \in A$ by $[a, b] = \{x \in A \mid a \leq x \leq b\}$. Assume that each closed interval $[a, b]$, $a, b \in A$, has a finite number of elements. Then $(A, <)$ is similar to the set $\boldsymbol{Z}$ of all integers in the usual ordering.

4.11 Let $(A, <)$ be a dense linearly ordered set. Show that for all $a, b \in A$, $a < b$, the closed interval $[a, b]$, as defined in Exercise 4.10, has infinitely many elements.

4.12 Show that all countable dense linearly ordered sets with both endpoints are similar.

4.13 Let $(\boldsymbol{Q}, <)$ be the set of all rational numbers in the usual ordering. Find subsets of $\boldsymbol{Q}$ similar to

(a) the sum of two copies of $(\boldsymbol{N}, <)$;
(b) the sum of $(\boldsymbol{N}, <)$ and $(\boldsymbol{N}, <^{-1})$;
(c) the lexicographic product of $(\boldsymbol{N}, <)$ with $(\boldsymbol{N}, <)$.

5. Complete Linear Orderings

We have seen in the preceding section that the usual ordering $<$ of the set $\boldsymbol{Q}$ of all rational numbers is universal among countable linear orderings (Theorem 4.10). However, when arithmetic operations on $\boldsymbol{Q}$ are considered, it becomes apparent that something is missing. For example, there is no rational number x such that $x^2 = 2$ (Exercise 5.1). Another example of this phenomenon appears when one considers decimal representations of rational numbers. Every rational number has a decimal expansion that is either finite (e.g., $1/4 = 0.25$) or infinite but periodic from some place on (e.g., $1/6 = 0.1666\cdots$) (see Section 10.1). Although it is possible to write down decimal expansions $0.a_1a_2a_3\cdots$ where $\langle a_i \rangle_{i=1}^{\infty}$ is an arbitrary sequence of integers between 0 and 9, unless the sequence is finite or eventually periodic, there is no rational number x such that $x = 0.a_1a_2a_3\cdots$. (As a specific example, consider $0.1010010001\cdots$.) It is clear from these considerations that the ordered set $(\boldsymbol{Q}, <)$ has gaps.

The notion of a gap can be formulated in terms involving only the linear ordering.

5.1 Definition Let $(P, <)$ be a linearly ordered set. A *gap* is a pair (A, B) of sets such that
(a) A and B are nonempty disjoint subsets of P and $A \cup B = P$.
(b) If $a \in A$ and $b \in B$, then $a < b$.
(c) A does not have a greatest element and B does not have a least element.

For example, let $B = \{x \in \boldsymbol{Q} \mid x > 0 \text{ and } x^2 > 2\}$ and $A = \boldsymbol{Q} - B = \{x \in \boldsymbol{Q} \mid x \leq 0 \text{ or } (x > 0 \text{ and } x^2 < 2)\}$. It is not difficult to check that (A, B) is a gap in $\boldsymbol{Q}$ (Exercise 5.2). Similarly, an infinite decimal expansion which is not eventually periodic gives rise to a gap (Exercise 5.3).

At this point, we ask the reader to recall the notions of upper and lower bound, and supremum and infimum, that were introduced in Chapter 2. We call a nonempty subset of a linearly ordered set P *bounded* if it has both lower and upper bounds. A set is *bounded from above* (*from below*) if it has an upper (lower) bound.

Let (A, B) be a gap in a linearly ordered set. The set A is bounded from above because any $b \in B$ is its upper bound. We claim that A does not have a supremum. For if c were a supremum of A then either c would be the greatest element of A or the least element of B, as one can easily verify.

On the other hand, let S be a nonempty set, bounded from above. Let

$$A = \{x \mid x \leq s \text{ for some } s \in S\},$$
$$B = \{x \mid x > s \text{ for every } s \in S\}.$$

The reader is asked to verify that the pair (A, B) satisfies the first two properties (a) and (b) in the definition of a gap. Now assume that S does not have a supremum. Then (A, B) is a gap, because the greatest element of A, or the least element of B, would be the supremum of S.

We notice that the existence of gaps is closely related to the existence, or rather nonexistence, of suprema of bounded sets. Thus we are led to the definition of a *complete* ordered set.

5.2 Definition Let $(P, <)$ be a dense linearly ordered set. P is *complete* if every nonempty $S \subseteq P$ bounded from above has a supremum. Note that $(P, <)$ is complete if and only if it does not have any gaps.

As we have seen, dense linearly ordered sets are not necessarily complete. However, every dense linearly ordered set can be *completed* by "filling the gaps," and the result is essentially uniquely determined. This is the content of the following important theorem.

5.3 Theorem *Let $(P, <)$ be a dense linearly ordered set without endpoints. Then there exists a complete linearly ordered set $(C, \prec)$ such that*

(a) $P \subseteq C$.

(b) If $p, q \in P$, then $p < q$ if and only if $p \prec q$ ($\prec$ coincides with $<$ on P).

(c) P is dense in C, i.e., for any $p, q \in P$ such that $p < q$, there is $c \in C$ with $p \prec c \prec q$.

(d) C does not have endpoints.

Moreover, this complete linearly ordered set $(C, \prec)$ is unique up to isomorphism over P. In other words, if $(C^*, \prec^*)$ is a complete linearly ordered set which satisfies (a)–(d), then there is an isomorphism h between $(C, \prec)$ and $(C^*, \prec^*)$ such that $h(x) = x$ for each $x \in P$. The linearly ordered set $(C, \prec)$ is called the completion *of $(P, <)$.*

As is usual in theorems of this type, the uniqueness part of the theorem is easier to prove. For that reason, we do it first.

Proof of Uniqueness of Completion. Let $(C, \prec)$ and $(C^*, \prec^*)$ be two complete linearly ordered sets satisfying (a)–(d). We show that there is an isomorphism h of C onto C^* such that $h(x) = x$ for each $x \in P$.

If $c \in C$, let $S_c = \{p \in P \mid p \preccurlyeq c\}$. Similarly, let $S_{c^*} = \{p \in P \mid p \preccurlyeq^* c^*\}$ for $c^* \in C^*$. If S is a nonempty subset of P bounded from above, let $\sup S$ be the supremum of S in $(C, \prec)$ and $\sup^* S$ be the supremum of S in $(C^*, \prec^*)$. Note that $\sup S_c = c$, $\sup^* S_{c^*} = c^*$.

We define the mapping h as follows: $h(c) = \sup^* S_c$.

Clearly, h is a mapping of C into C^*; we have to show that the mapping is onto and that

(a) If $c \prec d$, then $h(c) \prec^* h(d)$.

(b) $h(x) = x$ for each $x \in P$.

To show that h is onto, let $c^* \in C^*$ be arbitrary. Then $c^* = \sup^* S_{c^*}$, and if we let $c = \sup S_{c^*}$ then $S_c = S_{c^*}$ and $c^* = h(c)$. If $c \prec d$, then (since P

is dense in C) there is $p \in P$ such that $c \prec p \prec d$. It is readily seen that $\sup^* S_c \prec^* p \prec^* \sup^* S_d$ and hence $h(c) \prec^* h(d)$. From this we conclude that h is an isomorphism using Lemma 5.18 from Chapter 2. Finally, if $x \in P$, then $x = \sup S_x = \sup^* S_x$ and so $h(x) = x$. □

To prove the existence of a completion, we introduce the notion of a Dedekind cut.

5.4 Definition A *cut* is a pair (A, B) of sets such that
(a) A and B are disjoint nonempty subsets of P and $A \cup B = P$.
(b) If $a \in A$ and $b \in B$, then $a < b$.

We recall that a cut is a gap if, in addition, A does not have a greatest element and B does not have a least element. Notice that since P is dense, it is not possible that both A has a greatest element and B has a least element. Thus the remaining possibilities are that either B has a least element and A does not have a greatest element, or A has a greatest element and B does not have a least element. In either case, the supremum of A exists: In the first case, the supremum is the least element of B, and in the other case, the supremum is the greatest element of A. Hence, we consider only the first case and disregard the cuts where A has the greatest element.

5.5 Definition A cut (A, B) is a *Dedekind cut* if A does not have a greatest element.

We have two types of Dedekind cuts (A, B):
(a) Those where $B = \{x \in P \mid x \geq p\}$ for some $p \in P$; we denote $(A, B) = [p]$.
(b) Gaps.
Now we consider the set C of all Dedekind cuts (A, B) in $(P, <)$ and order C as follows:

$$(A, B) \preccurlyeq (A', B') \quad \text{if and only if} \quad A \subseteq A'.$$

We leave it to the reader to verify that $(C, \preccurlyeq)$ is a linearly ordered set.

If $p, q \in P$ are such that $p < q$, then we have $[p] \prec [q]$. Thus the linearly ordered set $(P', \prec)$, where $P' = \{[p] \mid p \in P\}$, is isomorphic to $(P, <)$. We intend to show that $(C, \prec)$ is a completion of $(P', \prec)$. Since $(P, <)$ and $(P', \prec)$ are isomorphic, it follows that $(P, <)$ has a completion.

It suffices to prove
(c′) P' is dense in $(C, \prec)$,
(d′) C does not have endpoints,
and, of course,
(e) $(C, \prec)$ is complete.

To show that P' is dense in C, let $c, d \in C$ be such that $c \prec d$; i.e., $c = (A, B)$, $d = (A', B')$, and $A \subset A'$. Let $p \in P$ be such that $p \in A'$ and $p \notin A$. Moreover, we can assume that p is not the least element of B. Then $(A, B) \prec [p] \prec (A', B')$ and hence P' is dense in C. [This also shows that $(C, \prec)$ is a densely ordered set.]

Similarly, if $(A,B) \in C$, then there is $p \in B$ that is not the least element of B, and we have $(A,B) \prec [p]$. Hence C does not have a greatest element. For analogous reasons, it does not have a least element.

To show that C is complete, let S be a nonempty subset of C, bounded from above. Therefore, there is $(A_0, B_0) \in C$ such that $A \subseteq A_0$ whenever $(A,B) \in S$. To find the supremum of S, let

$$A_S = \bigcup\{A \mid (A,B) \in S\} \quad \text{and} \quad B_S = P - A_S = \bigcap\{B \mid (A,B) \in S\}.$$

It is easy to verify that (A_S, B_S) is a cut. (Note that B_S is nonempty because $B_0 \subseteq B_S$.) In fact, (A_S, B_S) is a Dedekind cut: A_S does not have a greatest element since none of the A's does.

Since $A_S \supseteq A$ for each $(A,B) \in S$, (A_S, B_S) is an upper bound of S; let us show that (A_S, B_S) is the least upper bound of S. If $(\overline{A}, \overline{B})$ is any upper bound of S, then $A \subseteq \overline{A}$ for all $(A,B) \in S$, and, so, $A_S = \bigcup\{A \mid (A,B) \in S\} \subseteq \overline{A}$; hence $(A_S, B_S) \preccurlyeq (\overline{A}, \overline{B})$. Thus (A_S, B_S) is the supremum of S. □

Theorem 5.3 is now proved. In particular, the ordered set $(\boldsymbol{Q}, <)$ of rationals has a unique completion (up to isomorphism); this is the ordered set of real numbers. As the ordering of real numbers coincides with $<$ on $\boldsymbol{Q}$, it is customary to use the symbol $<$ (rather than $\prec$) for it as well.

5.6 Definition The completion of $(\boldsymbol{Q}, <)$ is denoted $(\boldsymbol{R}, <)$; the elements of $\boldsymbol{R}$ are the *real numbers*.

We get immediately the following characterization of $(\boldsymbol{R}, <)$.

5.7 Theorem *$(\boldsymbol{R}, <)$ is the unique (up to isomorphism) complete linearly ordered set without endpoints that has a countable subset dense in it.*

Proof. Let $(C, \prec)$ be a complete linearly ordered set without endpoints, and let P be a countable subset of C dense in C. Then $(P, \prec)$ is isomorphic to $(\boldsymbol{Q}, <)$ and by the uniqueness of completion (Theorem 5.3), $(C, \prec)$ is isomorphic to the completion of $(\boldsymbol{Q}, <)$, i.e., to $(\boldsymbol{R}, <)$. □

It remains to define algebraic operations on $\boldsymbol{R}$ and show that they satisfy the usual laws of algebra and that on rational numbers they agree with their previous definitions. As this topic is of more interest to algebra and real analysis than set theory, we do not pursue it here. The reader interested in learning how the arithmetic of real numbers can be rigorously established can at this point read Section 10.2.

Exercises

5.1 Prove that there is no $x \in \boldsymbol{Q}$ for which $x^2 = 2$. [*Hint*: Write $x = p/q$ where $p, q \in \boldsymbol{Z}$ are relatively prime, and use $p^2 = 2q^2$ to show that 2 has to divide both p and q.]

5.2 Show that (A, B), where $A = \{x \in \boldsymbol{Q} \mid x \leq 0 \text{ or } (x > 0 \text{ and } x^2 < 2)\}$, $B = \{x \in \boldsymbol{Q} \mid x > 0 \text{ and } x^2 > 2\}$, is a gap in $(\boldsymbol{Q}, <)$. [*Hint*: To prove that A does not have a greatest element, given $x > 0$, $x^2 < 2$, find a rational $\varepsilon > 0$ such that $(x + \varepsilon)^2 < 2$. It suffices to take $\varepsilon < x$ such that $x^2 + 3x\varepsilon < 2$.]

5.3 Let $0.a_1a_2a_3 \cdots$ be an infinite, but not periodic, decimal expansion. Let $A = \{x \in \boldsymbol{Q} \mid x \leq 0.a_1a_2 \cdots a_k \text{ for some } k \in \boldsymbol{N} - \{0\}\}$, $B = \{x \in \boldsymbol{Q} \mid x \geq 0.a_1a_2 \cdots a_k \text{ for all } k \in \boldsymbol{N} - \{0\}\}$. Show that (A, B) is a gap in $(\boldsymbol{Q}, <)$.

5.4 Show that a dense linearly ordered set $(P, <)$ is complete if and only if every nonempty $S \subseteq P$ bounded from below has an infimum.

5.5 Let D be dense in $(P, <)$, and let E be dense in $(D, <)$. Show that E is dense in $(P, <)$.

5.6 Let F be the set of all rational numbers that have a decimal expansion with only a finite number of nonzero digits. Show that F is dense in $\boldsymbol{Q}$.

5.7 Let D (the *dyadic rationals*) be the set of all numbers $m/2^n$ where m is an integer and n is a natural number. Show that D is dense in $\boldsymbol{Q}$.

5.8 Prove that the set $\boldsymbol{R} - \boldsymbol{Q}$ of all irrational numbers is dense in $\boldsymbol{R}$. [*Hint*: Given $a < b$, let $x = (a+b)/2$ if it is irrational, $x = (a+b)/\sqrt{2}$ otherwise. Use Exercise 5.1.]

6. Uncountable Sets

All infinite sets whose cardinalities we have determined up to this point turned out to be countable. Naturally, a question arises whether perhaps all infinite sets are countable. If it were so, this book might end with the preceding section. It was a great discovery of Georg Cantor that uncountable sets, in fact, exist. This discovery provided an impetus for the development of set theory and became a source of its depth and richness.

6.1 Theorem *The set $\boldsymbol{R}$ of all real numbers is uncountable.*

Proof. $(\boldsymbol{R}, <)$ is a dense linear ordering without endpoints. If $\boldsymbol{R}$ were countable, $(\boldsymbol{R}, <)$ would be isomorphic to $(\boldsymbol{Q}, <)$ by Theorem 4.9. But this is not possible because $(\boldsymbol{R}, <)$ is complete and $(\boldsymbol{Q}, <)$ is not. □

The proof above relies on the theory of linear orderings developed in Section 4. Cantor's original proof used his famous "diagonalization argument." We give it below.

Cantor's Proof of Theorem 6.1. Assume that $\boldsymbol{R}$ is countable, i.e., $\boldsymbol{R}$ is the range of some infinite sequence $\langle r_n \rangle_{n=1}^{\infty}$. Let $a_0^{(n)}.a_1^{(n)}a_2^{(n)}a_3^{(n)} \cdots$ be the decimal expansion of r_n. (We assume that a decimal expansion does not contain only the digit 9 from some place on, so each real number has a unique decimal expansion. See Section 10.1.) Let $b_n = 1$ if $a_n^{(n)} = 0$, $b_n = 0$ otherwise; and let r be the real number whose decimal expansion is $0.b_1b_2b_3\cdots$. We have $b_n \neq a_n^{(n)}$, hence $r \neq r_n$, for all $n = 1, 2, 3, \ldots$, a contradiction. □

The combinatorial heart of the diagonal argument (quite similar to Russell's Paradox, which is of later origin) becomes even clearer in the next theorem.

6.2 Theorem *The set of all sets of natural numbers is uncountable; in fact,* $|\mathcal{P}(\boldsymbol{N})| > |\boldsymbol{N}|$.

Proof. The function $f : \boldsymbol{N} \to \mathcal{P}(\boldsymbol{N})$ defined by $f(n) = \{n\}$ is one-to-one, so $|\boldsymbol{N}| \leq |\mathcal{P}(\boldsymbol{N})|$. We prove that for every sequence $\langle S_n \mid n \in \boldsymbol{N}\rangle$ of subsets of $\boldsymbol{N}$ there is some $S \subseteq \boldsymbol{N}$ such that $S \neq S_n$ for all $n \in \boldsymbol{N}$. This shows that there is no mapping of $\boldsymbol{N}$ onto $\mathcal{P}(\boldsymbol{N})$, and hence $|\boldsymbol{N}| < |\mathcal{P}(\boldsymbol{N})|$.

We define the set $S \subseteq \boldsymbol{N}$ as follows: $S = \{n \in \boldsymbol{N} \mid n \notin S_n\}$. The number n is used to distinguish S from S_n: If $n \in S_n$, then $n \notin S$, and if $n \notin S_n$, then $n \in S$. In either case, $S \neq S_n$, as required. □

Detailed study of uncountable sets is the subject of Chapter 5 (and subsequent chapters). Here we only prove that the set $2^{\boldsymbol{N}} = \{0,1\}^{\boldsymbol{N}}$ of all infinite sequences of 0's and 1's is also uncountable, and, in fact, has the same cardinality as $\mathcal{P}(\boldsymbol{N})$ and $\boldsymbol{R}$.

6.3 Theorem $|\mathcal{P}(\boldsymbol{N})| = |2^{\boldsymbol{N}}| = |\boldsymbol{R}|$.

Proof. For each $S \subseteq \boldsymbol{N}$ define the *characteristic function* of S, $\chi_S : \boldsymbol{N} \to \{0,1\}$, as follows:

$$\chi_S(n) = \begin{cases} 0 & \text{if } n \in S; \\ 1 & \text{if } n \notin S. \end{cases}$$

It is easy to check that the correspondence between sets and their characteristic functions is a one-to-one mapping of $\mathcal{P}(\boldsymbol{N})$ onto $\{0,1\}^{\boldsymbol{N}}$.

To complete the proof, we show that $|\boldsymbol{R}| \leq |\mathcal{P}(\boldsymbol{N})|$ and also $|2^{\boldsymbol{N}}| \leq |\boldsymbol{R}|$ and use the Cantor-Bernstein Theorem.

(a) We have constructed real numbers as cuts in the set $\boldsymbol{Q}$ of all rational numbers. The function that assigns to each real number $r = (A, B)$ the set $A \subseteq \boldsymbol{Q}$ is a one-to-one mapping of $\boldsymbol{R}$ into $\mathcal{P}(\boldsymbol{Q})$. Therefore $|\boldsymbol{R}| \leq |\mathcal{P}(\boldsymbol{Q})|$. As $|\boldsymbol{Q}| = |\boldsymbol{N}|$, we have $|\mathcal{P}(\boldsymbol{Q})| = |\mathcal{P}(\boldsymbol{N})|$ (Exercise 6.3). Hence $|\boldsymbol{R}| \leq |\mathcal{P}(\boldsymbol{N})|$.

(b) To prove $|2^{\boldsymbol{N}}| \leq |\boldsymbol{R}|$ we use the decimal representation of real numbers. The function that assigns to each infinite sequence $\langle a_n\rangle_{n=0}^{\infty}$ of 0's and 1's the unique real number whose decimal expansion is $0.a_0a_1a_2\cdots$ is a one-to-one mapping of $2^{\boldsymbol{N}}$ into $\boldsymbol{R}$. Therefore we have $|2^{\boldsymbol{N}}| \leq |\boldsymbol{R}|$.

□

We introduced $\aleph_0$ as a notation for the cardinal of $\boldsymbol{N}$. Due to Theorem 6.3, the cardinal number of $\boldsymbol{R}$ is usually denoted $2^{\aleph_0}$. The set $\boldsymbol{R}$ of all real numbers is also referred to as "the continuum"; for this reason, $2^{\aleph_0}$ is called the *cardinality of the continuum.* In this notation, Theorem 6.2 says that $\aleph_0 < 2^{\aleph_0}$.

Exercises

6.1 Use the diagonal argument to show that $\boldsymbol{N}^{\boldsymbol{N}}$ is uncountable. [*Hint*: Consider $\langle a_n \mid n \in \boldsymbol{N} \rangle$ where $a_n = \langle a_{nk} \mid k \in \boldsymbol{N} \rangle$. Define $d \in \boldsymbol{N}^{\boldsymbol{N}}$ by $d_n = a_{nn} + 1$.]

6.2 Show that $|\boldsymbol{N}^{\boldsymbol{N}}| = 2^{\aleph_0}$. [*Hint*: $2^{\boldsymbol{N}} \subseteq \boldsymbol{N}^{\boldsymbol{N}} \subseteq \mathcal{P}(\boldsymbol{N} \times \boldsymbol{N})$.]

6.3 Show that $|A| = |B|$ implies $|\mathcal{P}(A)| = |\mathcal{P}(B)|$.

Chapter 5

Cardinal Numbers

1. Cardinal Arithmetic

Cardinal numbers have been introduced in Chapter 4. The present chapter is devoted to the study of their general properties, with the particular emphasis on the cardinality of the continuum, $2^{\aleph_0}$. In this section we set out to define arithmetic operations (addition, multiplication, and exponentiation) on cardinal numbers and investigate the properties of these operations.

To define the *sum* $\kappa+\lambda$ of two cardinals, we use the analogy with finite sets: If a set A has a elements, a set B has b elements, and if A and B are disjoint, then $A \cup B$ has $a+b$ elements.

1.1 Definition $\kappa + \lambda = |A \cup B|$ where $|A| = \kappa$, $|B| = \lambda$, and $A \cap B = \emptyset$.

In order to make this definition legitimate, we have to show that $\kappa + \lambda$ does not depend on the choice of the sets A and B. This is the content of the following lemma.

1.2 Lemma *If A, B, A', B' are such that $|A| = |A'|$, $|B| = |B'|$, and $A \cap B = \emptyset = A' \cap B'$, then $|A \cup B| = |A' \cup B'|$.*

Proof. Let f and g be, respectively, a one-to-one mapping of A onto A' and of B onto B'. Then $f \cup g$ is a one-to-one mapping of $A \cup B$ onto $A' \cup B'$. □

Not only does addition of cardinals coincide with the ordinary addition of numbers in case of finite cardinals, but many of the usual laws of addition remain valid. For example, addition of cardinal numbers is commutative and associative:

(a) $\kappa + \lambda = \lambda + \kappa$.
(b) $\kappa + (\lambda + \mu) = (\kappa + \lambda) + \mu$.

These laws follow directly from the definition. Similarly, the following inequalities are easily established:

(c) $\kappa \le \kappa + \lambda$.
(d) If $\kappa_1 \le \kappa_2$ and $\lambda_1 \le \lambda_2$, then $\kappa_1 + \lambda_1 \le \kappa_2 + \lambda_2$.
However, not all laws of addition of numbers hold also for addition of cardinals. In particular, strict inequalities in formulas are rare in case of infinite cardinals and, as is discussed later (König's Theorem), those that hold are quite difficult to establish. As an example, take the simple fact that if $n \neq 0$, then $n + n > n$. If κ is infinite, then this is no longer true: We have seen that $\aleph_0 + \aleph_0 = \aleph_0$ [see 3.16(b) in Chapter 4], and the Axiom of Choice implies that $\kappa + \kappa = \kappa$ for *every* infinite κ.

The multiplication of cardinals is again motivated by properties of multiplication of numbers. If A and B are sets of a and b elements, respectively, then the product $A \times B$ has $a \cdot b$ elements.

1.3 Definition $\kappa \cdot \lambda = |A \times B|$, where $|A| = \kappa$ and $|B| = \lambda$.

The legitimacy of this definition follows from Lemma 1.4.

1.4 Lemma *If A, B, A', B' are such that $|A| = |A'|$, $|B| = |B'|$, then $|A \times B| = |A' \times B'|$.*

Proof. Let $f : A \to A'$, $g : B \to B'$ be mappings. We define $h : A \times B \to A' \times B'$ as follows:

$$h(a, b) = (f(a), g(b)).$$

Clearly, if f and g are one-to-one and onto, so is h. □

Again, multiplication has some expected properties; in particular, it is commutative and associative. Moreover, the distributive law holds.
(e) $\kappa \cdot \lambda = \lambda \cdot \kappa$.
(f) $\kappa \cdot (\lambda \cdot \mu) = (\kappa \cdot \lambda) \cdot \mu$.
(g) $\kappa \cdot (\lambda + \mu) = \kappa \cdot \lambda + \kappa \cdot \mu$.
The last property is a consequence of the equality

$$A \times (B \cup C) = (A \times B) \cup (A \times C),$$

that holds for any sets A, B, and C. We also have:
(h) $\kappa \le \kappa \cdot \lambda$ if $\lambda > 0$.
(i) If $\kappa_1 \le \kappa_2$ and $\lambda_1 \le \lambda_2$, then $\kappa_1 \cdot \lambda_1 \le \kappa_2 \cdot \lambda_2$.
To make the analogy between multiplication of cardinals and multiplication of numbers even more complete, let us prove
(j) $\kappa + \kappa = 2 \cdot \kappa$.

Proof. If $|A| = \kappa$, then $2 \cdot \kappa$ is the cardinal of $\{0, 1\} \times A$. We note that $\{0, 1\} \times A = (\{0\} \times A) \cup (\{1\} \times A)$, that $|\{0\} \times A| = |\{1\} \times A| = \kappa$, and that the two summands are disjoint. Hence $2 \cdot \kappa = \kappa + \kappa$. □

As a consequence of (j), we have
(k) $\kappa + \kappa \le \kappa \cdot \kappa$, whenever $\kappa \ge 2$.

As in the case of addition, multiplication of infinite cardinals has some properties different from those valid for finite numbers. For example, $\aleph_0 \cdot \aleph_0 = \aleph_0$ [see 3.16(b) in Chapter 4]. (And the Axiom of Choice implies that $\kappa \cdot \kappa = \kappa$ for all infinite cardinals.)

To define exponentiation of cardinal numbers, we note that if A and B are finite sets, with a and b elements respectively, then a^b is the number of all functions from B to A.

1.5 Definition $\kappa^\lambda = |A^B|$, where $|A| = \kappa$ and $|B| = \lambda$.

The definition of κ^λ does not depend on the choice of A and B.

1.6 Lemma *If* $|A| = |A'|$ *and* $|B| = |B'|$, *then* $|A^B| = |A'^{B'}|$.

Proof. Let $f : A \to A'$ and $g : B \to B'$ be one-to-one and onto. Let $F : A^B \to A'^{B'}$ be defined as follows: If $k \in A^B$, let $F(k) = h$, where $h \in A'^{B'}$ is such that $h(g(b)) = f(k(b))$ for all $b \in B$, i.e., $h = f \circ k \circ g^{-1}$.

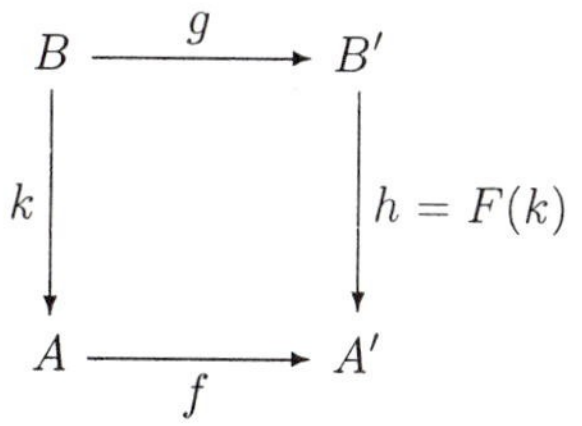

Then F is one-to-one and maps A^B onto $A'^{B'}$. □

It is easily seen from the definition of exponentiation that
(l) $\kappa \le \kappa^\lambda$ if $\lambda > 0$.
(m) $\lambda \le \kappa^\lambda$ if $\kappa > 1$.
(n) If $\kappa_1 \le \kappa_2$ and $\lambda_1 \le \lambda_2$, then $\kappa_1^{\lambda_1} \le \kappa_2^{\lambda_2}$.
We also have
(o) $\kappa \cdot \kappa = \kappa^2$.
To prove (o), it suffices to have a one-to-one correspondence between $A \times A$, the set of all pairs (a, b) with $a, b \in A$, and the set of all functions from $\{0, 1\}$ into A. This correspondence has been established in Section 5 of Chapter 3.

The next theorem gives further properties of exponentiation.

1.7 Theorem
(a) $\kappa^{\lambda+\mu} = \kappa^\lambda \cdot \kappa^\mu$.
(b) $(\kappa^\lambda)^\mu = \kappa^{\lambda\cdot\mu}$.
(c) $(\kappa \cdot \lambda)^\mu = \kappa^\mu \cdot \lambda^\mu$.

Proof. Let $\kappa = |K|$, $\lambda = |L|$, $\mu = |M|$. To show (a), assume that L and M are disjoint. We construct a one-to-one mapping F of $K^L \times K^M$ onto $K^{L\cup M}$.

If $(f,g) \in K^L \times K^M$, we let $F(f,g) = f \cup g$. We note that $f \cup g$ is a function, in fact a member of $K^{L \cup M}$, and every $h \in K^{L \cup M}$ is equal to $F(f,g)$ for some $(f,g) \in K^L \times K^M$ (namely, $f = h \restriction L$, $g = h \restriction M$). It is easily seen that F is one-to-one.

To prove (b), we look for a one-to-one map F of $K^{L \times M}$ onto $(K^L)^M$. A typical element of $K^{L \times M}$ is a function $f : L \times M \to K$. We let F assign to f the function $g : M \to K^L$ defined as follows: for all $m \in M$, $g(m) = h \in K^L$ where $h(l) = f(l,m)$ (for all $l \in L$). We leave it to the reader to verify that F is one-to-one and onto.

For a proof of (c) we need a one-to-one mapping F of $K^M \times L^M$ onto $(K \times L)^M$. For each $(f_1, f_2) \in K^M \times L^M$, let $F(f_1, f_2) = g : M \to (K \times L)$ where $g(m) = (f_1(m), f_2(m))$, for all $m \in M$. It is routine to check that F is one-to-one and onto. □

We now have a collection of useful general properties of cardinals, but the only specific cardinal numbers encountered so far are natural numbers, $\aleph_0$, and $2^{\aleph_0}$. We show next that, in fact, there exist many other cardinalities. The fundamental result in this direction is the general Cantor's Theorem (see Theorem 6.2 in Chapter 4 for a special case).

1.8 Cantor's Theorem *$|X| < |\mathcal{P}(X)|$, for every set X.*

Proof. The proof is a straightforward generalization of the proof of Theorem 6.2 in Chapter 4. Its heart is an abstract form of the diagonalization argument.

First, the function $f : X \to \mathcal{P}(X)$ defined by $f(x) = \{x\}$ is clearly one-to-one, and so $|X| \leq |\mathcal{P}(X)|$. It remains to be proven that there is no mapping of X onto $\mathcal{P}(X)$. So let f be a mapping of X into $\mathcal{P}(X)$. Consider the set $S = \{x \in X \mid x \notin f(x)\}$. We claim that S is not in the range of f. Suppose that $S = f(z)$ for some $z \in X$. By definition of S, $z \in S$ if and only if $z \notin f(z)$; so we have $z \in S$ if and only if $z \notin S$, a contradiction. This shows that f is not onto $\mathcal{P}(X)$, and the proof of $|X| < |\mathcal{P}(X)|$ is complete. □

The first half of Theorem 6.3 of Chapter 4 also generalizes to arbitrary sets.

1.9 Theorem *$|\mathcal{P}(X)| = 2^{|X|}$, for every set X.*

Proof. Replace $\boldsymbol{N}$ by X in the proof of $|\mathcal{P}(\boldsymbol{N})| = 2^{|\boldsymbol{N}|}$ in Theorem 6.3 in Chapter 4. □

Cantor's Theorem can now be restated in terms of cardinal numbers as follows:

$$\kappa < 2^\kappa \quad \text{for every cardinal number } \kappa.$$

We conclude this section with an observation that for any set of cardinal numbers, there exists a cardinal number greater than all of them.

1.10 Corollary *For any system of sets S there is a set Y such that $|Y| > |X|$ holds for all $X \in S$.*

Proof. Let $Y = \mathcal{P}(\bigcup S)$. By Cantor's Theorem, $|Y| > |\bigcup S|$, and clearly $|\bigcup S| \geq |X|$ for all $X \in S$ (because $X \subseteq \bigcup S$ if $X \in S$). □

Exercises

1.1 Prove properties (a)–(n) of cardinal arithmetic stated in the text of this section.

1.2 Show that $\kappa^0 = 1$ and $\kappa^1 = \kappa$ for all κ.

1.3 Show that $1^\kappa = 1$ for all κ and $0^\kappa = 0$ for all $\kappa > 0$.

1.4 Prove that $\kappa^\kappa \leq 2^{\kappa \cdot \kappa}$.

1.5 If $|A| \leq |B|$ and if $A \neq \emptyset$, then there is a mapping of B onto A. We later show, with the help of the Axiom of Choice, that the converse is also true: If there is a mapping of B onto A, then $|A| \leq |B|$.

1.6 If there is a mapping of B onto A, then $2^{|A|} \leq 2^{|B|}$. [*Hint*: Given g mapping B onto A, let $f(X) = g^{-1}[X]$, for all $X \subseteq A$.]

1.7 Use Cantor's Theorem to show that the "set of all sets" does not exist.

1.8 Let X be a set and let f be a one-to-one mapping of X into itself such that $f[X] \subset X$. Then X is infinite.

Call a set X *Dedekind infinite* if there is a one-to-one mapping of X onto its proper subset. A *Dedekind finite* set is a set that is not Dedekind infinite. The remaining exercises investigate properties of Dedekind finite and Dedekind infinite sets.

1.9 Every countable set is Dedekind infinite.

1.10 if X contains a countable subset, then X is Dedekind infinite.

1.11 If X is Dedekind infinite, then it contains a countable subset. [*Hint*: Let $x \in X - f[X]$; define $x_0 = x$, $x_1 = f(x_0)$, ..., $x_{n+1} = f(x_n)$, ... The set $\{x_n \mid n \in N\}$ is countable.]

Thus Dedekind infinite sets are exactly those that have a countable subset. Later, using the Axiom of Choice, we show that *every* infinite set has a countable subset; thus Dedekind infinite = infinite.

1.12 If A and B are Dedekind finite, then $A \cup B$ is Dedekind finite. [*Hint*: Use Exercise 1.11.]

1.13 If A and B are Dedekind finite, then $A \times B$ is Dedekind finite. [*Hint*: Use Exercise 1.11.]

1.14 If A is infinite, then $\mathcal{P}(\mathcal{P}(A))$ is Dedekind infinite. [*Hint*: For each $n \in N$, let $S_n = \{X \subset A \mid |X| = n\}$. The set $\{S_n \mid n \in N\}$ is a countable subset of $\mathcal{P}(\mathcal{P}(A))$.]

2. The Cardinality of the Continuum

We are by now well acquainted with the properties of the cardinal number $\aleph_0$, the cardinality of countable sets. We summarize them here for reference, using the concepts of cardinal arithmetic introduced in the preceding section.

(a) $\kappa < \aleph_0$ if and only if $\kappa \in \boldsymbol{N}$.
(b) $n + \aleph_0 = \aleph_0 + \aleph_0 = \aleph_0$ $(n \in \boldsymbol{N})$.
(c) $n \cdot \aleph_0 = \aleph_0 \cdot \aleph_0 = \aleph_0$ $(n \in \boldsymbol{N},\ n > 0)$.
(d) $\aleph_0^n = \aleph_0$ $(n \in \boldsymbol{N},\ n > 0)$.

In the present section we study the second most important infinite cardinal number, the cardinality of the continuum, $2^{\aleph_0}$. To begin with, we recall that $2^{\aleph_0}$ is indeed the cardinality of the set $\boldsymbol{R}$ of all real numbers.

2.1 Theorem $|\boldsymbol{R}| = 2^{\aleph_0}$.

Proof. This is the second half of Theorem 6.3 in Chapter 4. □

The next theorem summarizes arithmetic properties of the cardinality of the continuum.

2.2 Theorem
(a) $n + 2^{\aleph_0} = \aleph_0 + 2^{\aleph_0} = 2^{\aleph_0} + 2^{\aleph_0} = 2^{\aleph_0}$ $(n \in \boldsymbol{N})$.
(b) $n \cdot 2^{\aleph_0} = \aleph_0 \cdot 2^{\aleph_0} = 2^{\aleph_0} \cdot 2^{\aleph_0} = 2^{\aleph_0}$ $(n \in \boldsymbol{N},\ n > 0)$.
(c) $(2^{\aleph_0})^n = (2^{\aleph_0})^{\aleph_0} = n^{\aleph_0} = \aleph_0^{\aleph_0} = 2^{\aleph_0}$ $(n \in \boldsymbol{N},\ n > 0)$.

Proof.
(a) This follows from the obvious sequence of inequalities

$$2^{\aleph_0} \leq n + 2^{\aleph_0} \leq \aleph_0 + 2^{\aleph_0} \leq 2^{\aleph_0} + 2^{\aleph_0} = 2 \cdot 2^{\aleph_0} = 2^{1+\aleph_0} = 2^{\aleph_0}$$

by the Cantor-Bernstein Theorem.
(b) Similarly, we have

$$2^{\aleph_0} \leq n \cdot 2^{\aleph_0} \leq \aleph_0 \cdot 2^{\aleph_0} \leq 2^{\aleph_0} \cdot 2^{\aleph_0} = 2^{\aleph_0+\aleph_0} = 2^{\aleph_0}.$$

(c) We have both

$$2^{\aleph_0} \leq (2^{\aleph_0})^n \leq (2^{\aleph_0})^{\aleph_0} = 2^{\aleph_0^2} = 2^{\aleph_0}$$

and

$$2^{\aleph_0} \leq n^{\aleph_0} \leq \aleph_0^{\aleph_0} \leq (2^{\aleph_0})^{\aleph_0} = 2^{\aleph_0^2} = 2^{\aleph_0}.$$

□

It is interesting to notice that Theorem 2.2, although an easy corollary of laws of cardinal arithmetic and the Cantor-Bernstein Theorem, has some rather unexpected consequences. For example, $2^{\aleph_0} \cdot 2^{\aleph_0} = 2^{\aleph_0}$ means that $|\boldsymbol{R} \times \boldsymbol{R}| = |\boldsymbol{R}|$; however, the set $\boldsymbol{R} \times \boldsymbol{R}$ of all pairs of real numbers is in a one-to-one correspondence with the set of all points in the plane (via a cartesian coordinate system). Thus we see that there exists a one-to-one mapping of a straight line

$\boldsymbol{R}$ onto a plane $\boldsymbol{R} \times \boldsymbol{R}$ (and similarly, onto a three-dimensional space $\boldsymbol{R} \times \boldsymbol{R} \times \boldsymbol{R}$, etc.). These results (due to Cantor) astonished his contemporaries; they seem rather counterintuitive, and the reader may find it helpful to actually construct such a mapping (see Exercise 2.7). The next theorem shows that several important sets have the cardinality of the continuum.

2.3 Theorem

(a) The set of all points in the n-dimensional space $\boldsymbol{R}^n$ has cardinality $2^{\aleph_0}$.
(b) The set of all complex numbers has cardinality $2^{\aleph_0}$.
(c) The set of all infinite sequences of natural numbers has cardinality $2^{\aleph_0}$.
(d) The set of all infinite sequences of real numbers has cardinality $2^{\aleph_0}$.

Proof.

(a) $|\boldsymbol{R}^n| = (2^{\aleph_0})^n$ by definition of cardinal exponentiation; $(2^{\aleph_0})^n = 2^{\aleph_0}$ by Theorem 2.2(c).
(b) Complex numbers are represented by pairs of reals (see Exercise 2.6 in Chapter 10), so the cardinality of the set of all complex numbers is $|\boldsymbol{R} \times \boldsymbol{R}| = (2^{\aleph_0})^2 = 2^{\aleph_0}$.
(c) The set of all infinite sequences of natural numbers is $\boldsymbol{N}^{\boldsymbol{N}}$ and $|\boldsymbol{N}^{\boldsymbol{N}}| = \aleph_0^{\aleph_0} = 2^{\aleph_0}$.
(d) $|\boldsymbol{R}^{\boldsymbol{N}}| = (2^{\aleph_0})^{\aleph_0} = 2^{\aleph_0}$.

□

The next theorem helps to establish further results of this sort.

2.4 Theorem *If A is a countable subset of B and $|B| = 2^{\aleph_0}$, then $|B - A| = 2^{\aleph_0}$.*

(Here we remark that using the Axiom of Choice, we are able to show in general that if $|A| < |B|$, then $|B - A| = |B|$.)

Proof. We can assume without loss of generality that $B = \boldsymbol{R} \times \boldsymbol{R}$.

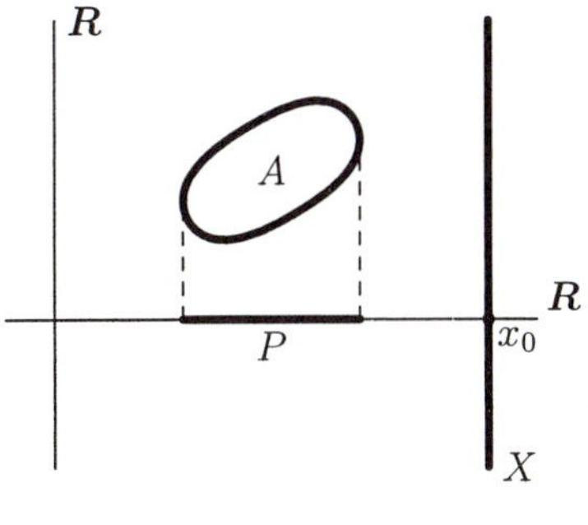

Let $P = \operatorname{dom} A$:

$$P = \{x \in \boldsymbol{R} \mid (x, y) \in A \text{ for some } y\}.$$

Since $|A| = \aleph_0$, we have $|P| \leq \aleph_0$. Thus there is $x_0 \in \boldsymbol{R}$ such that $x_0 \notin P$. Consequently, the set $X = \{x_0\} \times \boldsymbol{R}$ is disjoint from A, so $X \subseteq (\boldsymbol{R} \times \boldsymbol{R}) - A$. Clearly, $|X| = |\boldsymbol{R}| = 2^{\aleph_0}$, and we have $|(\boldsymbol{R} \times \boldsymbol{R}) - A| \geq 2^{\aleph_0}$. □

2.5 Theorem

(a) The set of all irrational numbers has cardinality $2^{\aleph_0}$.
(b) The set of all infinite sets of natural numbers has cardinality $2^{\aleph_0}$.
(c) The set of all one-to-one mappings of $\boldsymbol{N}$ *onto* $\boldsymbol{N}$ *has cardinality* $2^{\aleph_0}$.

Proof.

(a) The set of all rationals $\boldsymbol{Q}$ is countable, hence the set $\boldsymbol{R} - \boldsymbol{Q}$ of all irrational numbers has cardinality $2^{\aleph_0}$ by Theorem 2.4.

(b) The set of all subsets of $\boldsymbol{N}$, $\mathcal{P}(\boldsymbol{N})$, has cardinality $2^{\aleph_0}$, and the set of all finite subsets of $\boldsymbol{N}$ is countable (see Corollary 3.11 in Chapter 4), hence the set of all infinite subsets of $\boldsymbol{N}$ has the cardinality of the continuum.

(c) Let P be the set of all one-to-one mappings of $\boldsymbol{N}$ onto $\boldsymbol{N}$; as $P \subseteq \boldsymbol{N}^{\boldsymbol{N}}$, clearly $|P| \leq 2^{\aleph_0}$. Let E and O, respectively, be the sets of all even and odd natural numbers. If $X \subseteq E$ is infinite, define a mapping $f_X : \boldsymbol{N} \to \boldsymbol{N}$ as follows:

$$f_X(2k) = \text{the } k\text{th element of } X \ (k \in \boldsymbol{N});$$
$$f_X(2k+1) = \text{the } k\text{th element of } \boldsymbol{N} - X \ (k \in \boldsymbol{N}).$$

Notice that $\boldsymbol{N} - X \supseteq O$ is infinite, so f_X is a one-to-one mapping of $\boldsymbol{N}$ onto $\boldsymbol{N}$. Moreover, it is easy to show that $X_1 \neq X_2$ implies $f_{X_1} \neq f_{X_2}$. We thus have a one-to-one correspondence between infinite subsets of E and certain elements of P. Since there are $2^{\aleph_0}$ infinite subsets of E by Theorem 2.5(b), we get $|P| \geq 2^{\aleph_0}$ as needed.

□

Yet other similar results are provided by the next theorem. The definitions and basic properties of open sets and continuous functions can be found in Section 3, Chapter 10.

2.6 Theorem

(a) The set of all continuous functions on $\boldsymbol{R}$ *to* $\boldsymbol{R}$ *has cardinality* $2^{\aleph_0}$.
(b) The set of all open sets of reals has cardinality $2^{\aleph_0}$.

Proof.

(a) We use the fact (proved in Theorem 3.11, Chapter 10) that every continuous function on $\boldsymbol{R}$ is determined by its values on a dense set, in particular by its values at rational arguments: If f and g are two continuous functions on $\boldsymbol{R}$, and if $f(q) = g(q)$ for every rational number q, then $f = g$. Thus let C be the set of all continuous real-valued functions on $\boldsymbol{R}$. Let F be a mapping of C into $\boldsymbol{R}^{\boldsymbol{Q}}$ defined by $F(f) = f \restriction \boldsymbol{Q}$. By the fact above, F is one-to-one, so $|C| \leq |\boldsymbol{R}^{\boldsymbol{Q}}| = (2^{\aleph_0})^{\aleph_0} = 2^{\aleph_0}$. On the other hand, clearly $|C| \geq 2^{\aleph_0}$ (consider the constant functions).

(b) Every open set is a union of a system of open intervals with rational endpoints (see Lemma 3.14 in Chapter 10). There are $\aleph_0$ open intervals with rational endpoints (each such interval is determined by an ordered pair of rationals), and hence $2^{\aleph_0}$ such systems. This shows that there are at most $2^{\aleph_0}$ open sets. On the other hand, if $a, b \in \boldsymbol{R}$, $a \neq b$, then $(a, \infty) \neq (b, \infty)$, so there are at least $2^{\aleph_0}$ open sets.

□

The results of this section demonstrate the importance of the cardinal number $2^{\aleph_0}$. It should not be surprising that the problem of determining the magnitude of $2^{\aleph_0}$ is of fundamental significance. We know that $2^{\aleph_0}$ is greater than $\aleph_0$, but how much greater? Cantor conjectured that $2^{\aleph_0}$ is the next cardinal number after $\aleph_0$. This is the famous Continuum Hypothesis.

The Continuum Hypothesis There is no uncountable cardinal number κ such that $\kappa < 2^{\aleph_0}$.

In words, the Continuum Hypothesis asserts that every set of real numbers is either finite or countable, or else it is equipotent to the set of all real numbers. There are no cardinalities in between. In 1900, David Hilbert included the Continuum Problem in his famous list of open problems in mathematics (as Problem 1). It is still not fully resolved today. In 1939, Kurt Gödel showed that the Continuum Hypothesis is *consistent* with the axioms of set theory. That is, using the axioms of Zermelo-Fraenkel set theory (including the Axiom of Choice), one cannot refute the Continuum Hypothesis. In 1963, Paul Cohen proved that the Continuum Hypothesis is *independent* of the axioms. This means that one cannot prove the Continuum Hypothesis from the axioms. We discuss these questions in more detail in Chapter 15.

We conclude this section with an example of a set that has cardinality greater than the continuum.

2.7 Lemma *The set of all real-valued functions on real numbers has cardinality* $2^{2^{\aleph_0}} > 2^{\aleph_0}$.

Proof. The cardinal number of $\boldsymbol{R}^{\boldsymbol{R}}$ is $(2^{\aleph_0})^{2^{\aleph_0}} = 2^{\aleph_0 \cdot 2^{\aleph_0}} = 2^{2^{\aleph_0}}$. □

Exercises

2.1 Prove that the set of all finite sets of reals has cardinality $2^{\aleph_0}$. We remark here that the set of all countable sets of reals also has cardinality $2^{\aleph_0}$, but the proof of this requires the Axiom of Choice.

2.2 A real number x is *algebraic* if it is a solution of some equation

$$(*) \qquad a_n x^n + a_{n-1} x^{n-1} + \cdots + a_1 x + a_0 = 0,$$

where $a_0, \ldots, a_n$ are integers. If x is not algebraic, it is called *transcendental.* Show that the set of all algebraic numbers is countable and hence the set of all transcendental numbers has cardinality $2^{\aleph_0}$.

2.3 If a linearly ordered set P has a countable dense subset, then $|P| \leq 2^{\aleph_0}$.

2.4 The set of all closed subsets of reals has cardinality $2^{\aleph_0}$.

2.5 Show that, for $n > 0$, $n \cdot 2^{2^{\aleph_0}} = \aleph_0 \cdot 2^{2^{\aleph_0}} = 2^{\aleph_0} \cdot 2^{2^{\aleph_0}} = 2^{2^{\aleph_0}} \cdot 2^{2^{\aleph_0}} = (2^{2^{\aleph_0}})^n = (2^{2^{\aleph_0}})^{\aleph_0} = (2^{2^{\aleph_0}})^{2^{\aleph_0}} = 2^{2^{\aleph_0}}$.

2.6 The cardinality of the set of all discontinuous functions is $2^{2^{\aleph_0}}$. [*Hint*: Using Exercise 2.5, show that $|\boldsymbol{R}^{\boldsymbol{R}} - C| = 2^{2^{\aleph_0}}$ whenever $|C| \leq 2^{\aleph_0}$.]

2.7 Construct a one-to-one mapping of $\boldsymbol{R} \times \boldsymbol{R}$ onto $\boldsymbol{R}$. [*Hint*: If $a, b \in [0, 1]$ have decimal expansions $0.a_1a_2a_3 \cdots$ and $0.b_1b_2b_3 \cdots$, map the ordered pair (a, b) onto $0.a_1b_1a_2b_2a_3b_3 \cdots \in [0, 1]$. Make adjustments to avoid sequences where the digit 9 appears from some place onward.]

Chapter 6

Ordinal Numbers

1. Well-Ordered Sets

When we introduced natural numbers, we were motivated by the need to formalize the process of "counting": the natural numbers start with 0 and are generated by successively increasing the number by one unit: 0, 1, 2, 3, ..., and so on. We defined the operation of *successor* by $S(x) = x \cup \{x\}$ and introduced natural numbers as elements of the smallest set containing 0 and closed under S.

It is desirable to be able to continue the process of counting beyond natural numbers. The idea is that we can imagine an infinite number ω that comes "after" all natural numbers and then continue the counting process into the transfinite: ω, $\omega + 1$, $(\omega + 1) + 1$, and so on.

In the present chapter we formalize the process of transfinite counting, and introduce *ordinal* numbers as a generalization of natural numbers. As is usual in any meaningful generalization, the resulting concept shares many features of natural numbers. Most important, the theorems on induction and recursion are generalized to theorems on transfinite induction and transfinite recursion.

As a starting point, we use the fact that each natural number is identified with the set of all smaller natural numbers: $n = \{m \in \boldsymbol{N} \mid m < n\}$. (See Exercise 2.6 in Chapter 3.) By analogy, we let ω, the least transfinite number, to be the set $\boldsymbol{N}$ of all natural numbers: $\omega = \boldsymbol{N} = \{0, 1, 2, \ldots\}$.

It is easy to continue the process after this "limit" step is made: The operation of successor can be used to produce numbers following ω in the same way we used it to produce numbers following 0:

$$S(\omega) = \omega \cup \{\omega\} = \{0, 1, 2, \ldots, \omega\},$$
$$S(S(\omega)) = S(\omega) \cup \{S(\omega)\} = \{0, 1, 2, \ldots, \omega, S(\omega)\}, \quad \text{etc.}$$

We use the suggestive notation

$$S(\omega) = \omega + 1, \quad S(S(\omega)) = (\omega + 1) + 1 = \omega + 2, \quad \text{etc.}$$

In this fashion, we can generate greater and greater "numbers": ω, $\omega+1$, $\omega+2, \ldots, \omega+n, \ldots$, for all $n \in \boldsymbol{N}$. A number following all $\omega+n$ can again be conceived of as a set of all smaller numbers:

$$\omega \cdot 2 = \omega + \omega = \{0, 1, 2, \ldots, \omega, \omega+1, \omega+2, \ldots\}.$$

The reader may wish to introduce still greater numbers; e.g.,

$$\begin{aligned}
\omega \cdot 2 + 1 &= \omega + \omega + 1 = \{0, 1, 2, \ldots, \omega, \omega+1, \omega+2, \ldots, \omega+\omega\}, \\
\omega \cdot 3 &= \omega + \omega + \omega \\
&= \{0, 1, 2, \ldots, \omega, \omega+1, \omega+2, \ldots, \omega+\omega, \omega+\omega+1, \ldots\}, \\
\omega \cdot \omega &= \{0, 1, 2, \ldots, \omega, \omega+1, \ldots, \omega \cdot 2, \omega \cdot 2 + 1, \ldots, \omega \cdot 3, \ldots, \omega \cdot 4, \ldots\}.
\end{aligned}$$

The sets we generate behave very much like natural numbers, in this respect: they are linearly ordered by $\in$, and every nonempty subset has a least element. We called linear orderings with this property *well-orderings* (see Definition 2.3 in Chapter 3). As the concept of well-ordering is, along with cardinality, one of the most important ideas in abstract set theory, let us recall the definition:

1.1 Definition A set W is *well-ordered* by the relation $<$ if
(a) $(W, <)$ is a linearly ordered set.
(b) Every nonempty subset of W has a least element.

The sets generated above are all examples of sets well-ordered by the relation $\in$. Later in this chapter we show that all well-orderings can be represented by such sets and we introduce ordinal numbers to serve as order types of well-ordered sets. More examples of well-ordered sets can be found in the exercises at the end of this section.

The fundamental property of well-orderings is that they can be compared by their "lengths." The precise meaning of this is given in Theorem 1.3.

Let $(L, <)$ be a linearly ordered set. A set $S \subseteq L$ is called an *initial segment* of L if S is a proper subset of L (i.e., $S \neq L$) and if for every $a \in S$, all $x < a$ are also elements of S. (For instance, both the set of all negative reals and the set of all nonpositive reals are initial segments of the set of all real numbers.)

1.2 Lemma *If $(W, <)$ is a well-ordered set and if S is an initial segment of $(W, <)$, then there exists $a \in W$ such that $S = \{x \in W \mid x < a\}$.*

Proof. Let $X = W - S$ be the complement of S. As S is a proper subset of W, X is nonempty, and so it has a least element in the well-ordering $<$. Let a be the least element of X. If $x < a$, then x cannot belong to X, as a is its least member, so x belongs to S. If $x \geq a$, then x cannot be in S because otherwise a would also be in S as S is an initial segment. Thus $S = \{x \in W \mid x < a\}$. □

If a is an element of a well-ordered set $(W, <)$, we call the set

$$W[a] = \{x \in W \mid x < a\}$$

the *initial segment of* W *given by* a. (Note that if a is the least element of W, then $W[a]$ is empty.) By Lemma 1.2, each initial segment of a well-ordered set is of the form $W[a]$ for some $a \in W$. Of course, $W[a]$ is also well-ordered by $<$ (more precisely, by $< \cap W[a]^2$); we usually do not mention the well-ordering relation explicitly when it is understood from the context.

1.3 Theorem *If* $(W_1, <_1)$ *and* $(W_2, <_2)$ *are well-ordered sets, then exactly one of the following holds:*
(a) either W_1 *and* W_2 *are isomorphic, or*
(b) W_1 *is isomorphic to an initial segment of* W_2*, or*
(c) W_2 *is isomorphic to an initial segment of* W_1*.*
In each case, the isomorphism is unique.

This theorem provides the method of comparison for well-orderings mentioned above: we say that W_1 has *smaller order type* than W_2 if W_1 is isomorphic to $W_2[a]$ for some $a \in W_2$.

Before we prove Theorem 1.3, we prove the following lemma and state some of its corollaries. A function f on a linearly ordered set $(L, <)$ into L is *increasing* if $x_1 < x_2$ implies $f(x_1) < f(x_2)$. Note that an increasing function is one-to-one, and is an isomorphism of $(L, <)$ and $(\operatorname{ran} f, <)$.

1.4 Lemma *If* $(W, <)$ *is a well-ordered set and if* $f : W \to W$ *is an increasing function, then* $f(x) \geq x$ *for all* $x \in W$*.*

Proof. If the set $X = \{x \in W \mid f(x) < x\}$ is nonempty, it has a least element a. But then $f(a) < a$, and $f(f(a)) < f(a)$ because f is increasing. This means that $f(a) \in X$, which is a contradiction because a is least in X. □

1.5 Corollary
(a) No well-ordered set is isomorphic to an initial segment of itself.
(b) Each well-ordered set has only one automorphism, the identity.
(c) If W_1 *and* W_2 *are isomorphic well-ordered sets, then the isomorphism between* W_1 *and* W_2 *is unique.*

Proof.
(a) Assume that f is an isomorphism between W and $W[a]$ for some $a \in W$. Then $f(a) \in W[a]$ and therefore $f(a) < a$, contrary to the lemma, as f is an increasing function.
(b) Let f be an automorphism of W. Both f and f^{-1} are increasing functions and so for all $x \in W$, $f(x) \geq x$ and $f^{-1}(x) \geq x$, therefore $x \geq f(x)$. It follows that $f(x) = x$ for all $x \in W$.
(c) Let f and g be isomorphisms between W_1 and W_2. Then $f \circ g^{-1}$ is an automorphism of W_2 and hence is the identity map. It follows that $f = g$. □

1.6 *Proof of Theorem 1.3* Let W_1 and W_2 be well-ordered sets. It is a consequence of Lemma 1.4 that the three cases (a), (b), and (c) are mutually exclusive: For example, if W_1 were isomorphic to $W_2[a_2]$ for some $a_2 \in W_2$ and at the same time W_2 were isomorphic to $W_1[a_1]$ for some $a_1 \in W_1$, then the composition of the two isomorphisms would be an isomorphism of a well-ordered set onto its own initial segment.

Also, the uniqueness of the isomorphism in each case follows from Corollary 1.5; thus, we only have to show that one of the three cases (a), (b), and (c) always holds. We define a set of pairs $f \subseteq W_1 \times W_2$ and show that either f or f^{-1} is an isomorphism attesting to (a), (b), or (c). Let

$$f = \{(x, y) \in W_1 \times W_2 \mid W_1[x] \text{ is isomorphic to } W_2[y]\}.$$

First, it follows from Corollary 1.5 (a) that f is a one-to-one function: If $W_1[x]$ is isomorphic both to $W_2[y]$ and to $W_2[y']$, then $y = y'$ because otherwise $W_2[y]$ would be an initial segment of $W_2[y']$ (or vice versa) while they are isomorphic, and that is impossible. Hence $(x, y) \in f$ and $(x, y') \in f$ imply $y = y'$. A similar argument shows that $(x, y) \in f$ and $(x', y) \in f$ imply $x = x'$.

Second, $x < x'$ implies $f(x) < f(x')$: If h is the isomorphism between $W_1[x']$ and $W_2[f(x')]$, then the restriction $h \restriction W_1[x]$ is an isomorphism between $W_1[x]$ and $W_2[h(x)]$, so $f(x) = h(x)$ and $f(x) < f(x')$.

Hence f is an isomorphism between its domain, a subset of W_1, and its range, a subset of W_2. If the domain of f is W_1 and the range of f is W_2, then W_1 is isomorphic to W_2. We show now that if the domain of f is not all of W_1 then it is its initial segment, and the range of f is all of W_2. (This is enough to complete the proof as the remaining case is obtained by interchanging the role of W_1 and W_2.)

So assume that $\operatorname{dom} f \neq W_1$. We note that the set $S = \operatorname{dom} f$ is an initial segment of W_1: If $x \in S$ and $z < x$, let h be the isomorphism between $W_1[x]$ and $W_2[f(x)]$; then $h \restriction W_1[z]$ is an isomorphism between $W_1[z]$ and $W_2[h(z)]$, so $z \in S$. To show that the set $T = \operatorname{ran} f = W_2$, we assume otherwise and, by a similar argument as above, show that T is an initial segment of W_2. But then $\operatorname{dom} f = W_1[a]$ for some $a \in W_1$, and $\operatorname{ran} f = W_2[b]$ for some $b \in W_2$. In other words, f is an isomorphism between $W_1[a]$ and $W_2[b]$. This means, by the definition of f, that $(a, b) \in f$, so $a \in \operatorname{dom} f = W_1[a]$, that is, $a < a$, a contradiction. □

Exercises

1.1 Give an example of a linearly ordered set $(L, <)$ and an initial segment S of L which is not of the form $\{x \mid x < a\}$, for any $a \in L$.

1.2 $\omega + 1$ is not isomorphic to ω (in the well-ordering by $\in$).

1.3 There exist $2^{\aleph_0}$ well-orderings of the set of all natural numbers.

1.4 For every infinite subset A of $\boldsymbol{N}$, $(A, <)$ is isomorphic to $(\boldsymbol{N}, <)$.

1.5 Let $(W_1, <_1)$ and $(W_2, <_2)$ be disjoint well-ordered sets, each isomorphic to $(\boldsymbol{N}, <)$. Show that the sum of the two linearly ordered sets (as defined in Lemma 4.5 in Chapter 4) is a well-ordering, and is isomorphic to the ordinal number $\omega + \omega = \{0, 1, 2, \dots, \omega, \omega + 1, \omega + 2, \dots\}$.

1.6 Show that the lexicographic product $(\boldsymbol{N} \times \boldsymbol{N}, <)$ (see Lemma 4.6 in Chapter 4) is isomorphic to $\omega \cdot \omega$.

1.7 Let $(W, <)$ be a well-ordered set, and let $a \notin W$. Extend $<$ to $W' = W \cup \{a\}$ by making a greater than all $x \in W$. Then W has smaller order type than W'.

1.8 The sets $W = \boldsymbol{N} \times \{0, 1\}$ and $W' = \{0, 1\} \times \boldsymbol{N}$, ordered lexicographically, are nonisomorphic well-ordered sets. (See the remark following Theorem 4.7 in Chapter 4.)

2. Ordinal Numbers

In Chapter 3 we introduced the natural numbers to represent both the cardinality and the order type of finite sets and used them to prove theorems on induction and recursion. We now generalize this definition by introducing *ordinal numbers.*

Ordinal numbers continue the procedure of generating larger numbers into transfinite. As was the case with natural numbers, ordinal numbers are defined in such a way that each is well-ordered by the $\in$ relation. Moreover, the collection of all ordinal numbers (which as we shall see is not a set) is itself well-ordered by $\in$, and contains the natural numbers as an initial segment. And most significantly, ordinal numbers are representatives for *all* well-ordered sets: every well-ordered set is isomorphic to an ordinal number. Thus ordinal numbers can be viewed as *order types* of well-ordered sets.

2.1 Definition A set T is *transitive* if every element of T is a subset of T.

In other words, a transitive set has the property that $u \in v \in T$ implies $u \in T$. For more on transitive sets we refer the reader to the exercises in this section, and to Chapter 14.

2.2 Definition A set α is an *ordinal number* if
(a) α is transitive.
(b) α is well-ordered by $\in_\alpha$.

It is a standard practice to use lowercase Greek letters to denote ordinal numbers. Also, the term *ordinal* is often used for *ordinal number.*

For every natural number m, if $k \in l \in m$ (i.e., $k < l < m$), then $k \in m$. Hence every natural number is a transitive set. Also, every natural number is well-ordered by the $\in$ relation (because every $n \in \boldsymbol{N}$ is a subset of $\boldsymbol{N}$ and $\boldsymbol{N}$ is well-ordered by $\in$). Thus

2.3 Theorem *Every natural number is an ordinal.*

The set $\boldsymbol{N}$ of all natural numbers is easily seen to be transitive, and is also well-ordered by $\in$. Thus $\boldsymbol{N}$ is an ordinal number.

2.4 Definition $\omega = \boldsymbol{N}$.

We have just given the set $\boldsymbol{N}$ a new name ω.

2.5 Lemma *If α is an ordinal number, then $S(\alpha)$ is also an ordinal number.*

Proof. $S(\alpha) = \alpha \cup \{\alpha\}$ is a transitive set. Moreover, $\alpha \cup \{\alpha\}$ is well-ordered by $\in$, α being its greatest element, and $\alpha \subset \alpha \cup \{\alpha\}$ being the initial segment given by α. So $S(\alpha)$ is an ordinal number. □

We denote the successor of α by $\alpha + 1$:

$$\alpha + 1 = S(\alpha) = \alpha \cup \{\alpha\}.$$

An ordinal number α is called a *successor ordinal* if $\alpha = \beta + 1$ for some β. Otherwise, it is called a *limit ordinal.* For all ordinals α and β, we define $\alpha < \beta$ if and only if $\alpha \in \beta$, thus extending the definition of ordering of natural numbers from Chapter 3. The next theorem shows that $<$ does indeed have all the properties of a linear ordering, in fact, of a well-ordering.

2.6 Theorem *Let α, β, and γ be ordinal numbers.*
(a) *If $\alpha < \beta$ and $\beta < \gamma$, then $\alpha < \gamma$.*
(b) *$\alpha < \beta$ and $\beta < \alpha$ cannot both hold.*
(c) *Either $\alpha < \beta$ or $\alpha = \beta$ or $\beta < \alpha$ holds.*
(d) *Every nonempty set of ordinal numbers has a $<$-least element. Consequently, every set of ordinal numbers is well-ordered by $<$.*
(e) *For every set of ordinal numbers X, there is an ordinal number $\alpha \notin X$. (In other words, "the set of all ordinal numbers" does not exist.)*

The proof uses the following lemmas.

2.7 Lemma *If α is an ordinal number, then $\alpha \notin \alpha$.*

None of the axioms of set theory we have considered excludes the existence of sets X such that $X \in X$. However, the sets which arise in mathematical practice do not have this peculiar property.

Proof. If $\alpha \in \alpha$ then the linearly ordered set $(\alpha, \in_\alpha)$ has an element $x = \alpha$ such that $x \in x$, contrary to asymmetry of $\in_\alpha$. □

2.8 Lemma *Every element of an ordinal number is an ordinal number.*

Proof. Let α be an ordinal and let $x \in \alpha$. First we prove that x is transitive. Let u and v be such that $u \in v \in x$; we wish to show that $u \in x$. Since α is transitive and $x \in \alpha$, we have $v \in \alpha$ and therefore, also $u \in \alpha$. Thus u, v, and x are all elements of α and $u \in v \in x$. Since $\in_\alpha$ linearly orders α, we conclude that $u \in x$.

Second, we prove that $\in_x$ is a well-ordering of x. But by transitivity of α we have $x \subseteq \alpha$ and therefore, the relation $\in_x$ is a restriction of the relation $\in_\alpha$. Since $\in_\alpha$ is a well-ordering, so is $\in_x$. □

2.9 Lemma *If α and β are ordinal numbers such that $\alpha \subset \beta$, then $\alpha \in \beta$.*

Proof. Let $\alpha \subset \beta$. Then $\beta - \alpha$ is a nonempty subset of β, and hence has a least element γ in the ordering $\in_\beta$. Notice that $\gamma \subseteq \alpha$: If not, then any $\delta \in \gamma - \alpha$ would be an element of $\beta - \alpha$ smaller than γ (by transitivity of β). The proof is complete if we show that $\alpha \subseteq \gamma$ (and so $\alpha = \gamma \in \beta$).

Let $\delta \in \alpha$; we show that $\delta \in \gamma$. If not, then either $\gamma \in \delta$ or $\gamma = \delta$ (both δ and γ belong to β, which is linearly ordered by $\in$). But this implies that $\gamma \in \alpha$, since α is transitive. That contradicts the choice of $\gamma \in \beta - \alpha$. □

Proof of Theorem 2.6

(a) If $\alpha \in \beta$ and $\beta \in \gamma$, then $\alpha \in \gamma$ because γ is transitive.

(b) Assume that $\alpha \in \beta$ and $\beta \in \alpha$. By transitivity, $\alpha \in \alpha$, contradicting Lemma 2.7.

(c) If α and β are ordinals, $\alpha \cap \beta$ is also an ordinal (check properties (a) and (b) from the definition) and $\alpha \cap \beta \subseteq \alpha$, $\alpha \cap \beta \subseteq \beta$. If $\alpha \cap \beta = \alpha$, then $\alpha \subseteq \beta$ and so $\alpha \in \beta$ or $\alpha = \beta$ by Lemma 2.9. Similarly, $\alpha \cap \beta = \beta$ implies $\beta \in \alpha$ or $\beta = \alpha$. The only remaining case, $\alpha \cap \beta \subset \alpha$ and $\alpha \cap \beta \subset \beta$, cannot occur, because then $\alpha \cap \beta \in \alpha$ and $\alpha \cap \beta \in \beta$, leading to $\alpha \cap \beta \in \alpha \cap \beta$ and a contradiction with Lemma 2.7.

(d) Let A be a nonempty set of ordinals. Take $\alpha \in A$, and consider the set $\alpha \cap A$. If $\alpha \cap A = \emptyset$, α is the least element of A. If $\alpha \cap A \neq \emptyset$, $\alpha \cap A \subseteq \alpha$ has a least element β in the ordering $\in_\alpha$. Then β is the least element of A in the ordering $<$.

(e) Let X be a set of ordinal numbers. Since all elements of X are transitive sets, $\bigcup X$ is also a transitive set (see Exercise 2.5). It immediately follows from part (d) of this theorem that $\in$ well-orders $\bigcup X$; consequently, $\bigcup X$ is an ordinal number. Now let $\alpha = S(\bigcup X)$; α is an ordinal number and $\alpha \notin X$. [Otherwise, we get $\alpha \subseteq \bigcup X$ and, by Lemma 2.9, either $\alpha = \bigcup X$ or $\alpha \in \bigcup X$. In both cases, $\alpha \in S(\bigcup X) = \alpha$, contradicting Lemma 2.7.]

□

The ordinal number $\bigcup X$ used in the proof of (e) is called the *supremum* of X and is denoted $\sup X$. This is justified by observing that $\bigcup X$ is the least ordinal greater than or equal to all elements of X:

(a) If $\alpha \in X$, then $\alpha \subseteq \bigcup X$, so $\alpha \leq \bigcup X$.

(b) If $\alpha \leq \gamma$ for all $\alpha \in X$, then $\alpha \subseteq \gamma$ for all $\alpha \in X$ and so $\bigcup X \subseteq \gamma$, i.e., $\bigcup X \leq \gamma$.

If the set X has a greatest element β in the ordering $<$, then $\sup X = \beta$. Otherwise, $\sup X > \gamma$ for all $\gamma \in X$ (and it is the least such ordinal). We see that every set of ordinals has a supremum (in the ordering $<$).

The last theorem of this section restates the fact that ordinals are a generalization of the natural numbers:

2.10 Theorem *The natural numbers are exactly the finite ordinal numbers.*

Proof. We already know (from Theorem 2.3) that every natural number is an ordinal, and of course, every natural number is a finite set. So we only have to prove that all ordinals that are not natural numbers are infinite sets. If α is an ordinal and $\alpha \notin \boldsymbol{N}$, then by Theorem 2.6(b) it must be the case that $\alpha \geq \omega$ (because $\alpha \not< \omega$), so $\alpha \supseteq \omega$ because α is transitive. So α has an infinite subset and hence is infinite. □

Every ordinal is a well-ordered set, under the well-ordering $\in$. If α and β are distinct ordinals, then they are not isomorphic as well-ordered sets because one is an initial segment of the other. We also prove that *any* well-ordered set is isomorphic to an ordinal number. This, however, requires an introduction of another axiom, and is done in the next section.

One final comment. Lemma 2.8 establishes that each ordinal number α has the property that

$$\alpha = \{\beta \mid \beta \text{ is an ordinal and } \beta < \alpha\}.$$

If we view α as a set of ordinals, then if α is a successor, say $\beta + 1$, then it has a greatest element, namely β. If α is a limit ordinal, then it does not have a greatest element, and $\alpha = \sup\{\beta \mid \beta < \alpha\}$.

Note also that by definition, 0 is a limit ordinal, and $\sup \emptyset = 0$.

Exercises

2.1 A set X is transitive if and only if $X \subseteq \mathcal{P}(X)$.

2.2 A set X is transitive if and only if $\bigcup X \subseteq X$.

2.3 Are the following sets transitive?
- (a) $\{\emptyset, \{\emptyset\}, \{\{\emptyset\}\}\}$,
- (b) $\{\emptyset, \{\emptyset\}, \{\{\emptyset\}\}, \{\emptyset, \{\emptyset\}\}\}$,
- (c) $\{\emptyset, \{\{\emptyset\}\}\}$.

2.4 Which of the following statements are true?
- (a) If X and Y are transitive, then $X \cup Y$ is transitive.
- (b) If X and Y are transitive, then $X \cap Y$ is transitive.
- (c) If $X \in Y$ and Y is transitive, then X is transitive.
- (d) If $X \subseteq Y$ and Y is transitive, then X is transitive.
- (e) If Y is transitive and $S \subseteq \mathcal{P}(Y)$, then $Y \cup S$ is transitive.

2.5 If every $X \in S$ is transitive, then $\bigcup S$ is transitive.

2.6 An ordinal α is a natural number if and only if every nonempty subset of α has a greatest element.

2.7 If a set of ordinals X does not have a greatest element, then $\sup X$ is a limit ordinal.

2.8 If X is a nonempty set of ordinals, then $\bigcap X$ is an ordinal. Moreover, $\bigcap X$ is the least element of X.

3. The Axiom of Replacement

As we indicated at the end of Section 2, well-ordered sets can be represented by ordinal numbers. The following theorem states precisely what we mean by "representation."

3.1 Theorem *Every well-ordered set is isomorphic to a unique ordinal number.*

We give a proof of this theorem below. Although the reader may find the proof entirely acceptable, it nevertheless has a deficiency: it uses an assumption which, however plausible, does not follow from the axioms that we have introduced so far. For that reason it is necessary to introduce an additional axiom.

Proof. Let $(W, <)$ be a well-ordered set. Let A be the set of all those $a \in W$ for which $W[a]$ is isomorphic to some ordinal number. As no two distinct ordinals can be isomorphic (one is an initial segment of the other), this ordinal number is uniquely determined, and we denote it by α_a.

Now suppose that there exists a set S such that $S = \{\alpha_a \mid a \in A\}$. The set S is well-ordered by $\in$ as it is a set of ordinals. It is also transitive, because if $\gamma \in \alpha_a \in S$, let φ be the isomorphism between $W[a]$ and α_a and let $c = \varphi^{-1}(\gamma)$; it is easy to see that $\varphi \restriction c$ is an isomorphism between $W[c]$ and γ and so $\gamma \in S$.

Therefore, S is an ordinal number, $S = \alpha$.

A similar argument shows that $a \in A$, $b < a$ imply $b \in A$: let φ be the isomorphism of $W[a]$ and α_a. Then $\varphi \restriction W[b]$ is an isomorphism of $W[b]$ and an initial segment I of α_a. By Lemma 1.2, there exists $\beta < \alpha_a$ such that $I = \{\gamma \in \alpha_a \mid \gamma < \beta\} = \beta$; i.e., $\beta = \alpha_b$. This shows that $b \in A$ and $\alpha_b < \alpha_a$. We conclude that either $A = W$ or $A = W[c]$ for some $c \in W$ (Lemma 1.2 again).

We now define a function $f : A \to S = \alpha$ by $f(a) = \alpha_a$. From the definition of S and the fact that $b < a$ implies $\alpha_b < \alpha_a$ it is obvious that f is an isomorphism of $(A, <)$ and α. If $A = W[c]$, we would thus have $c \in A$, a contradiction. Therefore $A = W$, and f is an isomorphism of $(W, <)$ and the ordinal α.

This would complete the proof of Theorem 3.1, *if* we were justified to make the assumption that the set S exists. The elements of S are certainly well specified, but there is no reason why the existence of such a set should formally follow from the axioms that were considered so far.

To further illustrate the problem involved, consider two other examples:

To construct a sequence

$$\langle \emptyset, \{\emptyset\}, \{\{\emptyset\}\}, \{\{\{\emptyset\}\}\}, \ldots \rangle,$$

we might define

$$a_0 = \emptyset,$$
$$a_{n+1} = \{a_n\} \quad \text{for all } n \in \boldsymbol{N},$$

following the general pattern of recursive definitions. The difficulty here is that to apply the Recursion Theorem we need a set A, *given in advance*, such that $g : \boldsymbol{N} \times A \to A$ defined by $g(n, x) = \{x\}$ can be used to compute the $(n+1)$st term of the sequence from its nth term. But it is not obvious how to prove from our axioms that any set A such that

$$\emptyset \in A, \quad \{\emptyset\} \in A, \quad \{\{\emptyset\}\} \in A, \quad \{\{\{\emptyset\}\}\} \in A, \ldots$$

exists. It seems as if the definition of A itself required recursion.

Let us consider another example. In Chapter 3, we have postulated existence of ω; from it, the sets $\omega + 1 = \omega \cup \{\omega\}$, $\omega + 2 = (\omega + 1) \cup \{\omega + 1\}$, etc., can easily be obtained by repeated use of operations union and unordered pair. In the introductory remarks in Section 1, we "defined" $\omega + \omega$ as the union of ω and the set of all $\omega + n$ for all $n \in \omega$, and passed over the question of existence of this set. Although intuitively it is hardly any more questionable than the existence of ω, the existence of $\omega + \omega$ cannot be proved from the axioms we accepted so far. We know that, to each $n \in \omega$, there corresponds a unique set $\omega + n$; as yet, we do not have any axiom that would allow us to collect all these $\omega + n$ into one set. The next axiom schema removes this shortcoming.

The Axiom Schema of Replacement Let $\mathbf{P}(x, y)$ be a property such that for every x there is a unique y for which $\mathbf{P}(x, y)$ holds.

For every set A, there is a set B such that, for every $x \in A$, there is $y \in B$ for which $\mathbf{P}(x, y)$ holds.

We hope that the following comments provide additional motivation for the schema.

3.2 Let $\boldsymbol{F}$ be the operation defined by the property $\mathbf{P}$; that is, let $\boldsymbol{F}(x)$ denote the unique y for which $\mathbf{P}(x, y)$. (See Section 2 of Chapter 1.) The corresponding Axiom of Replacement can then be stated as follows:

For every set A there is a set B such that for all $x \in A$, $\boldsymbol{F}(x) \in B$.

Of course, B may also contain elements not of the form $\boldsymbol{F}(x)$ for any $x \in A$; however, an application of the Axiom Schema of Comprehension shows that

$$\begin{aligned}\{y \in B \mid y = \boldsymbol{F}(x) \text{ for some } x \in A\} &= \{y \in B \mid \mathbf{P}(x, y) \text{ holds for some } x \in A\} \\ &= \{y \mid \mathbf{P}(x, y) \text{ holds for some } x \in A\}\end{aligned}$$

exists. We call this set the *image* of A by $\boldsymbol{F}$ and denote it $\{\boldsymbol{F}(x) \mid x \in A\}$ or simply $\boldsymbol{F}[A]$.

3.3 Further intuitive justification for the Axiom Schema of Replacement can be given by comparing it with the Axiom Schema of Comprehension. The latter allows us to go through elements of a given set A, check for each $x \in A$ whether or not it has the property $\mathbf{P}(x)$, and collect those x which do into a set. In an entirely analogous way, the Axiom Schema of Replacement allows us to go through elements of A, take for each $x \in A$ the corresponding unique y having the property $\mathbf{P}(x, y)$, and collect all such y into a set. It is intuitively obvious that the set $\boldsymbol{F}[A]$ is "no larger than" the set A. In contrast, all known examples of "paradoxical sets" are "large," on the order of the "set of all sets."

3.4 Let $\boldsymbol{F}$ be again the operation defined by $\mathbf{P}$, as in (3.2). The Axiom of Replacement then implies that the operation $\boldsymbol{F}$ on elements of a given set A can be represented, "replaced," by a function, i.e., a set of ordered pairs. Precisely:

For every set A, there is a function f such that $\operatorname{dom} f = A$ and $f(x) = \boldsymbol{F}(x)$ for all $x \in A$.

We simply let $f = \{(x, y) \in A \times B \mid \mathbf{P}(x, y)\}$, where B is the set provided by the Axiom of Replacement. We use notation $\boldsymbol{F} \restriction A$ for this uniquely determined function f. Notice that $\operatorname{ran}(\boldsymbol{F} \restriction A) = \boldsymbol{F}[A]$.

We can now complete the *proof of Theorem 3.1*:

We have concluded earlier that in order to prove the theorem, we only have to guarantee the existence of the set $S = \{\alpha_a \mid a \in W\}$, where for each $a \in W$, α_a is the unique ordinal number isomorphic to $W[a]$.

Let $\mathbf{P}(x, y)$ be the property:

Either $x \in W$ and y is the unique ordinal isomorphic to $W[x]$, or $x \notin W$ and $y = \emptyset$.

Applying the Axiom of Replacement [with $\mathbf{P}(x, y)$ as above] we conclude that (for $A = W$) there exists a set B such that for all $a \in W$ there is $\alpha \in B$ for which $\mathbf{P}(a, \alpha)$ holds. Then we let

$$S = \{\alpha \in B \mid \mathbf{P}(a, \alpha) \text{ holds for some } a \in W\} = \boldsymbol{F}[W]$$

where $\boldsymbol{F}$ is the operation defined by $\mathbf{P}$. □

Using Theorem 3.1, we have:

3.5 Definition If W is a well-ordered set, then the *order type* of W is the unique ordinal number isomorphic to W.

And what about the examples mentioned above? It turns out that we need a more general Recursion Theorem. Compare the following theorem with the Recursion Theorem proved in Chapter 3:

3.6 The Recursion Theorem *Let $\boldsymbol{G}$ be an operation. For any set a there is a unique infinite sequence $\langle a_n \mid n \in \boldsymbol{N} \rangle$ such that*

(a) $a_0 = a$.

(b) $a_{n+1} = \boldsymbol{G}(a_n, n)$ for all $n \in \boldsymbol{N}$.

With this theorem, the existence of the sequence $\langle \emptyset, \{\emptyset\}, \{\{\emptyset\}\}, \ldots \rangle$ and of $\omega + \omega$ follows (see Exercise 3.2). We prove this Recursion Theorem, as well as the more general Transfinite Recursion Theorem, in the next section.

Exercises

3.1 Let $\mathbf{P}(x, y)$ be a property such that for every x there is at most one y for which $\mathbf{P}(x, y)$ holds. Then for every set A there is a set B such that, for all $x \in A$, if $\mathbf{P}(x, y)$ holds for some y, then $\mathbf{P}(x, y)$ holds for some $y \in B$.

3.2 Use Theorem 3.6 to prove the existence of
(a) The set $\{\emptyset, \{\emptyset\}, \{\{\emptyset\}\}, \{\{\{\emptyset\}\}\}, \ldots\}$.
(b) The set $\{\boldsymbol{N}, \mathcal{P}(\boldsymbol{N}), \mathcal{P}(\mathcal{P}(\boldsymbol{N})), \ldots\}$.
(c) The set $\omega + \omega = \omega \cup \{\omega, \omega + 1, (\omega + 1) + 1, \ldots\}$.

3.3 Use Theorem 3.6 to define

$$V_0 = \emptyset;$$
$$V_{n+1} = \mathcal{P}(V_n) \quad (n \in \omega);$$
$$V_\omega = \bigcup_{n \in \omega} V_n.$$

3.4 (a) Every $x \in V_\omega$ is finite.
(b) V_ω is transitive.
(c) V_ω is an inductive set.
The elements of V_ω are called *hereditarily finite sets*.

3.5 (a) If $x \in V_\omega$ and $y \in V_\omega$, then $\{x, y\} \in V_\omega$.
(b) If $X \in V_\omega$, then $\bigcup X \in V_\omega$ and $\mathcal{P}(X) \in V_\omega$.
(c) If $A \in V_\omega$ and f is a function on A such that $f(x) \in V_\omega$ for each $x \in A$, then $f[X] \in V_\omega$.
(d) If X is a finite subset of V_ω, then $X \in V_\omega$.

4. Transfinite Induction and Recursion

The Induction Principle and the Recursion Theorem are the main tools for proving theorems about natural numbers and for constructing functions with domain $\boldsymbol{N}$. We used them both extensively in the previous chapters. In this section, we show how these results generalize to ordinal numbers.

4.1 The Transfinite Induction Principle *Let $\mathbf{P}(x)$ be a property (possibly with parameters). Assume that, for all ordinal numbers α:*

(4.2) *If $\mathbf{P}(\beta)$ holds for all $\beta < \alpha$, then $\mathbf{P}(\alpha)$.*

Then $\mathbf{P}(\alpha)$ holds for all ordinals α.

Proof. Suppose that some ordinal number γ fails to have property $\mathbf{P}$, and let S be the set of all ordinal numbers $\beta \leq \gamma$ that do not have property $\mathbf{P}$. The set S has a least element α. Since every $\beta < \alpha$ has property $\mathbf{P}$, it follows by (4.2) that $\mathbf{P}(\alpha)$ holds, a contradiction. □

It is sometimes convenient to use the Transfinite Induction Principle in a form which resembles more closely the usual formulation of the Induction Principle for $\boldsymbol{N}$. We do it by treating successor and limit ordinals separately.

4.3 The Second Version of the Transfinite Induction Principle *Let $\mathbf{P}(x)$ be a property. Assume that*
(a) $\mathbf{P}(0)$ holds.
(b) $\mathbf{P}(\alpha)$ implies $\mathbf{P}(\alpha+1)$ for all ordinals α.
(c) For all limit ordinals $\alpha \neq 0$, if $\mathbf{P}(\beta)$ holds for all $\beta < \alpha$, then $\mathbf{P}(\alpha)$ holds.
Then $\mathbf{P}(\alpha)$ holds for all ordinals α.

Proof. It suffices to show that the assumptions (a), (b), and (c) imply (4.2). So let α be an ordinal such that $\mathbf{P}(\beta)$ for all $\beta < \alpha$. If $\alpha = 0$, then $\mathbf{P}(\alpha)$ holds by (a). If α is a successor, i.e., if there is $\beta < \alpha$ such that $\alpha = \beta + 1$, we know that $\mathbf{P}(\beta)$ holds, so $\mathbf{P}(\alpha)$ holds by (b). If $\alpha \neq 0$ is limit, we have $\mathbf{P}(\alpha)$ by (c). □

We proceed to generalize the Recursion Theorem. Functions whose domain is an ordinal α are called *transfinite sequences of length α*.

4.4 Theorem *Let Ω be an ordinal number, A a set, and $S = \bigcup_{\alpha<\Omega} A^\alpha$ the set of all transfinite sequences of elements of A of length less than Ω. Let $g : S \to A$ be a function. Then there exists a unique function $f : \Omega \to A$ such that*

$$f(\alpha) = g(f \restriction \alpha) \quad \textit{for all } \alpha < \Omega.$$

The reader might try to prove this theorem in a way entirely analogous to the proof of the Recursion Theorem in Chapter 3. We do not go into the details since this theorem follows from the subsequent more general Transfinite Recursion Theorem.

If ϑ is an ordinal and f is a transfinite sequence of length ϑ, we use the notation

$$f = \langle a_\alpha \mid \alpha < \vartheta \rangle.$$

Theorem 4.4 states that if g is a function on the set of all transfinite sequences of elements of A of length less than Ω with values in A, then there is a transfinite sequence $\langle a_\alpha \mid \alpha < \Omega \rangle$ such that for all $\alpha < \Omega$, $a_\alpha = g(\langle a_\xi \mid \xi < \alpha \rangle)$.

4.5 The Transfinite Recursion Theorem *Let $\boldsymbol{G}$ be an operation; then the property $\mathbf{P}$ stated in (4.6) defines an operation $\boldsymbol{F}$ such that $\boldsymbol{F}(\alpha) = \boldsymbol{G}(\boldsymbol{F} \restriction \alpha)$ for all ordinals α.*

Proof. We call t *a computation of length α based on $\boldsymbol{G}$* if t is a function, $\operatorname{dom} t = \alpha + 1$ and for all $\beta \leq \alpha$, $t(\beta) = \boldsymbol{G}(t \restriction \beta)$.

Let $\mathbf{P}(x, y)$ be the property

(4.6)

$$\begin{cases} x \text{ is an ordinal number and } y = t(x) \text{ for some computation } t \text{ of length } x \\ \quad \text{based on } \boldsymbol{G}, \\ \text{or } x \text{ is not an ordinal number and } y = \emptyset. \end{cases}$$

We prove first that $\mathbf{P}$ defines an operation.

We have to show that for each x there is a unique y such that $\mathbf{P}(x, y)$. This is obvious if x is not an ordinal. To prove it for ordinals, it suffices to show, by transfinite induction: For every ordinal α there is a unique computation of length α.

We make the inductive assumption that for all $\beta < \alpha$ there is a unique computation of length β, and endeavor to prove the existence and uniqueness of a computation of length α.

Existence: According to the Axiom Schema of Replacement applied to the property "y is a computation of length x" and the set α, there is a set

$$T = \{t \mid t \text{ is a computation of length } \beta \text{ for some } \beta < \alpha\}.$$

Moreover, the inductive assumption implies that for every $\beta < \alpha$ there is a unique $t \in T$ such that the length of t is β.

T is a system of functions; let $\bar{t} = \bigcup T$. Finally, let $\tau = \bar{t} \cup \{(\alpha, \boldsymbol{G}(\bar{t}))\}$. We prove that τ is a computation of length α.

4.7 Claim *τ is a function and* $\operatorname{dom} \tau = \alpha + 1$.

We see immediately that $\operatorname{dom} \bar{t} = \bigcup_{t \in T} \operatorname{dom} t = \bigcup_{\beta \in \alpha} (\beta + 1) = \alpha$; consequently, $\operatorname{dom} \tau = \operatorname{dom} \bar{t} \cup \{\alpha\} = \alpha + 1$.

Next, since $\alpha \notin \operatorname{dom} \bar{t}$, it is enough to prove that $\bar{t}$ is a function. This follows from the fact that T is a compatible system of functions.

Indeed, let t_1 and $t_2 \in T$ be arbitrary, and let $\operatorname{dom} t_1 = \beta_1$, $\operatorname{dom} t_2 = \beta_2$. Assume that, e.g., $\beta_1 \leq \beta_2$; then $\beta_1 \subseteq \beta_2$, and it suffices to show that $t_1(\gamma) = t_2(\gamma)$ for all $\gamma < \beta_1$. We do that by transfinite induction. So assume that $\gamma < \beta_1$ and $t_1(\delta) = t_2(\delta)$ for all $\delta < \gamma$. Then $t_1 \restriction \gamma = t_2 \restriction \gamma$, and we have $t_1(\gamma) = \boldsymbol{G}(t_1 \restriction \gamma) = \boldsymbol{G}(t_2 \restriction \gamma) = t_2(\gamma)$. We conclude that $t_1(\gamma) = t_2(\gamma)$ for all $\gamma < \beta_1$. This completes the proof of Claim 4.7.

4.8 Claim $\tau(\beta) = \boldsymbol{G}(\tau \restriction \beta)$ *for all* $\beta \leq \alpha$.

This is clear if $\beta = \alpha$, as $\tau(\alpha) = \boldsymbol{G}(\bar{t}) = \boldsymbol{G}(\tau \restriction \alpha)$. If $\beta < \alpha$, pick $t \in T$ such that $\beta \in \operatorname{dom} t$. We have $\tau(\beta) = t(\beta) = \boldsymbol{G}(t \restriction \beta) = \boldsymbol{G}(\tau \restriction \beta)$ because t is a computation, and $t \subseteq \tau$.

Claims 4.7 and 4.8 together prove that τ is a computation of length α.

Uniqueness: Let σ be another computation of length α; we prove $\tau = \sigma$. As τ and σ are functions and $\operatorname{dom} \tau = \alpha + 1 = \operatorname{dom} \sigma$, it suffices to prove by transfinite induction that $\tau(\gamma) = \sigma(\gamma)$ for all $\gamma \leq \alpha$.

Assume that $\tau(\delta) = \sigma(\delta)$ for all $\delta < \gamma$. Then $\tau(\gamma) = \boldsymbol{G}(\tau \restriction \gamma) = \boldsymbol{G}(\sigma \restriction \gamma) = \sigma(\gamma)$. The assertion follows.

This concludes the proof that the property $\mathbf{P}$ defines an operation $\boldsymbol{F}$. Notice that for any computation t, $\boldsymbol{F} \restriction \operatorname{dom} t = t$. This is because for any $\beta \in \operatorname{dom} t$, $t_\beta = t \restriction (\beta + 1)$ is obviously a computation of length β, and so, by the definition of $\boldsymbol{F}$, $\boldsymbol{F}(\beta) = t_\beta(\beta) = t(\beta)$.

To prove that $\boldsymbol{F}(\alpha) = \boldsymbol{G}(\boldsymbol{F} \restriction \alpha)$ for all α, let t be the unique computation of length α; we have $\boldsymbol{F}(\alpha) = t(\alpha) = \boldsymbol{G}(t \restriction \alpha) = \boldsymbol{G}(\boldsymbol{F} \restriction \alpha)$. □

We need again a parametric version of the Transfinite Recursion Theorem. If $\boldsymbol{F}(z, x)$ is an operation in two variables, we write $\boldsymbol{F}_z(x)$ in place of $\boldsymbol{F}(z, x)$.

Notice that, for any fixed z, $\boldsymbol{F}_z$ is an operation in one variable. If $\boldsymbol{F}$ is defined by $\mathbf{Q}(z, x, y)$, the notations $\boldsymbol{F}_z[A]$ and $\boldsymbol{F}_z \restriction A$ thus have meaning:

$$\boldsymbol{F}_z[A] = \{y \mid \mathbf{Q}(z, x, y) \text{ for some } x \in A\};$$
$$\boldsymbol{F}_z \restriction A = \{(x, y) \mid \mathbf{Q}(z, x, y) \text{ for some } x \in A\}.$$

We can now state a parametric version of Theorem 4.5.

4.9 The Transfinite Recursion Theorem, Parametric Version *Let $\boldsymbol{G}$ be an operation. The property $\mathbf{Q}$ stated in (4.10) defines an operation $\boldsymbol{F}$ such that $\boldsymbol{F}(z, \alpha) = \boldsymbol{G}(z, \boldsymbol{F}_z \restriction \alpha)$ for all ordinals α and all sets z.*

Proof. Call t a computation of length α *based on $\boldsymbol{G}$ and z* if t is a function, $\operatorname{dom} t = \alpha + 1$, and, for all $\beta \le \alpha$, $t(\beta) = \boldsymbol{G}(z, t \restriction \beta)$.

Let $\mathbf{Q}(z, x, y)$ be the property

(4.10)
$$\begin{cases} x \text{ is an ordinal number and } y = t(x) \text{ for some computation } t \text{ of length } x \\ \quad \text{based on } \boldsymbol{G} \text{ and } z, \text{ or} \\ x \text{ is not an ordinal number and } y = \emptyset. \end{cases}$$

Then carry z as a parameter through the rest of the proof in an obvious way. □

It is sometimes necessary to distinguish between successor ordinals and limit ordinals in our constructions. It is convenient to reformulate the Transfinite Recursion Theorem with this distinction in mind.

4.11 Theorem *Let $\boldsymbol{G}_1$, $\boldsymbol{G}_2$, and $\boldsymbol{G}_3$ be operations, and let $\boldsymbol{G}$ be the operation defined in (4.12) below. Then the property $\mathbf{P}$ stated in (4.6) (based on $\boldsymbol{G}$) defines an operation $\boldsymbol{F}$ such that*

$$\begin{aligned} \boldsymbol{F}(0) &= \boldsymbol{G}_1(\emptyset), \\ \boldsymbol{F}(\alpha + 1) &= \boldsymbol{G}_2(\boldsymbol{F}(\alpha)) \quad \textit{for all } \alpha, \\ \boldsymbol{F}(\alpha) &= \boldsymbol{G}_3(\boldsymbol{F} \restriction \alpha) \quad \textit{for all limit } \alpha \neq 0. \end{aligned}$$

Proof. Define an operation $\boldsymbol{G}$ by

(4.12)

$$\begin{cases} \boldsymbol{G}(x) = y \text{ if and only if either} \\ \text{(a) } x = \emptyset \text{ and } y = \boldsymbol{G}_1(\emptyset), \text{ or} \\ \text{(b) } x \text{ is a function, } \operatorname{dom} x = \alpha + 1 \text{ for an ordinal } \alpha \text{ and } y = \boldsymbol{G}_2(x(\alpha)), \text{ or} \\ \text{(c) } x \text{ is a function, } \operatorname{dom} x = \alpha \text{ for a limit ordinal } \alpha \neq 0 \text{ and } y = \boldsymbol{G}_3(x), \text{ or} \\ \text{(d) } x \text{ is none of the above and } y = \emptyset. \end{cases}$$

Let **P** be the property stated in (4.6) in the proof of the Transfinite Recursion Theorem (based on $\boldsymbol{G}$). The operation $\boldsymbol{F}$ defined by **P** then satisfies $\boldsymbol{F}(\alpha) = \boldsymbol{G}(\boldsymbol{F} \restriction \alpha)$ for all α. Using our definition of $\boldsymbol{G}$, it is easy to verify that $\boldsymbol{F}$ has the required properties. □

A parametric version of Theorem 4.11 is straightforward and we leave it to the reader.

We conclude this section with the proofs of Theorems 3.6 and 4.4.

Proof of Theorem 3.6. Let $\boldsymbol{G}$ be an operation. We want to find, for every set a, a sequence $\langle a_n \mid n \in \omega \rangle$ such that $a_0 = a$ and $a_{n+1} = \boldsymbol{G}(a_n, n)$ for all $n \in \boldsymbol{N}$.

By the parametric version of the Transfinite Recursion Theorem 4.11, there is an operation $\boldsymbol{F}$ such that $\boldsymbol{F}(0) = a$ and $\boldsymbol{F}(n+1) = \boldsymbol{G}(\boldsymbol{F}(n), n)$ for all $n \in \boldsymbol{N}$. Now we apply the Axiom of Replacement: There exists a sequence $\langle a_n \mid n \in \omega \rangle$ that is equal to $\boldsymbol{F} \restriction \omega$, and the Theorem follows. □

Proof of Theorem 4.4. Define an operation $\boldsymbol{G}$ by

$$\boldsymbol{G}(t) = \begin{cases} g(t) & \text{if } t \in S; \\ \emptyset & \text{otherwise.} \end{cases}$$

The Transfinite Recursion Theorem provides an operation $\boldsymbol{F}$ such that $\boldsymbol{F}(\alpha) = \boldsymbol{G}(\boldsymbol{F} \restriction \alpha)$ holds for all ordinals α. Let $f = \boldsymbol{F} \restriction \Omega$. □

Exercises

4.1 Prove a more general Transfinite Recursion Theorem (Double Recursion Theorem): Let $\boldsymbol{G}$ be an operation in two variables. Then there is an operation $\boldsymbol{F}$ such that $\boldsymbol{F}(\beta, \alpha) = \boldsymbol{G}(\boldsymbol{F} \restriction (\beta \times \alpha))$ for all ordinals β and α. [*Hint*: Computations are functions on $(\beta + 1) \times (\alpha + 1)$.]

4.2 Using the Recursion Theorem 4.9 show that there is a binary operation $\boldsymbol{F}$ such that

(a) $\boldsymbol{F}(x, 1) = 0$ for all x.

(b) $\boldsymbol{F}(x, n+1) = 0$ if and only if there exist y and z such that $x = (y, z)$ and $\boldsymbol{F}(y, n) = 0$.

We say that x is an *n-tuple* (where $n \in \omega$, $n > 0$) if $\boldsymbol{F}(x, n) = 0$. Prove that this definition of n-tuples coincides with the one given in Exercise 5.17 in Chapter 3.

5. Ordinal Arithmetic

We now use the Transfinite Recursion Theorem of the preceding section to define addition, multiplication, and exponentiation of ordinal numbers. These definitions are straightforward generalizations of the corresponding definitions for natural numbers.

5.1 Definition — Addition of Ordinal Numbers *For all ordinals* β,

(a) $\beta + 0 = \beta$.

(b) $\beta + (\alpha + 1) = (\beta + \alpha) + 1$ *for all* α.

(c) $\beta + \alpha = \sup\{\beta + \gamma \mid \gamma < \alpha\}$ *for all limit* $\alpha \neq 0$.

If we let $\alpha = 0$ in (b), we have the equality $\beta + 1 = \beta + 1$; the left-hand side denotes the sum of ordinal numbers β and 1 while the right-hand side is the successor of β.

To see how Definition 5.1 conforms with the formal version of the Transfinite Recursion Theorem, let us consider operations $\boldsymbol{G}_1$, $\boldsymbol{G}_2$, and $\boldsymbol{G}_3$ where $\boldsymbol{G}_1(z,x) = z$, $\boldsymbol{G}_2(z,x) = x + 1$, and $\boldsymbol{G}_3(z,x) = \sup(\operatorname{ran} x)$ if x is a function (and $\boldsymbol{G}_3(z,x) = 0$ otherwise).

The parametric form of Theorem 4.11 then provides an operation $\boldsymbol{F}$ such that for all z

(5.2)

$$\begin{cases} \boldsymbol{F}(z,0) = \boldsymbol{G}_1(z,0) = z. \\ \boldsymbol{F}(z,\alpha+1) = \boldsymbol{G}_2(z,\boldsymbol{F}_z(\alpha)) = \boldsymbol{F}(z,\alpha) + 1 \text{ for all } \alpha. \\ \boldsymbol{F}(z,\alpha) = \boldsymbol{G}_3(z,\boldsymbol{F}_z \restriction \alpha) = \sup(\operatorname{ran}(\boldsymbol{F}_z \restriction \alpha)) = \sup(\{\boldsymbol{F}(z,\gamma) \mid \gamma < \alpha\}) \\ \text{for limit } \alpha \neq 0. \end{cases}$$

If β and α are ordinals, then we write $\beta + \alpha$ instead of $\boldsymbol{F}(\beta, \alpha)$ and see that the conditions (5.2) are exactly the clauses (a), (b), and (c) from Definition 5.1.

In the subsequent applications of the Transfinite Recursion Theorem, we use the abbreviated form as in Definition 5.1, without explicitly formulating the defining operations $\boldsymbol{G}_1$, $\boldsymbol{G}_2$, and $\boldsymbol{G}_3$ to conform with Theorem 4.11.

One consequence of (5.1) is that for all β,

$$(\beta + 1) + 1 = \beta + 2,$$
$$(\beta + 2) + 1 = \beta + 3,$$

etc. Also, we have (if $\alpha = \beta = \omega$):

$$\omega + \omega = \sup\{\omega + n \mid n < \omega\},$$

and similarly,

$$(\omega + \omega) + \omega = \sup\{(\omega + \omega) + n \mid n < \omega\}.$$

In contrast to these examples, let us consider the sum $m + \omega$ for $m < \omega$. We have $m + \omega = \sup\{m + n \mid n < \omega\} = \omega$, because, if m is a natural number,

$m + n$ is also a natural number. We see that $m + \omega \neq \omega + m$; the addition of ordinals is not commutative. One should also notice that, while $1 \neq 2$, we have $1 + \omega = 2 + \omega$. Thus cancellations on the right in equations and inequalities are not allowed. However, addition of ordinal numbers is associative and allows left cancellations. (Lemma 5.4.)

In Chapter 4 we defined the sum of linearly ordered sets. We now prove that the ordinal sum defined in 5.1 agrees with the earlier more general definition.

5.3 Theorem *Let $(W_1, <_1)$ and $(W_2, <_2)$ be well-ordered sets, isomorphic to ordinals α_1 and α_2, respectively, and let $(W, <)$ be the sum of $(W_1, <_1)$ and $(W_2, <_2)$. Then $(W, <)$ is isomorphic to the ordinal $\alpha_1 + \alpha_2$.*

Proof. We assume that W_1 and W_2 are disjoint, that $W = W_1 \cup W_2$, and that each element in W_1 precedes in $<$ each element of W_2, while $<$ agrees with $<_1$ and with $<_2$ on both W_1 and W_2. We prove the theorem by induction on α_2.

If $\alpha_2 = 0$ then $W_2 = \emptyset$, $W = W_1$, and $\alpha_1 + \alpha_2 = \alpha_1$.

If $\alpha_2 = \beta + 1$ then W_2 has a greatest element a, and $W[a]$ is isomorphic to $\alpha_1 + \beta$; the isomorphism extends to an isomorphism between W and $\alpha_1 + \alpha_2 = (\alpha_1 + \beta) + 1$.

Let α_2 be a limit ordinal. For each $\beta < \alpha_2$ there is an isomorphism f_β of $\alpha_1 + \beta$ onto $W[a_\beta]$ where $a_\beta \in W_2$; moreover, f_β is unique, a_β is the βth element of W_2, and if $\beta < \gamma$ then $f_\beta \subset f_\gamma$. Let $f = \bigcup_{\beta<\alpha_2} f_\beta$. As $\alpha_1 + \alpha_2 = \bigcup_{\beta<\alpha_2}(\alpha_1 + \beta)$, it follows that f is an isomorphism of $\alpha_1 + \alpha_2$ onto W. □

5.4 Lemma

(a) *If α_1, α_2, and β are ordinals, then $\alpha_1 < \alpha_2$ if and only if $\beta + \alpha_1 < \beta + \alpha_2$.*
(b) *For all ordinals α_1, α_2, and β, $\beta + \alpha_1 = \beta + \alpha_2$ if and only if $\alpha_1 = \alpha_2$.*
(c) *$(\alpha + \beta) + \gamma = \alpha + (\beta + \gamma)$ for all ordinals α, β, and γ.*

Proof.

(a) We first use transfinite induction on α_2 to show that $\alpha_1 < \alpha_2$ implies $\beta + \alpha_1 < \beta + \alpha_2$. Let us then assume that α_2 is an ordinal greater than α_1 and that $\alpha_1 < \delta$ implies $\beta + \alpha_1 < \beta + \delta$ for all $\delta < \alpha_2$. If α_2 is a successor ordinal, then $\alpha_2 = \delta + 1$ where $\delta \geq \alpha_1$. By the inductive assumption in case $\delta > \alpha_1$, and trivially in case $\delta = \alpha_1$, we obtain $\beta + \alpha_1 \leq \beta + \delta < (\beta + \delta) + 1 = \beta + (\delta + 1) = \beta + \alpha_2$. If α_2 is a limit ordinal, then $\alpha_1 + 1 < \alpha_2$ and we have $\beta + \alpha_1 < (\beta + \alpha_1) + 1 = \beta + (\alpha_1 + 1) \leq \sup\{\beta + \delta \mid \delta < \alpha_2\} = \beta + \alpha_2$
To prove the converse we assume that $\beta + \alpha_1 < \beta + \alpha_2$. If $\alpha_2 < \alpha_1$, the implication already proved would show $\beta + \alpha_2 < \beta + \alpha_1$. Since $\alpha_2 = \alpha_1$ is also impossible (it implies $\beta + \alpha_2 = \beta + \alpha_1$), we conclude from linearity of $<$ that $\alpha_1 < \alpha_2$.
(b) This follows immediately from (a): If $\alpha_1 \neq \alpha_2$, then either $\alpha_1 < \alpha_2$ or $\alpha_2 < \alpha_1$ and, consequently, either $\beta + \alpha_1 < \beta + \alpha_2$ or $\beta + \alpha_2 < \beta + \alpha_1$. If $\alpha_1 = \alpha_2$, then $\beta + \alpha_1 = \beta + \alpha_2$ holds trivially.

(c) We proceed by transfinite induction on γ. If $\gamma = 0$, then $(\alpha + \beta) + 0 = \alpha + \beta = \alpha + (\beta + 0)$. Let us assume that the equality holds for γ, and let us prove it for $\gamma + 1$:

$$\begin{aligned}(\alpha + \beta) + (\gamma + 1) &= [(\alpha + \beta) + \gamma] + 1 = [\alpha + (\beta + \gamma)] + 1 \\ &= \alpha + [(\beta + \gamma) + 1] = \alpha + [\beta + (\gamma + 1)]\end{aligned}$$

(we have used the inductive assumption in the second step, and the second clause from the definition of addition in the other steps).
Finally, let γ be a limit ordinal, $\gamma \neq 0$. Then $(\alpha+\beta)+\gamma = \sup\{(\alpha+\beta)+\delta \mid \delta < \gamma\} = \sup\{\alpha + (\beta + \delta) \mid \delta < \gamma\}$. We observe that $\sup\{\beta + \delta \mid \delta < \gamma\} = \beta + \gamma$ (the third clause in the definition of addition) and that $\beta + \gamma$ is a limit ordinal (if $\xi < \beta + \gamma$ then $\xi \leq \beta + \delta$ for some $\delta < \gamma$ and so $\xi + 1 \leq (\beta + \delta) + 1 = \beta + (\delta + 1) < \beta + \gamma$ because γ is limit). It remains to notice that $\sup\{\alpha + (\beta + \delta) \mid \delta < \gamma\} = \sup\{\alpha + \xi \mid \xi < \beta + \gamma\}$ (because $\beta + \gamma = \sup\{\beta + \delta \mid \delta < \gamma\}$), and so we have $(\alpha + \beta) + \gamma = \sup\{\alpha + \xi \mid \xi < \beta + \gamma\} = \alpha + (\beta + \gamma)$, again by the third clause in Definition 5.1.

□

The following lemma shows that it is possible to define subtraction of ordinal numbers:

5.5 Lemma *If $\alpha \leq \beta$ then there is a unique ordinal number ξ such that $\alpha+\xi = \beta$.*

Proof. As α is an initial segment of the well-ordered set β (or $\alpha = \beta$), Theorem 5.3 implies that $\beta = \alpha + \xi$ where ξ is the order type of the set $\beta - \alpha = \{\nu \mid \alpha \leq \nu < \beta\}$. By Lemma 5.4(b), the ordinal ξ is unique. □

Next, we give a definition of ordinal multiplication:

5.6 Definition — Multiplication of Ordinal Numbers *For all ordinals β,*

(a) $\beta \cdot 0 = 0$.
(b) $\beta \cdot (\alpha + 1) = \beta \cdot \alpha + \beta$ *for all* α.
(c) $\beta \cdot \alpha = \sup\{\beta \cdot \gamma \mid \gamma < \alpha\}$ *for all limit* $\alpha \neq 0$.

5.7 Examples

(a) $\beta \cdot 1 = \beta \cdot (0 + 1) = \beta \cdot 0 + \beta = 0 + \beta = \beta$.
(b) $\beta \cdot 2 = \beta \cdot (1 + 1) = \beta \cdot 1 + \beta = \beta + \beta$; in particular, $\omega \cdot 2 = \omega + \omega$.
(c) $\beta \cdot 3 = \beta \cdot (2 + 1) = \beta \cdot 2 + \beta = \beta + \beta + \beta$, etc.
(d) $\beta \cdot \omega = \sup\{\beta \cdot n \mid n \in \omega\} = \sup\{\beta, \beta + \beta, \beta + \beta + \beta, \dots\}$.
(e) $1 \cdot \alpha = \alpha$ for all α, but this requires an inductive proof:
$1 \cdot 0 = 0$; $1 \cdot (\alpha + 1) = 1 \cdot \alpha + 1 = \alpha + 1$;
$1 \cdot \alpha = \sup\{1 \cdot \gamma \mid \gamma < \alpha\} = \sup\{\gamma \mid \gamma < \alpha\} = \alpha$ if α is limit, $\alpha \neq 0$.
(f) $2 \cdot \omega = \sup\{2 \cdot n \mid n \in \omega\} = \omega$. Since $\omega \cdot 2 = \omega + \omega \neq \omega$, we conclude that multiplication of ordinals is generally not commutative.

For more properties of ordinal products, see Exercises 5.1, 5.2 and 5.7.

Ordinal multiplication as defined in 5.6 also agrees with the general definition of products of linearly ordered sets as defined in Chapter 4:

5.8 Theorem *Let α and β be ordinal numbers. Both the lexicographic and the antilexicographic orderings of the product $\alpha \times \beta$ are well-orderings. The order type of the antilexicographic ordering of $\alpha \times \beta$ is $\alpha \cdot \beta$, while the lexicographic ordering of $\alpha \times \beta$ has order type $\beta \cdot \alpha$.*

Proof. Let $\prec$ denote the antilexicographic ordering of $\alpha \times \beta$. We define an isomorphism between $(\alpha \times \beta, \prec)$ and $\alpha \cdot \beta$ as follows: for $\xi < \alpha$ and $\eta < \beta$, let $f(\xi, \eta) = \alpha \cdot \eta + \xi$. The range of f is the set $\{\alpha \cdot \eta + \xi \mid \eta < \beta \text{ and } \xi < \alpha\} = \alpha \cdot \beta$ and f is an isomorphism (we leave the details — proved by induction — to the reader; see also Exercises 5.1, 5.2, 5.7, and 5.8). □

5.9 Definition — Exponentiation of Ordinal Numbers *For all β,*
(a) $\beta^0 = 1$.
(b) $\beta^{\alpha+1} = \beta^\alpha \cdot \beta$ for all α.
(c) $\beta^\alpha = \sup\{\beta^\gamma \mid \gamma < \alpha\}$ for all limit $\alpha \neq 0$.

5.10 Example
(a) $\beta^1 = \beta$, $\beta^2 = \beta \cdot \beta$, $\beta^3 = \beta^2 \cdot \beta = \beta \cdot \beta \cdot \beta$, etc.
(b) $\beta^\omega = \sup\{\beta^n \mid n \in \omega\}$; in particular,
$1^\omega = 1$,
$2^\omega = \omega$, $3^\omega = \omega$, ..., $n^\omega = \omega$ for any $n \in \omega$,
$\omega^\omega = \sup\{\omega^n \mid n \in \omega\} > \omega$.

It should be pointed out that ordinal arithmetic differs substantially from arithmetic of cardinals. Thus, for instance, $2^\omega = \omega$ and ω^ω are countable ordinals, while $2^{\aleph_0} = \aleph_0^{\aleph_0}$ is uncountable.

One can use arithmetic operations to generate larger and larger ordinals: $0, 1, 2, 3, \ldots, \omega, \omega+1, \omega+2, \ldots, \omega \cdot 2, \omega \cdot 2 + 1, \ldots, \omega \cdot 3, \ldots, \omega \cdot 4, \ldots,$ $\omega \cdot \omega = \omega^2, \omega^2 + 1, \ldots, \omega^2 \cdot 2, \ldots, \omega^3, \ldots, \omega^4, \ldots, \omega^\omega, \omega^\omega + 1, \ldots, \omega^\omega \cdot 2,$ $\ldots, \omega^\omega \cdot \omega = \omega^{\omega+1}, \ldots, \omega^{\omega^2}, \ldots, \omega^{\omega^3}, \ldots, \omega^{\omega^\omega}, \ldots, \omega^{\omega^{\omega^\omega}}, \ldots$

The process can easily be continued. We define $\varepsilon = \sup\{\omega, \omega^\omega, \omega^{\omega^\omega}, \omega^{\omega^{\omega^\omega}}, \ldots\}$. One can then form $\varepsilon + 1$, $\varepsilon + \omega$, ε^ω, ε^ε, $\varepsilon^{\varepsilon^\varepsilon}$, etc.

Exercises

5.1 Prove the associative law $(\alpha \cdot \beta) \cdot \gamma = \alpha \cdot (\beta \cdot \gamma)$.
5.2 Prove the distributive law $\alpha \cdot (\beta + \gamma) = \alpha \cdot \beta + \alpha \cdot \gamma$.
5.3 Simplify:
(a) $(\omega + 1) + \omega$.
(b) $\omega + \omega^2$.
(c) $(\omega + 1) \cdot \omega^2$.

5.4 For every ordinal α, there is a unique limit ordinal β and a unique natural number n such that $\alpha = \beta + n$. [*Hint*: $\beta = \sup\{\gamma \leq \alpha \mid \gamma \text{ is limit}\}$.]

5.5 Let $\alpha \leq \beta$. The equation $\xi + \alpha = \beta$ may have 0, 1, or infinitely many solutions.

5.6 Find the least $\alpha > \omega$ such that $\xi + \alpha = \alpha$ for all $\xi < \alpha$.

5.7 (a) If α_1, α_2, and β are ordinals and $\beta \neq 0$, then $\alpha_1 < \alpha_2$ if and only if $\beta \cdot \alpha_1 < \beta \cdot \alpha_2$.
(b) For all ordinals α_1, α_2, and $\beta \neq 0$, $\beta \cdot \alpha_1 = \beta \cdot \alpha_2$ if and only if $\alpha_1 = \alpha_2$.

5.8 Let α, β, and γ be ordinals, and let $\alpha < \beta$. Then
(a) $\alpha + \gamma \leq \beta + \gamma$,
(b) $\alpha \cdot \gamma \leq \beta \cdot \gamma$
and $\leq$ cannot be replaced by $<$ in either inequality.

5.9 Show that the following rules do not hold for all ordinals α, β, and γ:
(a) If $\alpha + \gamma = \beta + \gamma$, then $\alpha = \beta$.
(b) If $\gamma > 0$ and $\alpha \cdot \gamma = \beta \cdot \gamma$, then $\alpha = \beta$.
(c) $(\beta + \gamma) \cdot \alpha = \beta \cdot \alpha + \gamma \cdot \alpha$.

5.10 An ordinal α is a limit ordinal if and only if $\alpha = \omega \cdot \beta$ for some β.

5.11 Find a set A of rational numbers such that $(A, \leq_Q)$ is isomorphic to $(\alpha, \leq)$ where
(a) $\alpha = \omega + 1$,
(b) $\alpha = \omega \cdot 2$,
(c) $\alpha = \omega \cdot 3$,
(d) $\alpha = \omega^\omega$,
(e) $\alpha = \varepsilon$.
[*Hint*: $\{n - 1/m \mid m, n \in \boldsymbol{N} - \{0\}\}$ is isomorphic to ω^2, etc.]

5.12 Show that $(\omega \cdot 2)^2 \neq \omega^2 \cdot 2^2$.

5.13 (a) $\alpha^{\beta+\gamma} = \alpha^\beta \cdot \alpha^\gamma$.
(b) $(\alpha^\beta)^\gamma = \alpha^{\beta \cdot \gamma}$.

5.14 (a) If $\alpha \leq \beta$ then $\alpha^\gamma \leq \beta^\gamma$.
(b) If $\alpha > 1$ and if $\beta < \gamma$, then $\alpha^\beta < \alpha^\gamma$.

5.15 Find the least ξ such that
(a) $\omega + \xi = \xi$.
(b) $\omega \cdot \xi = \xi$, $\xi \neq 0$.
(c) $\omega^\xi = \xi$.
[*Hint for part (a)*: Let $\xi_0 = 0$, $\xi_{n+1} = \omega + \xi_n$, $\xi = \sup\{\xi_n \mid n \in \omega\}$.]

5.16 (Characterization of Ordinal Exponentiation) Let α and β be ordinals. For $f : \beta \to \alpha$, let $s(f) = \{\xi < \beta \mid f(\xi) \neq 0\}$. Let $S(\beta, \alpha) = \{f \mid f : \beta \to \alpha \text{ and } s(f) \text{ is finite}\}$. Define $\prec$ on $S(\beta, \alpha)$ as follows: $f \prec g$ if and only if there is $\xi_0 < \beta$ such that $f(\xi_0) < g(\xi_0)$ and $f(\xi) = g(\xi)$ for all $\xi > \xi_0$. Show that $(S(\beta, \alpha), \prec)$ is isomorphic to $(\alpha^\beta, <)$.

6. The Normal Form

Using exponentiation, one can represent ordinal numbers in a way similar to decimal expansion of integers. Ordinal numbers can be expressed uniquely in *normal form*, to be made precise in the theorem below. We apply the normal form to prove an interesting result about so-called Goodstein sequences of integers.

First we observe that the ordinal functions $\alpha+\beta$, $\alpha\cdot\beta$, and α^β are *continuous* in the second variable: If γ is a limit ordinal and $\beta = \sup_{\nu<\gamma}\beta_\nu$, then

$$(6.1) \qquad \alpha+\beta = \sup_{\nu<\gamma}(\alpha+\beta_\nu), \quad \alpha\cdot\beta = \sup_{\nu<\gamma}(\alpha\cdot\beta_\nu), \quad \alpha^\beta = \sup_{\nu<\gamma}\alpha^{\beta_\nu}.$$

This follows directly from Definitions 5.1, 5.6, and 5.9. As a consequence, we have:

6.2 Lemma
(a) If $0<\alpha\le\gamma$ then there is a greatest ordinal β such that $\alpha\cdot\beta\le\gamma$.
(b) If $1<\alpha\le\gamma$ then there is a greatest ordinal β such that $\alpha^\beta\le\gamma$.

Proof. Since $\alpha\cdot(\gamma+1)\ge\gamma+1>\gamma$, there exists a δ such that $\alpha\cdot\delta>\gamma$. Similarly, because $\alpha^{\gamma+1}\ge\gamma+1>\gamma$, there is a δ with $\alpha^\delta>\gamma$. The least δ such that $\alpha\cdot\delta>\gamma$ (or that $\alpha^\delta>\gamma$) must be a successor ordinal because of (6.1), say $\delta=\beta+1$. Then β is greatest such that $\alpha\cdot\beta\le\gamma$ (respectively, $\alpha^\beta\le\gamma$). □

The following lemma is the analogue of division of integers:

6.3 Lemma *If γ is an arbitrary ordinal and if $\alpha\ne0$, then there exists a unique ordinal β and a unique $\rho<\alpha$ such that $\gamma=\alpha\cdot\beta+\rho$.*

Proof. Let β be the greatest ordinal such that $\alpha\cdot\beta\le\gamma$ (if $\alpha>\gamma$ then $\beta=0$), and let ρ be the unique ρ (by Lemma 5.5) such that $\alpha\cdot\beta+\rho=\gamma$. The ordinal ρ is less than α, because otherwise we would have $\alpha\cdot(\beta+1)=\alpha\cdot\beta+\alpha\le\alpha\cdot\beta+\rho=\gamma$, contrary to the maximality of β.

To prove uniqueness, let $\gamma=\alpha\cdot\beta_1+\rho_1=\alpha\cdot\beta_2+\rho_2$ with $\rho_1,\rho_2<\alpha$. Assume that $\beta_1<\beta_2$. Then $\beta_1+1\le\beta_2$ and we have $\alpha\cdot\beta_1+(\alpha+\rho_2)=\alpha\cdot(\beta_1+1)+\rho_2\le\alpha\cdot\beta_2+\rho_2=\alpha\cdot\beta_1+\rho_1$, and by Lemma 5.4(a), $\rho_1\ge\alpha+\rho_2\ge\alpha$, a contradiction. Thus $\beta_1=\beta_2$, and $\rho_1=\rho_2$ follows by Lemma 5.5. □

The normal form is analogous to decimal expansion of integers, with the base for exponentiation being the ordinal ω:

6.4 Theorem *Every ordinal $\alpha>0$ can be expressed uniquely as*

$$\alpha=\omega^{\beta_1}\cdot k_1+\omega^{\beta_2}\cdot k_2+\cdots+\omega^{\beta_n}\cdot k_n,$$

where $\beta_1>\beta_2>\cdots>\beta_n$, and $k_1>0,\ k_2>0,\ \ldots,\ k_n>0$ are finite.

We remark that it is possible to have $\alpha = \omega^\alpha$, see Exercise 6.1.

Proof. We first prove the existence of the normal form, by induction on α. The ordinal $\alpha = 1$ can be expressed as $1 = \omega^0 \cdot 1$.

Now let $\alpha > 0$ be arbitrary. By Lemma 6.2(b) there exists a greatest β such that $\omega^\beta \leq \alpha$ (if $\alpha < \omega$ then $\beta = 0$). Then by Lemma 6.3 there exist unique δ and ρ such that $\rho < \omega^\beta$ and $\alpha = \omega^\beta \cdot \delta + \rho$. As $\omega^\beta \leq \alpha$, we have $\delta > 0$ and $\rho < \alpha$. We claim that δ is finite. If δ were infinite, then $\alpha \geq \omega^\beta \cdot \delta \geq \omega^\beta \cdot \omega = \omega^{\beta+1}$, contradicting the maximality of β. Thus let $\beta_1 = \beta$ and $k_1 = \delta$.

If $\rho = 0$ then $\alpha = \omega^{\beta_1} \cdot k_1$ is in normal form. If $\rho > 0$ then by the induction hypothesis,

$$\rho = \omega^{\beta_2} \cdot k_2 + \cdots + \omega^{\beta_n} \cdot k_n$$

for some $\beta_2 > \cdots > \beta_n$ and finite $k_2, \ldots, k_n > 0$. As $\rho < \omega^{\beta_1}$, we have $\omega^{\beta_2} \leq \rho < \omega^{\beta_1}$ and so $\beta_1 > \beta_2$. It follows that $\alpha = \omega^{\beta_1} \cdot k_1 + \omega^{\beta_2} \cdot k_2 + \cdots + \omega^{\beta_n} \cdot k_n$ is expressed in normal form.

To prove uniqueness, we first observe that if $\beta < \gamma$, then $\omega^\beta \cdot k < \omega^\gamma$ for every finite k: this is because $\omega^\beta \cdot k < \omega^\beta \cdot \omega = \omega^{\beta+1} \leq \omega^\gamma$. From this it easily follows that if $\alpha = \omega^{\beta_1} \cdot k_1 + \cdots + \omega^{\beta_n} \cdot k_n$ is in normal form and $\gamma > \beta_1$, then $\alpha < \omega^\gamma$.

We prove the uniqueness of normal form by induction on α. For $\alpha = 1$, the expansion $1 = \omega^0 \cdot 1$ is clearly unique. So let $\alpha = \omega^{\beta_1} \cdot k_1 + \cdots \omega^{\beta_n} \cdot k_n = \omega^{\gamma_1} \cdot \ell_1 + \cdots + \omega^{\gamma_m} \cdot \ell_m$. The preceding observation implies that $\beta_1 = \gamma_1$. If we let $\delta = \omega^{\beta_1} = \omega^{\gamma_1}$, $\rho = \omega^{\beta_2} \cdot k_2 + \cdots + \omega^{\beta_n} \cdot k_n$ and $\sigma = \omega^{\gamma_2} \cdot \ell_2 + \cdots + \omega^{\gamma_m} \cdot \ell_m$, we have $\alpha = \delta \cdot k_1 + \rho = \delta \cdot \ell_1 + \sigma$, and since $\rho < \delta$ and $\sigma < \delta$, Lemma 6.3 implies that $k_1 = \ell_1$ and $\rho = \sigma$. By the induction hypothesis, the normal form for ρ is unique, and so $m = n$, $\beta_2 = \gamma_2$, ..., $\beta_n = \gamma_n$, $k_2 = \ell_2$, ..., $k_n = \ell_n$. If follows that the normal form expansion for α is unique. □

We use the normal form to prove an interesting result on *Goodstein sequences.* Let us first recall that for every natural number $a \geq 2$, every natural number m can be written in base a, i.e., as a sum of powers of a:

$$m = a^{b_1} \cdot k_1 + \cdots + a^{b_n} \cdot k_n,$$

with $b_1 > \cdots > b_n$ and $0 < k_i < a$, $i = 1, \ldots, n$. For instance, the number 324 can be written as $4^4 + 4^3 + 4$ in base 4 and $7^2 \cdot 6 + 7 \cdot 4 + 2$ in base 7. A *weak Goodstein sequence* starting at $m > 0$ is a sequence $m_0, m_1, m_2, \ldots$ of natural numbers defined as follows:

First, let $m_0 = m$, and write m_0 in base 2:

$$m_0 = 2^{b_1} + \cdots + 2^{b_n}.$$

To obtain m_1, increase the base by 1 (from 2 to 3) and then subtract 1:

$$m_1 = 3^{b_1} + \cdots + 3^{b_n} - 1.$$

In general, to obtain m_{k+1} from m_k (as long as $m_k \neq 0$), write m_k in base $k+2$, increase the base by 1 (to $k+3$) and subtract 1. For example, the weak Goodstein sequence starting at $m = 21$ is as follows:

$$\begin{aligned}
m_0 &= 21 = 2^4 + 2^2 + 1 \\
m_1 &= 3^4 + 3^2 = 90 \\
m_2 &= 4^4 + 4^2 - 1 = 4^4 + 4 \cdot 3 + 3 = 271 \\
m_3 &= 5^4 + 5 \cdot 3 + 2 = 642 \\
m_4 &= 6^4 + 6 \cdot 3 + 1 = 1315 \\
m_5 &= 7^4 + 7 \cdot 3 = 2422 \\
m_6 &= 8^4 + 8 \cdot 2 + 7 = 4119 \\
m_7 &= 9^4 + 9 \cdot 2 + 6 = 6585 \\
m_8 &= 10^4 + 10 \cdot 2 + 5 = 10\,025 \\
&\text{etc.}
\end{aligned}$$

Even though weak Goodstein sequences increase rapidly at first, we have

6.5 Theorem *For each $m > 0$, the weak Goodstein sequence starting at m eventually terminates with $m_n = 0$ for some n.*

Proof. We use the normal form for ordinals. Let $m > 0$ and $m_0, m_1, m_2, \ldots$ be the weak Goodstein sequence starting at m. Its ath term is written in base $a + 2$:

$$m_a = (a+2)^{b_1} k_1 + \cdots + (a+2)^{b_n} k_n. \tag{6.6}$$

Consider the ordinal

$$\alpha_a = \omega^{b_1} \cdot k_1 + \cdots + \omega^{b_n} \cdot k_n$$

obtained by replacing base $a + 2$ by ω in (6.6). It is easily seen that $\alpha_0 > \alpha_1 > \alpha_2 > \cdots > \alpha_a > \cdots$ is a decreasing sequence of ordinals, necessarily finite. Therefore, there exists some n such that $\alpha_n = 0$. But clearly $m_a \leq \alpha_a$ for every $a = 0, 1, 2, \ldots, n$. Hence $m_n = 0$. □

Remark For the weak Goodstein sequence $m_0, m_1, m_2, \ldots$ starting at $m = 21$ displayed above, the corresponding ordinal numbers are $\omega^4 + \omega^2 + 1$, $\omega^4 + \omega^2$, $\omega^4 + \omega \cdot 3 + 3$, $\omega^4 + \omega \cdot 3 + 2$, $\omega^4 + \omega \cdot 3 + 1$, $\omega^4 + \omega \cdot 3$, $\omega^4 + \omega \cdot 2 + 7$, ...

We shall now outline an even stronger result. A number m is written in *pure base* $a \geq 2$ if it is first written in base a, then so are the exponents and the exponents of exponents, etc. For instance, the number 324 written in pure base 3 is $3^{3+2} + 3^{3+1}$.

The *Goodstein sequence* starting at $m > 0$ is a sequence $m_0, m_1, m_2, \ldots$ obtained as follows: Let $m_0 = m$ and write m_0 in pure base 2. To define m_1, replace each 2 by 3, and then subtract 1. In general, to get m_{k+1}, write m_k in pure base $k+2$, replace each $k+2$ by $k+3$, and subtract 1. For example, the Goodstein sequence starting at $m = 21$ is as follows:

$$m_0 = 21 = 2^{2^2} + 2^2 + 1$$
$$m_1 = 3^{3^3} + 3^3 \sim 7.6 \times 10^{12}$$
$$m_2 = 4^{4^4} + 4^4 - 1 = 4^{4^4} + 4^3 \cdot 3 + 4^2 \cdot 3 + 4 \cdot 3 + 3 \sim 1.3 \times 10^{154}$$
$$m_3 = 5^{5^5} + 5^3 \cdot 3 + 5^2 \cdot 3 + 5 \cdot 3 + 2 \sim 1.9 \times 10^{2184}$$
$$m_4 = 6^{6^6} + 6^3 \cdot 3 + 6^2 \cdot 3 + 6 \cdot 3 + 1 \sim 2.6 \times 10^{36\,305}$$

etc.

Goodstein sequences initially grow even more rapidly than weak Goodstein sequences. But still:

6.7 Theorem *For each $m > 0$, the Goodstein sequence starting at m eventually terminates with $m_n = 0$ for some n.*

Proof. Again, we define a (finite) sequence of ordinals $\alpha_0 > \alpha_1 > \cdots > \alpha_a > \cdots$ as follows: When m_a is written in pure base $a+2$, we get α_a by replacing each $a+2$ by ω. For instance, in the example above, the ordinals are $\omega^{\omega^\omega} + \omega^\omega + 1$, $\omega^{\omega^\omega} + \omega^\omega$, $\omega^{\omega^\omega} + \omega^3 \cdot 3 + \omega^2 \cdot 3 + \omega \cdot 3 + 3$, $\omega^{\omega^\omega} + \omega^3 \cdot 3 + \omega^2 \cdot 3 + \omega \cdot 3 + 2$, $\omega^{\omega^\omega} + \omega^3 \cdot 3 + \omega^2 \cdot 3 + \omega \cdot 3 + 1$, etc. The ordinals α_a are in normal form, and again, it can be shown that they form a (finite) decreasing sequence. Therefore $\alpha_n = 0$ for some n, and since $m_a \leq \alpha_a$ for all a, we have $m_n = 0$. □

Exercises

6.1 Show that $\omega^\varepsilon = \varepsilon$.

6.2 Find the first few terms of the Goodstein sequence starting at $m = 28$.

Chapter 7

Alephs

1. Initial Ordinals

In Chapter 5 we started the investigation of cardinalities of infinite sets. Although we proved several results involving the concept of $|X|$, the cardinality of a set X, we have not defined $|X|$ itself, except in the case when X is finite or countable.

In the present chapter we consider the question of finding "representatives" of cardinalities. Natural numbers play this role satisfactorily for finite sets. We generalized the concept of natural number and showed that the resulting ordinal numbers have many properties of natural numbers, in particular, inductive proofs and recursive constructions on them are possible. However, ordinal numbers do not represent cardinalities, instead, they represent types of well-orderings. Since any infinite set can be well-ordered in many different ways (if at all) (see Exercise 1.1), there are many ordinal numbers of the same cardinality; ω, $\omega + 1$, $\omega + 2$, ... , $\omega + \omega$, ... , $\omega \cdot \omega$, $\omega \cdot \omega + 1$, ... are all countable ordinal numbers; that is, $|\omega| = |\omega + 1| = |\omega + \omega| = \cdots = \aleph_0$. An explanation of the good behavior of ordinal numbers of finite cardinalities — that is, natural numbers — is provided by Theorem 4.3 in Chapter 4: All linear orderings of a finite set are isomorphic, and they are well-orderings. So for any finite set X, there is precisely one ordinal number n such that $|n| = |X|$. We have called this n the cardinal number of X.

In spite of these difficulties, it is now rather easy to get representatives for cardinalities of infinite (well-orderable) sets; we simply take the *least* ordinal number of any given cardinality as the representative of that cardinality.

1.1 Definition An ordinal number α is called an *initial ordinal* if it is not equipotent to any $\beta < \alpha$.

1.2 Example Every natural number is an initial ordinal. ω is an initial ordinal, because ω is not equipotent to any natural number. $\omega + 1$ is not initial, because $|\omega| = |\omega + 1|$. Similarly, none of $\omega + 2$, $\omega + 3$, $\omega + \omega$, $\omega \cdot \omega$, ω^ω, ... , is initial.

1.3 Theorem *Each well-orderable set X is equipotent to a unique initial ordinal number.*

Proof. By Theorem 3.1 in Chapter 6, X is equipotent to some ordinal α. Let α_0 be the least ordinal equipotent to X. Then α_0 is an initial ordinal, because $|\alpha_0| = |\beta|$ for some $\beta < \alpha_0$ would imply $|X| = |\beta|$, a contradiction.

If $\alpha_0 \neq \alpha_1$ are initial ordinals, they cannot be equipotent, because $|\alpha_0| = |\alpha_1|$ and, say, $\alpha_0 < \alpha_1$, would violate the fact that α_1 is initial. This proves the uniqueness. □

1.4 Definition If X is a well-orderable set, then the *cardinal number* of X, denoted $|X|$, is the unique initial ordinal equipotent to X. In particular, $|X| = \omega$ for any countable set X and $|X| = n$ for any finite set of n elements, in agreement with our previous definitions.

According to Theorem 1.3, the cardinal numbers of well-orderable sets are precisely the initial ordinal numbers. A natural question is whether there are any other initial ordinals besides the natural numbers and ω. The next theorem shows that there are arbitrarily large initial ordinals. Actually, we prove something more general. Let A be any set; A may not be well-orderable itself, but it certainly has some well-orderable subsets; for example, all finite subsets of A are well-orderable.

1.5 Definition For any A, let $h(A)$ be the least ordinal number which is not equipotent to any subset of A. $h(A)$ is called the *Hartogs number* of A.

By definition, $h(A)$ is the least ordinal α such that $|\alpha| \not\leq |A|$.

1.6 Lemma *For any A, $h(A)$ is an initial ordinal number.*

Proof. Assume that $|\beta| = |h(A)|$ for some $\beta < h(A)$. Then β is equipotent to a subset of A, and β is equipotent to $h(A)$. We conclude that $h(A)$ is equipotent to a subset of A, i.e., $h(A) < h(A)$, a contradiction. □

So far, we have evaded the main difficulty: How do we know that the Hartogs number of A exists? If all infinite ordinals were countable, $h(\omega)$ would consist of all ordinals!

1.7 Lemma *The Hartogs number of A exists for all A.*

Proof. By Theorem 3.1 in Chapter 6, for every well-ordered set (W, R) where $W \subseteq A$, there is a unique ordinal α such that $(\alpha, <)$ is isomorphic to (W, R). The Axiom Schema of Replacement implies that there exists a set H such that, for every well-ordering $R \in \mathcal{P}(A \times A)$, the ordinal α isomorphic to it is in H. We claim that H contains all ordinals equipotent to a subset of A. Indeed, if f is a one-to-one function mapping α into A, we set $W = \operatorname{ran} f$ and $R = \{(f(\beta), f(\gamma)) \mid \beta < \gamma < \alpha\}$. $R \subseteq A \times A$ is then a well-ordering isomorphic to α (by the isomorphism f). These considerations show that $h(A) = \{\alpha \in H \mid \alpha$ is an ordinal equipotent to a subset of $A\}$, and justify the existence of $h(A)$ by the Axiom Schema of Comprehension. □

The preceding developments allow us to define a "scale" of larger and larger initial ordinal numbers by transfinite recursion.

1.8 Definition

$$\begin{aligned} \omega_0 &= \omega; \\ \omega_{\alpha+1} &= h(\omega_\alpha) \quad \text{for all } \alpha; \\ \omega_\alpha &= \sup\{\omega_\beta \mid \beta < \alpha\} \quad \text{if } \alpha \text{ is a limit ordinal, } \alpha \neq 0. \end{aligned}$$

The remark following Definition 1.5 shows that $|\omega_{\alpha+1}| > |\omega_\alpha|$, for each α, and so $|\omega_\alpha| < |\omega_\beta|$ whenever $\alpha < \beta$.

1.9 Theorem
(a) ω_α is an infinite initial ordinal number for each α.
(b) If Ω is an infinite initial ordinal number, then $\Omega = \omega_\alpha$ for some α.

Proof.
(a) The proof is by induction on α. The only nontrivial case is when α is a limit ordinal. Suppose that $|\omega_\alpha| = |\gamma|$ for some $\gamma < \omega_\alpha$; then there is $\beta < \alpha$ such that $\gamma \leq \omega_\beta$ (by the definition of supremum). But this implies $|\omega_\alpha| = |\gamma| \leq |\omega_\beta| \leq |\omega_\alpha|$ and yields a contradiction.
(b) First, an easy induction show that $\alpha \leq \omega_\alpha$ for all α. Therefore, for every infinite initial ordinal Ω, there is an ordinal α such that $\Omega < \omega_\alpha$ (for example, $\alpha = \Omega + 1$). Thus it suffices to prove the following claim: For every infinite initial ordinal $\Omega < \omega_\alpha$, there is some $\gamma < \alpha$ such that $\Omega = \omega_\gamma$. The proof proceeds by induction on α. The claim is trivially true for $\alpha = 0$. If $\alpha = \beta + 1$, $\Omega < \omega_\alpha = h(\omega_\beta)$ implies that $|\Omega| \leq |\omega_\beta|$, so either $\Omega = \omega_\beta$ and we can let $\gamma = \beta$, or $\Omega < \omega_\beta$ and existence of $\gamma < \beta < \alpha$ follows from the inductive assumption. If α is a limit ordinal, $\Omega < \omega_\alpha = \sup\{\omega_\beta \mid \beta < \alpha\}$ implies that $\Omega < \omega_\beta$ for some $\beta < \alpha$. The inductive assumption again guarantees the existence of some $\gamma < \beta$ such that $\Omega = \omega_\gamma$.

□

The conclusion of this section is that every well-orderable set is equipotent to a unique initial ordinal number and that infinite initial ordinal numbers form a transfinite sequence ω_α with α ranging over all ordinal numbers. Infinite initial ordinals are, by definition, the cardinalities of infinite well-orderable sets. It is customary to call these cardinal numbers *alephs*; thus we define

$$\aleph_\alpha = \omega_\alpha$$

for each α.

The cardinal number of a well-orderable set is thus either a natural number or an aleph. In particular, $|\boldsymbol{N}| = \aleph_0$ in agreement with our previous usage. The reader should also note that the ordering of cardinal numbers by size defined in Chapter 4 agrees with the ordering of natural numbers and alephs as ordinals by $<$ (i.e., $\in$): If $|X| = \aleph_\alpha$ and $|Y| = \aleph_\beta$, then $|X| < |Y|$ holds if and only if $\aleph_\alpha < \aleph_\beta$ (i.e., $\omega_\alpha \in \omega_\beta$) and a similar equivalence holds if one or both of $|X|$ and $|Y|$ are natural numbers.

In Chapter 5 we defined addition, multiplication, and exponentiation of cardinal numbers; these operations agree with the corresponding ordinal addition, multiplication, and exponentiation, as defined in Chapter 6, as long as the ordinals involved are natural numbers but they may differ for infinite ordinals. For example, $\omega_0 + \omega_0 \neq \omega_0$ if $+$ stands for the ordinal addition, but $\omega_0 + \omega_0 = \omega_0$ if $+$ stands for the cardinal addition. The addition of cardinal numbers is commutative, but the addition of ordinal numbers is not. To avoid confusion, we employ the convention of using the ω-symbolism when the ordinal operations are involved, and the *aleph*-symbolism for the cardinal operations. Thus $\omega_0 + \omega_0$ and 2^{ω_0} indicate ordinal addition and exponentiation ($\omega_0 + \omega_0 = \sup\{\omega + n \mid n < \omega_0\} > \omega_0$, $2^{\omega_0} = \sup\{2^n \mid n < \omega_0\} = \omega_0$), while $\aleph_0 + \aleph_0$ and $2^{\aleph_0}$ are cardinal operations ($\aleph_0 + \aleph_0 = \aleph_0$, $2^{\aleph_0}$ is uncountable).

Exercises

1.1 If X is an infinite well-orderable set, then X has nonisomorphic well-orderings.

1.2 If α and β are at most countable ordinals then $\alpha + \beta$, $\alpha \cdot \beta$, and α^β are at most countable. [*Hint*: Use the representation of ordinal operations from Theorems 5.3 and 5.8 and Exercise 5.16 in Chapter 6. Another possibility is a proof by transfinite induction.]

1.3 For any set A, there is a mapping of $\mathcal{P}(A \times A)$ onto $h(A)$. [*Hint*: Define $f(R)$ = the ordinal isomorphic to R, if $R \subseteq A \times A$ is a well-ordering of its field; $f(R) = 0$ otherwise.]

1.4 $|A| < |A| + h(A)$ for all A.

1.5 $|h(A)| < |\mathcal{P}(\mathcal{P}(A \times A))|$ for all A. [*Hint*: Prove that $|\mathcal{P}(h(A))| \leq |\mathcal{P}(\mathcal{P}(A \times A))|$ by assigning to each $X \in \mathcal{P}(h(A))$ the set of all well-orderings $R \subseteq A \times A$ for which the ordinal isomorphic to R belongs to X.]

1.6 Let $h^*(A)$ be the least ordinal α such that there exists no function with domain A and range α. Prove:

(a) If $\alpha \geq h^*(A)$, then there is no function with domain A and range α.

(b) $h^*(A)$ is an initial ordinal.

(c) $h(A) \leq h^*(A)$.

(d) If A is well-orderable, then $h(A) = h^*(A)$.

(e) $h^*(A)$ exists for all A.

[*Hint for part (e)*: Show that $\alpha \in h^*(A)$ if and only if $\alpha = 0$ or $\alpha =$ the ordinal isomorphic to R, where R is a well-ordering of some partition of A into equivalence classes.]

2. Addition and Multiplication of Alephs

Let us recall the definitions of cardinal addition and multiplication: Let κ and λ be cardinal numbers. We have defined $\kappa + \lambda$ as the cardinality of the set $X \cup Y$, where $|X| = \kappa$, $|Y| = \lambda$, and X and Y are disjoint:

$$|X| + |Y| = |X \cup Y| \quad \text{if } X \cap Y = \emptyset.$$

As we have shown, this definition does not depend on the choice of X and Y. The product $\kappa \cdot \lambda$ has been defined as the cardinality of the cartesian product $X \times Y$, where X and Y are any two sets of respective cardinalities κ and λ:

$$|X| \cdot |Y| = |X \times Y|;$$

and again, this definition is independent of the choice of X and Y. We verified that addition and multiplication of cardinal numbers satisfy various arithmetic laws such as commutativity, associativity, and distributivity:

$$\begin{aligned} \kappa + \lambda &= \lambda + \kappa, \\ \kappa \cdot \lambda &= \lambda \cdot \kappa, \\ \kappa + (\lambda + \mu) &= (\kappa + \lambda) + \mu, \\ \kappa \cdot (\lambda \cdot \mu) &= (\kappa \cdot \lambda) \cdot \mu, \\ \kappa \cdot (\lambda + \mu) &= \kappa \cdot \lambda + \kappa \cdot \mu. \end{aligned}$$

Also, if κ and λ are finite cardinals (i.e., natural numbers), then the operations $\kappa + \lambda$ and $\kappa \cdot \lambda$ coincide with the ordinary arithmetic operations.

The arithmetic of infinite numbers differs substantially from the arithmetic of finite numbers and in fact, the rules for addition and multiplication of alephs are very simple. For instance:

$$\aleph_0 + n = \aleph_0$$

for every natural number n. (If we add n elements to a countable set, the result is a countable set.) We even have

$$\aleph_0 + \aleph_0 = \aleph_0,$$

since, for example, we can view the set of all natural numbers as the union of two disjoint countable sets: the set of even numbers and the set of odd numbers. We also recall that

$$\aleph_0 \cdot \aleph_0 = \aleph_0.$$

(The set of all pairs of natural numbers is countable.) We now prove a general theorem that determines completely the result of addition and multiplication of alephs.

2.1 Theorem $\aleph_\alpha \cdot \aleph_\alpha = \aleph_\alpha$, *for every* α.

Before proving the theorem, let us look at its consequences for addition and multiplication of cardinal numbers.

2.2 Corollary *For every* α *and* β *such that* $\alpha \leq \beta$, *we have*

$$\aleph_\alpha \cdot \aleph_\beta = \aleph_\beta.$$

Also,

$$n \cdot \aleph_\alpha = \aleph_\alpha,$$

for every positive natural number n.

Proof. If $\alpha \leq \beta$, then on the one hand, we have $\aleph_\beta = 1 \cdot \aleph_\beta \leq \aleph_\alpha \cdot \aleph_\beta$, and on the other hand, Theorem 2.1 gives $\aleph_\alpha \cdot \aleph_\beta \leq \aleph_\beta \cdot \aleph_\beta = \aleph_\beta$. Thus by the Cantor-Bernstein Theorem $\aleph_\alpha \cdot \aleph_\beta = \aleph_\beta$.

The equality $n \cdot \aleph_\alpha = \aleph_\alpha$ is proved similarly. □

2.3 Corollary *For every* α *and* β *such that* $\alpha \leq \beta$, *we have*

$$\aleph_\alpha + \aleph_\beta = \aleph_\beta.$$

Also,

$$n + \aleph_\alpha = \aleph_\alpha$$

for all natural numbers n.

Proof. If $\alpha \leq \beta$, then $\aleph_\beta \leq \aleph_\alpha + \aleph_\beta \leq \aleph_\beta + \aleph_\beta = 2 \cdot \aleph_\beta = \aleph_\beta$, and the assertion follows. The second part is proved similarly. □

Proof of Theorem 2.1. We prove the theorem by transfinite induction. For every α, we construct a certain well-ordering $\prec$ of the set $\omega_\alpha \times \omega_\alpha$, and show, using the induction hypothesis $\aleph_\beta \cdot \aleph_\beta \leq \aleph_\beta$ for all $\beta < \alpha$, that the order-type of the well-ordered set $(\omega_\alpha \times \omega_\alpha, \prec)$ is at most ω_α. Then it follows that $\aleph_\alpha \cdot \aleph_\alpha \leq \aleph_\alpha$, and since $\aleph_\alpha \cdot \aleph_\alpha \geq \aleph_\alpha$, we have $\aleph_\alpha \cdot \aleph_\alpha = \aleph_\alpha$.

We construct the well-ordering $\prec$ of $\omega_\alpha \times \omega_\alpha$ uniformly for all ω_α; that is, we define a property $\prec$ of pairs of ordinals and show that $\prec$ well-orders $\omega_\alpha \times \omega_\alpha$ for every ω_α.

We let $(\alpha_1, \alpha_2) \prec (\beta_1, \beta_2)$ if and only if:
either $\max\{\alpha_1, \alpha_2\} < \max\{\beta_1, \beta_2\}$, or $\max\{\alpha_1, \alpha_2\} = \max\{\beta_1, \beta_2\}$ and $\alpha_1 < \beta_1$, or $\max\{\alpha_1, \alpha_2\} = \max\{\beta_1, \beta_2\}$, $\alpha_1 = \beta_1$ and $\alpha_2 < \beta_2$.

We now show that $\prec$ is a well-ordering (of any set of pairs of ordinals).

First, we show that $\prec$ is transitive. Let α_1, α_2, β_1, β_2, γ_1, γ_2 be such that $(\alpha_1, \alpha_2) \prec (\beta_1, \beta_2)$ and $(\beta_1, \beta_2) \prec (\gamma_1, \gamma_2)$. It follows from the definition of $\prec$ that $\max\{\alpha_1, \alpha_2\} \leq \max\{\beta_1, \beta_2\} \leq \max\{\gamma_1, \gamma_2\}$; hence $\max\{\alpha_1, \alpha_2\} \leq \max\{\gamma_1, \gamma_2\}$. If $\max\{\alpha_1, \alpha_2\} < \max\{\gamma_1, \gamma_2\}$ then $(\alpha_1, \alpha_2) \prec (\gamma_1, \gamma_2)$. Thus assume that $\max\{\alpha_1, \alpha_2\} = \max\{\beta_1, \beta_2\} = \max\{\gamma_1, \gamma_2\}$. Then we have $\alpha_1 \leq$

$\beta_1 \le \gamma_1$, and so $\alpha_1 \le \gamma_1$. If $\alpha_1 < \gamma_1$, then $(\alpha_1, \alpha_2) \prec (\gamma_1, \gamma_2)$; otherwise, we have $\alpha_1 = \beta_1 = \gamma_1$. In this last case, $\max\{\alpha_1, \alpha_2\} = \max\{\beta_1, \beta_2\} = \max\{\gamma_1, \gamma_2\}$, and $\alpha_1 = \beta_1 = \gamma_1$, so, necessarily, $\alpha_2 < \beta_2 < \gamma_2$, and it follows again that $(\alpha_1, \alpha_2) \prec (\gamma_1, \gamma_2)$.

Next we verify that for any $\alpha_1, \alpha_2, \beta_1, \beta_2$, either $(\alpha_1, \alpha_2) \prec (\beta_1, \beta_2)$ or $(\beta_1, \beta_2) \prec (\alpha_1, \alpha_2)$ or $(\alpha_1, \alpha_2) = (\beta_1, \beta_2)$ (and that these three cases are mutually exclusive). This follows directly from the definition: Given (α_1, α_2) and (β_1, β_2), we first compare the ordinals $\max\{\alpha_1, \alpha_2\}$ and $\max\{\beta_1, \beta_2\}$, then the ordinals α_1 and β_1, and last the ordinals α_2 and β_2.

We now show that $\prec$ is a well-ordering. Let X be a nonempty set of pairs of ordinals; we find the $\prec$-least element of X. Let δ be the least maximum of the pairs in X; i.e., let δ be the least element of the set $\{\max\{\alpha, \beta\} \mid (\alpha, \beta) \in X\}$. Further, let

$$Y = \{(\alpha, \beta) \in X \mid \max\{\alpha, \beta\} = \delta\}.$$

The set Y is a nonempty subset of X, and for every $(\alpha, \beta) \in Y$ we have $\max\{\alpha, \beta\} = \delta$; moreover, $\delta < \max\{\alpha', \beta'\}$ for any $(\alpha', \beta') \in X - Y$, and hence $(\alpha, \beta) \prec (\alpha', \beta')$ whenever $(\alpha, \beta) \in Y$ and $(\alpha', \beta') \in X - Y$. Therefore, the least element of Y, if it exists, is also the least element of X. Now let α_0 be the least ordinal in the set $\{\alpha \mid (\alpha, \beta) \in Y \text{ for some } \beta\}$, and let

$$Z = \{(\alpha, \beta) \in Y \mid \alpha = \alpha_0\}.$$

The set Z is a nonempty subset of Y, and we have $(\alpha, \beta) \prec (\alpha', \beta')$ whenever $(\alpha, \beta) \in Z$ and $(\alpha', \beta') \in Y - Z$.

Finally, let β_0 be the least ordinal in the set $\{\beta \mid (\alpha_0, \beta) \in Z\}$. Clearly, (α_0, β_0) is the least element of Z, and it follows that (α_0, β_0) is the least element of X.

Having shown that $\prec$ is a well-ordering of $\omega_\alpha \times \omega_\alpha$ for every α, we use this well-ordering to prove, by transfinite induction on α, that $|\omega_\alpha \times \omega_\alpha| \le \aleph_\alpha$, that is, $\aleph_\alpha \cdot \aleph_\alpha \le \aleph_\alpha$.

We have already proved that $\aleph_0 \cdot \aleph_0 = \aleph_0$, and so our assertion is true for $\alpha = 0$. So let $\alpha > 0$, and let us assume that $\aleph_\beta \cdot \aleph_\beta \le \aleph_\beta$ for all $\beta < \alpha$. We prove that $|\omega_\alpha \times \omega_\alpha| \le \aleph_\alpha$. If suffices to show that the order-type of the well-ordered set $(\omega_\alpha \times \omega_\alpha, \prec)$ is at most ω_α. If the order-type of $\omega_\alpha \times \omega_\alpha$ were greater than ω_α, then there would exist $(\alpha_1, \alpha_2) \in \omega_\alpha \times \omega_\alpha$ such that the cardinality of the set

$$X = \{(\xi_1, \xi_2) \in \omega_\alpha \times \omega_\alpha \mid (\xi_1, \xi_2) \prec (\alpha_1, \alpha_2)\}$$

is at least $\aleph_\alpha$. Thus it suffices to prove that for any $(\alpha_1, \alpha_2) \in \omega_\alpha \times \omega_\alpha$, we have $|X| < \aleph_\alpha$.

Let $\beta = \max\{\alpha_1, \alpha_2\} + 1$. Then $\beta \in \omega_\alpha$, and, for every $(\xi_1, \xi_2) \in X$, we have $\max\{\xi_1, \xi_2\} \le \max\{\alpha_1, \alpha_2\} < \beta$, so $\xi_1 \in \beta$ and $\xi_2 \in \beta$. In other words, $X \subseteq \beta \times \beta$.

Let $\gamma < \alpha$ be such that $|\beta| \le \aleph_\gamma$. Then $|X| \le |\beta \times \beta| = |\beta| \cdot |\beta| \le \aleph_\gamma \cdot \aleph_\gamma$, and by the induction hypothesis, $\aleph_\gamma \cdot \aleph_\gamma \le \aleph_\gamma$. Hence, $|X| \le \aleph_\gamma$, and so $|X| < \aleph_\alpha$ as claimed.

Now it follows that $|\omega_\alpha \times \omega_\alpha| \le \aleph_\alpha$. Thus we have proved, by transfinite induction on α, that $\aleph_\alpha \cdot \aleph_\alpha \le \aleph_\alpha$ for all α. Since $\aleph_\alpha \le \aleph_\alpha \cdot \aleph_\alpha$, we have $\aleph_\alpha \cdot \aleph_\alpha = \aleph_\alpha$, and the proof of Theorem 2.1 is complete. □

Exercises

2.1 Give a direct proof of $\aleph_\alpha + \aleph_\alpha = \aleph_\alpha$ by expressing ω_α as a disjoint union of two sets of cardinality $\aleph_\alpha$.

2.2 Give a direct proof of $n \cdot \aleph_\alpha = \aleph_\alpha$ by constructing a one-to-one mapping of ω_α onto $n \times \omega_\alpha$ (where n is a positive natural number).

2.3 Show that

(a) $\aleph_\alpha^n = \aleph_\alpha$, for all positive natural numbers n.

(b) $|[\aleph_\alpha]^n| = \aleph_\alpha$, where $[\aleph_\alpha]^n$ is the set of all n-element subsets of $\aleph_\alpha$, for all $n > 0$.

(c) $|[\aleph_\alpha]^{<\omega}| = \aleph_\alpha$, where $[\aleph_\alpha]^{<\omega}$ is the set of all finite subsets of $\aleph_\alpha$.

[*Hint*: Use Theorem 2.1 and induction; for (c), proceed as in the proof of Theorem 3.10 in Chapter 4, and use $\aleph_0 \cdot \aleph_\alpha = \aleph_\alpha$.]

2.4 If α and β are ordinals and $|\alpha| \le \aleph_\gamma$ and $|\beta| \le \aleph_\gamma$, then $|\alpha + \beta| \le \aleph_\gamma$, $|\alpha \cdot \beta| \le \aleph_\gamma$, $|\alpha^\beta| \le \aleph_\gamma$ (where $\alpha+\beta$, $\alpha \cdot \beta$, and α^β are *ordinal* operations).

2.5 If X is the image of ω_α by some function f, then $|X| \le \aleph_\alpha$. [*Hint*: Construct a one-to-one mapping g of X into ω_α by letting $g(x)$ = the least element of the inverse image of $\{x\}$ by f.]

2.6 If X is a subset of ω_α such that $|X| < \aleph_\alpha$, then $|\omega_\alpha - X| = \aleph_\alpha$.

Chapter 8

The Axiom of Choice

1. The Axiom of Choice and its Equivalents

In the preceding chapter, we left unanswered a fundamental question: Which sets can be well-ordered? Interestingly, the question was posed in this form only at a later stage in the development of set theory. Cantor considered it quite obvious that *every* set can be well-ordered. Here is a fairly intuitive "proof" of this "fact." In order to well-order a set A, it suffices to construct a one-to-one mapping of some ordinal λ onto A. We proceed by transfinite recursion. Let a be any set not in A. Define

$$f(0) = \begin{cases} \text{some element of } A & \text{if } A \neq \emptyset, \\ a & \text{otherwise,} \end{cases}$$

$$f(1) = \begin{cases} \text{some element of } A - \{f(0)\} & \text{if } A - \{f(0)\} \neq \emptyset, \\ a & \text{otherwise,} \end{cases}$$

etc. Generally,

$$f(\alpha) = \begin{cases} \text{some element of } A - \operatorname{ran}(f \restriction \alpha) & \text{if } A - \operatorname{ran}(f \restriction \alpha) \neq \emptyset, \\ a & \text{otherwise.} \end{cases}$$

Intuitively, f lists the elements of A, one by one, as long as they are available; when A is exhausted, f has value a.

We first notice that A does get exhausted at some stage $\lambda < h(A)$, the Hartogs number of A. The reason is that, for $\alpha < \beta$, if $f(\beta) \neq a$, then $f(\beta) \in A - \operatorname{ran}(f \restriction \beta)$, $f(\alpha) \in \operatorname{ran}(f \restriction \beta)$, and thus $f(\alpha) \neq f(\beta)$. If $f(\alpha) \neq a$ were to hold for all $\alpha < h(A)$, f would be a one-to-one mapping of $h(A)$ into A, contradicting the definition of $h(A)$ as the least ordinal which cannot be mapped into A by a one-to-one function.

Let λ be the least $\alpha < h(A)$ such that $f(\alpha) = a$. The considerations of the previous paragraph immediately show that $f \restriction \lambda$ is one-to-one. The "proof" is complete if we show that $\operatorname{ran}(f \upharpoonright \lambda) = A$. Clearly $\operatorname{ran}(f \upharpoonright \lambda) \subseteq A$; if $\operatorname{ran}(f \upharpoonright \lambda) \subset A$, $A - \operatorname{ran}(f \upharpoonright \lambda) \neq \emptyset$ and $f(\lambda) \neq a$, contradicting our definition of λ.

Our use of quotation marks indicates that something is wrong with this argument, but it may not be obvious precisely what it is. However, if one tries to justify this transfinite recursion by the Recursion Theorem, say in the form stated in Theorem 4.5 in Chapter 6, one discovers a need for a function G such that f can be defined by $f(\alpha) = G(f \upharpoonright \alpha)$. Such a function G should have the following properties:

$$\begin{aligned} &G(f \upharpoonright \alpha) \in A - \operatorname{ran}(f \upharpoonright \alpha) && \text{if } A - \operatorname{ran}(f \upharpoonright \alpha) \neq \emptyset, \\ &G(f \upharpoonright \alpha) = a && \text{otherwise.} \end{aligned}$$

If A were well-orderable, some such G could easily be defined; e.g.:

$$G(x) = \begin{cases} \text{the } \prec\text{-least element of } A - \operatorname{ran} x \text{ if } x \text{ is a function and } A - \operatorname{ran} x \neq \emptyset, \\ a \text{ otherwise,} \end{cases}$$

where $\prec$ is some well-ordering of A. But in the absence of well-orderings on A, no property which could be used to define such a function G is obvious.

More specifically, let S be a system of sets. A function g defined on S is called a *choice function* for S if $g(X) \in X$ for all nonempty $X \in S$.

If we now assume that there is a choice function g for $\mathcal{P}(A)$, we are able to fill the gap in the previous proof by defining

$$G(x) = \begin{cases} g(A - \operatorname{ran} x) & \text{if } x \text{ is a function and } A - \operatorname{ran} x \neq \emptyset, \\ a & \text{otherwise.} \end{cases}$$

We proved the difficult half of the following theorem, essentially due to Ernst Zermelo:

1.1 Theorem *A set A can be well-ordered if and only if the set $\mathcal{P}(A)$ of all subsets of A has a choice function.*

Proof. The proof of the second half is easy. If $\prec$ well-orders A, we define a choice function g on $\mathcal{P}(A)$ by

$$g(x) = \begin{cases} \text{the least element of } x \text{ in the well-ordering } \prec & \text{if } x \neq \emptyset, \\ \emptyset & \text{if } x = \emptyset. \end{cases}$$

□

The problem of well-ordering the set A is now reduced to an equivalent question, that of finding a choice function for $\mathcal{P}(A)$. We first notice Theorem 1.2.

1.2 Theorem *Every finite system of sets has a choice function.*

Proof. Proceed by induction. Let us assume that every system with n elements has a choice function, and let $|S| = n + 1$. Fix $X \in S$; the set $S - \{X\}$ has n elements, and, consequently, a choice function g_X. If $X = \emptyset$, $g = g_X \cup \{(X, \emptyset)\}$ is a choice function for S. If $X \neq \emptyset$, $g^x = g_X \cup \{(X, x)\}$ is a choice function for S (for any $x \in X$). □

The reader might be well advised to analyze the reasons why this proof cannot be generalized to show that every countable system of sets has a choice function. Also, while it is easy to find a choice function for $\mathcal{P}(\boldsymbol{N})$ or $\mathcal{P}(\boldsymbol{Q})$ (why?), no such function for $\mathcal{P}(\boldsymbol{R})$ suggests itself.

Although choice functions for infinite systems of sets of real numbers have been tacitly used by analysts at least since the end of the nineteenth century, it took some time to realize that the assumption of their existence is not entirely trivial. The following axiom was formulated by Zermelo in 1904.

Axiom of Choice There exists a choice function for every system of sets.

Sixty years later, in 1963, Paul Cohen showed that the Axiom of Choice cannot be proved from the axioms of Zermelo-Fraenkel set theory (more on this in Chapter 15). The Axiom of Choice is thus a new principle of set formation; it differs from the other set forming principles in that it is not *effective*. That is, the Axiom of Choice asserts that certain sets (the choice functions) exist without describing those sets as collections of objects having a particular property. Because of this, and because of some of its counterintuitive consequences (see Section 2), some mathematicians raised objections to its use.

We next examine several equivalent formulations of the Axiom of Choice and some of the consequences it has in set theory and in mathematics. Following that, at the end of Section 2, we resume discussion of its justification. To keep track of the use of the Axiom of Choice in this chapter, we denote the theorems whose proofs depend on it, and exercises in which it has to be used, by an asterisk. In later chapters, we use the Axiom of Choice without explicitly pointing it out each time.

1.3 Theorem *The following statements are equivalent:*

(a) (The Axiom of Choice) There exists a choice function for every system of sets.

(b) Every partition has a set of representatives.

(c) If $\langle X_i \mid i \in I\rangle$ is an indexed system of nonempty sets, then there is a function f such that $f(i) \in X_i$ for all $i \in I$.

We remind the reader that a partition of a set A is a system of mutually disjoint nonempty sets whose union equals A. We call $X \subseteq A$ a *set of representatives* for a partition S of A if, for every $C \in S$, $X \cap C$ has a unique element. (See Section 4 of Chapter 2 for these definitions.)

The statement (c) in Theorem 1.3 can be equivalently formulated as follows (compare with Exercise 5.10 in Chapter 3):
(d) If $X_i \neq \emptyset$ for all $i \in I$, then $\prod_{i \in I} X_i \neq \emptyset$.

Proof. *(a) implies (b).* Let f be a choice function for the partition S; then $X = \operatorname{ran} f$ is a set of representatives for S. Notice that for any $C \in S$, $f(C) \in X \cap C$, but $f(D) \notin X \cap C$ for $D \neq C$ [because $f(D) \in D$ and $D \cap C = \emptyset$]. So $X \cap C = \{f(C)\}$ for any $C \in S$.

(b) implies (c). Let $C_i = \{i\} \times X_i$. Since $i \neq i'$ implies $C_i \cap C_{i'} = \emptyset$, $S = \{C_i \mid i \in I\}$ is a partition. If f is a set of representatives for S, f is a set of ordered pairs, and for each $i \in I$, there is a unique x such that $(i, x) \in f \cap C_i$. But this means that f is a function on I and $f(i) \in X_i$ for all $i \in I$.

(c) implies (a). Let S be a system of sets; set $I = S - \{\emptyset\}$, $X_C = C$ for all $C \in I$. Then $\langle X_C \mid C \in I \rangle$ is an indexed system of nonempty sets. If $f \in \prod_{C \in I} X_C$, f is a choice function for S if $\emptyset \notin S$. If $\emptyset \in S$, then $f \cup \{(\emptyset, \emptyset)\}$ is a choice function for S. □

Theorem 1.3 gives several equivalent formulations of the Axiom of Choice. Many other statements are known to be equivalent to the Axiom, and we use some in the applications. The most frequently used equivalent is the Well-Ordering Theorem, which states that every set can be well-ordered (its equivalence with the Axiom follows from Theorem 1.1). Another frequently used version (Zorn's Lemma) is given in Theorem 1.13.

But first we present some consequences of the Axiom of Choice.

1.4 Theorem* *Every infinite set has a countable subset.*

Proof. Let A be an infinite set. A can be well-ordered, or equivalently, arranged in a transfinite one-to-one sequence $\langle a_\alpha \mid \alpha < \Omega \rangle$ whose length Ω is an infinite ordinal. The range $C = \{a_\alpha \mid \alpha < \omega\}$ of the initial segment $\langle a_\alpha \mid \alpha < \omega \rangle$ of this sequence is a countable subset of A. □

1.5 Theorem* *For every infinite set S there exists a unique aleph $\aleph_\alpha$ such that $|S| = \aleph_\alpha$.*

Proof. As S can be well-ordered, it is equipotent to some infinite ordinal, and hence to a unique initial ordinal number ω_α. □

Thus in the set theory with the Axiom of Choice, we can define for any set X its cardinal number $|X|$ as the initial ordinal equipotent to X. Sets X and Y are equipotent if and only if $|X|$ is the same ordinal as $|Y|$ (i.e., $|X| = |Y|$). Also, the ordering $<$ of cardinal numbers by size agrees with the ordering of ordinals by $\in$: $|X| < |Y|$ if and only if $|X| \in |Y|$. These considerations rigorously justify Assumption 1.7 in Chapter 4:

There are sets called *cardinals* with the property that for every set X there is a unique cardinal $|X|$, and sets X and Y are equipotent if and only if $|X|$ is equal to $|Y|$.

As $\in$ is a linear ordering (actually a well-ordering) on any set of ordinal numbers, we have the following theorem.

1.6 Theorem* *For any sets A and B either $|A| \leq |B|$ or $|B| \leq |A|$.*

1.7 Theorem* *The union of a countable collection of countable sets is countable.*

Proof. (Compare with Theorem 3.9 in Chapter 4.) Let S be a countable set whose every element is countable, and let $A = \bigcup S$. We show that A is countable.

As S is countable, there is a one-to-one sequence $\langle A_n \mid n \in \boldsymbol{N} \rangle$ such that $S = \{A_n \mid n \in \boldsymbol{N}\}$. For each $n \in \boldsymbol{N}$, the set A_n is countable, so there exists a sequence whose range is A_n.

By the Axiom of Choice, we can *choose* one such sequence for each n. [That is: For each n, let S_n be the set of all sequences whose range is A_n. Let F be a choice function on $\{S_n \mid n \in \boldsymbol{N}\}$, and let $s_n = F(S_n)$ for each n.]

Having chosen one $s_n = \langle a_n(k) \mid k \in \boldsymbol{N} \rangle$ for each n, we obtain a mapping f of $\boldsymbol{N} \times \boldsymbol{N}$ onto A by letting $f(n, k) = a_n(k)$. Since $\boldsymbol{N} \times \boldsymbol{N}$ is countable and A is its image under f, A is also countable. □

1.8 Corollary* *The set of all real numbers is not the union of countably many countable sets.*

Proof. The set $\boldsymbol{R}$ is uncountable. □

1.9 Corollary* *The ordinal ω_1 is not the supremum of a countable set of countable ordinals.*

Proof. If $\{\alpha_n \mid n \in \boldsymbol{N}\}$ is a set of countable ordinals, then its supremum

$$\alpha = \sup\{\alpha_n \mid n \in \boldsymbol{N}\} = \bigcup_{n \in \boldsymbol{N}} \alpha_n$$

is a countable set, and hence $\alpha < \omega_1$. □

1.10 Theorem* $2^{\aleph_0} \geq \aleph_1$.

Proof. This follows from Theorem 1.5 and the fact that $2^{\aleph_0} > \aleph_0$. □

As a result, the Continuum Hypothesis can be reformulated as a conjecture that

$$2^{\aleph_0} = \aleph_1,$$

the least uncountable cardinal number.

1.11 Theorem* *If f is a function and A is a set, then $|f[A]| \leq |A|$.*

Proof. For each $b \in f[A]$, let $X_b = f^{-1}(\{b\})$. Note that $X_b \neq \emptyset$ and $X_{b_1} \cap X_{b_2} = \emptyset$ if $b_1 \neq b_2$. Take $g \in \prod_{b \in f[A]} X_b$; then $g : f[A] \to A$ and $b_1 \neq b_2$ implies $g(b_1) \in X_{b_1}$, $g(b_1) \in X_{b_2}$, so $g(b_1) \neq g(b_2)$. We conclude that g is a one-to-one mapping of $f[A]$ into A, and consequently $|f[A]| \leq |A|$. □

1.12 Theorem* *If $|S| \leq \aleph_\alpha$ and, for all $A \in S$, $|A| \leq \aleph_\alpha$, then $|\bigcup S| \leq \aleph_\alpha$.*

Proof. This is a generalization of Theorem 1.7. We assume that $S \neq \emptyset$ and all $A \in S$ are nonempty, write $S = \{A_\nu \mid \nu < \aleph_\alpha\}$ and for each $\nu < \aleph_\alpha$, choose a transfinite sequence $A_\nu = \langle a_\nu(\kappa) \mid \kappa < \aleph_\alpha \rangle$ such that $A_\nu = \{a_\nu(\kappa) \mid \kappa < \aleph_\alpha\}$. (Cf. Exercise 1.9 in Chapter 4.) We define a mapping f on $\aleph_\alpha \times \aleph_\alpha$ onto $\bigcup S$ by $f(\nu, \kappa) = a_\nu(\kappa)$. By Theorem 1.11,

$$\left|\bigcup S\right| \leq |\aleph_\alpha \times \aleph_\alpha| = \aleph_\alpha$$

(the last step is Theorem 2.1 in Chapter 7). □

We next derive Zorn's Lemma, another important version of the Axiom of Choice.

1.13 Theorem *The following statements are equivalent:*

(a) (The Axiom of Choice) There exists a choice function for every system of sets.

(b) (The Well-Ordering Principle) Every set can be well-ordered.

(c) (Zorn's Lemma) If every chain in a partially ordered set has an upper bound, then the partially ordered set has a maximal element.

We remind the reader that a chain is a linearly ordered subset of a partially ordered set (see Section 5 of Chapter 2 for this definition, as well as the definitions of "ordered set," "upper bound," "maximal element," and other concepts related to orderings).

Proof. Equivalence of (a) and (b) follows immediately from Theorem 1.1; therefore, it is enough to show that (a) implies (c) and (c) implies (a).

(a) implies (c). Let $(A, \preccurlyeq)$ be a (partially) ordered set in which every chain has an upper bound. Our strategy is to search for a maximal element of $(A, \preccurlyeq)$ by constructing a $\preccurlyeq$-increasing transfinite sequence of elements of A.

We fix some $b \notin A$ and a choice function g for $\mathcal{P}(A)$, and define $\langle a_\alpha \mid \alpha < h(A) \rangle$ by transfinite recursion. Given $\langle a_\xi \mid \xi < \alpha \rangle$, we consider two cases. If $b \neq a_\xi$ for all $\xi < \alpha$ and $A_\alpha = \{a \in A \mid a_\xi \prec a \text{ holds for all } \xi < \alpha\} \neq \emptyset$, we let $a_\alpha = g(A_\alpha)$; otherwise we let $a_\alpha = b$.

We leave to the reader the easy task of justifying this definition by Theorem 4.4 in Chapter 7. We note that $a_\alpha = b$ for some $\alpha < h(A)$; otherwise, $\langle a_\xi \mid \xi < h(A) \rangle$ would be a one-to-one mapping of $h(A)$ into A. Let λ be the least α for which $a_\alpha = b$. Then the set $C = \{a_\xi \mid \xi < \lambda\}$ is a chain in $(A, \preccurlyeq)$ and so it has an upper bound $c \in A$. If $c \prec a$ for some $a \in A$, we have $a \in A_\lambda \neq \emptyset$ and $a_\lambda = g(A_\lambda) \neq b$, a contradiction. So c is a maximal element of A. (It is easy to see that, in fact, $\lambda = \beta + 1$ and $c = a_\beta$.)

(c) implies (a). It suffices if we show that every system of nonempty sets S has some choice function. Let F be the system of all functions f for which $\operatorname{dom} f \subseteq S$ and $f(X) \in X$ holds for any $X \in S$. The set F is ordered by inclusion $\subseteq$. Moreover, if F_0 is a linearly ordered subset of $(F, \subseteq)$ (i.e., either $f \subseteq g$ or $g \subseteq f$ holds for any $f, g \in F_0$), $f_0 = \bigcup F_0$ is a function. (See Theorem 3.12 in Chapter 2.) It is easy to check that $f_0 \in F$ and f_0 is an upper bound on F_0 in $(F, \subseteq)$.

The assumptions of Zorn's Lemma being satisfied, we conclude that $(F, \subseteq)$ has a maximal element $\overline{f}$. The proof is complete if we show that $\operatorname{dom} \overline{f} = S$. If not, select some $X \in S - \operatorname{dom} \overline{f}$ and $x \in X$. Clearly, $\overline{\overline{f}} = \overline{f} \cup \{(X, x)\} \in F$ and $\overline{\overline{f}} \supset \overline{f}$, contradicting the maximality of $\overline{f}$. □

We conclude this section with a theorem needed in Chapter 11.

1.14 Theorem* *If $(A, \preccurlyeq)$ is a linear ordering such that $|\{y \in A \mid y \preccurlyeq x\}| < \aleph_\gamma$ for all $x \in A$, then $|A| \leq \aleph_\gamma$.*

Proof. Exactly as in the proof of Zorn's Lemma, we construct an increasing sequence $\langle a_\xi \mid \xi < \lambda \rangle$ of elements of A for which $A_\lambda = \emptyset$, i.e., there is no $a \in A$ such that $a_\xi \prec a$ holds for all $\xi \in A$. As $(A, \preccurlyeq)$ is linearly ordered, this means that, for every $a \in A$, $a \preccurlyeq a_\xi$ for some $\xi \in A$. (We say that the sequence $\langle a_\xi \mid \xi < \lambda \rangle$ is *cofinal* in $(A, \preccurlyeq)$.)

We have $A = \bigcup_{\xi < \lambda} \{y \in A \mid y \preccurlyeq a_\xi\}$, $|\{y \in A \mid y \preccurlyeq a_\xi\}| < \aleph_\gamma$ by assumption, and $\lambda \leq \aleph_\gamma$ (otherwise, $\omega_\gamma < \lambda$ and $\langle a_\xi \mid \xi < \omega_\gamma \rangle$ is a one-to-one mapping of $\aleph_\gamma$ into $\{y \in A \mid y \preccurlyeq a_{\omega_\gamma}\}$, contradicting the assumption). By Theorem 1.12, $|A| \leq \aleph_\gamma$. □

Exercises

1.1 Prove: If a set A can be linearly ordered, then every system of finite subsets of A has a choice function. (It does not follow from the Zermelo-Fraenkel axioms that every set can be linearly ordered.)

1.2 If A can be well-ordered, then $\mathcal{P}(A)$ can be linearly ordered. [*Hint*: Let $<$ be a well-ordering of A; for $X, Y \subseteq A$ define $X \prec Y$ if and only if the $<$-least element of $X \triangle Y$ belongs to X.]

1.3* Let $(A, \leq)$ be an ordered set in which every chain has an upper bound. Then for every $a \in A$, there is a $\leq$-maximal element x of A such that $a \leq x$.

1.4 Prove that Zorn's Lemma is equivalent to the statement: For all $(A, \leq)$, the set of all chains of $(A, \leq)$ has an $\subseteq$-maximal element.

1.5 Prove that Zorn's Lemma is equivalent to the statement: If A is a system of sets such that, for each $B \subseteq A$ which is linearly ordered by $\subseteq$, $\bigcup B \in A$, then A has an $\subseteq$-maximal element.

1.6 A system of sets A has *finite character* if $X \in A$ if and only if every finite subset of X belongs to A. Prove that Zorn's Lemma is equivalent to the following (Tukey's Lemma): Every system of sets of finite character has an $\subseteq$-maximal element. [*Hint*: Use Exercise 1.5.]

1.7* Let E be a binary relation on a set A. Show that there exists a function $f : A \to A$ such that for all $x \in A$, $(x, f(x)) \in E$ if and only if there is some $y \in A$ such that $(x, y) \in E$.

1.8* Prove that every uncountable set has a subset of cardinality $\aleph_1$.

1.9* Every infinite set is equipotent to some of its proper subsets. Equivalently, Dedekind finite sets are precisely the finite sets.

1.10* Let $(A, <)$ be a linearly ordered set. A sequence $\langle a_n \mid n \in \omega \rangle$ of elements of A is *decreasing* if $a_{n+1} < a_n$ for all $n \in \omega$. Prove that $(A, <)$ is a well-ordering if and only if there exists no infinite decreasing sequence in A.

1.11* Prove the following distributive laws (see Exercise 3.13 in Chapter 2).

$$\bigcap_{t \in T} \Big(\bigcup_{s \in S} A_{t,s} \Big) = \bigcup_{f \in S^T} \Big(\bigcap_{t \in T} A_{t,f(t)} \Big).$$

$$\bigcup_{t \in T} \Big(\bigcap_{s \in S} A_{t,s} \Big) = \bigcap_{f \in S^T} \Big(\bigcup_{t \in T} A_{t,f(t)} \Big).$$

1.12* Prove that for every ordering $\preccurlyeq$ on A, there is a linear ordering $\leq$ on A such that $a \preccurlyeq b$ implies $a \leq b$ for all $a, b \in A$ (i.e., every partial ordering can be extended to a linear ordering).

1.13* (Principle of Dependent Choices) If R is a binary relation on $M \neq \emptyset$ such that for each $x \in M$ there is $y \in M$ for which xRy, then there is a sequence $\langle x_n \mid n \in \omega \rangle$ such that $x_n R x_{n+1}$ holds for all $n \in \omega$.

1.14 Assuming only the Principle of Dependent Choices, prove that every countable system of sets has a choice function (the Axiom of Countable Choice).

1.15 If every set is equipotent to an ordinal number, then the Axiom of Choice holds.

1.16 If for any sets A and B either $|A| \leq |B|$ or $|B| \leq |A|$, then the Axiom of Choice holds. [*Hint*: Compare A and $B = h(A)$.]

1.17* If B is an infinite set and A is a subset of B such that $|A| < |B|$, then $|B - A| = |B|$.

2. The Use of the Axiom of Choice in Mathematics

In this section we present several examples of the use of the Axiom of Choice in mathematics. We have chosen examples which illustrate the variety of roles the Axiom of Choice plays in mathematics, but which do not require extensive background outside of set theory. Other applications of the Axiom of Choice can be found in the exercises and in most textbooks of general topology, abstract algebra, and functional analysis. The examples are followed by some discussion of importance and soundness of the Axiom of Choice.

2.1 Example. Closure Points. According to the usual definition, a sequence of real numbers $\langle x_n \mid n \in \boldsymbol{N}\rangle$ *converges* to $a \in \boldsymbol{R}$ if for every positive real number ε there exists $n_\varepsilon \in \boldsymbol{N}$ such that $|x_n - a| < \varepsilon$ holds for all natural numbers $n \geq n_\varepsilon$. (In this example, $|x|$ denotes the absolute value of x, and not the cardinality of x.)

Let A be a set of real numbers. Advanced calculus textbooks commonly characterize closure points of A in either (or both) of the following ways:

(a) $a \in \boldsymbol{R}$ is a *closure point* of A if and only if there exists a sequence $\langle x_n \mid n \in \boldsymbol{N}\rangle$ with values in A, which converges to a.

(b) $a \in \boldsymbol{R}$ is a *closure point* of A if and only if for every positive real number ε there exists $x \in A$ such that $|x - a| < \varepsilon$.

It is then necessary to prove that (a) and (b) are equivalent.

(a) implies (b). Given $\varepsilon > 0$, there is $n_\varepsilon \in \boldsymbol{N}$ such that $|x_n - a| < \varepsilon$ for all $n \geq n_\varepsilon$. In particular, $|x_{n_\varepsilon} - a| < \varepsilon$ and $x_{n_\varepsilon} \in A$.

(b) implies (a). The usual proof proceeds as follows: Let $X_n = \{x \in A \mid |x - a| < 1/n\}$. By (b), $X_n \neq \emptyset$ for all $n \in \boldsymbol{N}$. Let $\langle x_n \mid n \in \boldsymbol{N}\rangle$ be a sequence such that $x_n \in X_n$ for all $n \in \boldsymbol{N}$. Then each $x_n \in A$ and $\langle x_n \mid n \in \boldsymbol{N}\rangle$ converges to a. □

The question usually passed over is, What reasons do we have to assume that any such sequence $\langle x_n \mid n \in \boldsymbol{N}\rangle$ exists? Notice that we do not give any property $\mathbf{P}(x, y)$ such that $\mathbf{P}(n, y)$ holds if and only if $y = x_n$ (for all $n \in \boldsymbol{N}$). Such a property can be exhibited in special cases (e.g., if A is open — see Exercise 2.1); however, it has been shown that the equivalence of (a) and (b) for all $A \subseteq \boldsymbol{R}$ cannot be proved from the axioms of Zermelo-Fraenkel set theory alone. Of course, if we do assume the Axiom of Choice, the fact that $X_n \neq \emptyset$ for all $n \in \boldsymbol{N}$ immediately implies that $\prod_{n \in \boldsymbol{N}} X_n \neq \emptyset$.

2.2 Example. Continuity of a Function. The standard definition of continuity of a real-valued function of a real variable is as follows:

(a) $f : \boldsymbol{R} \to \boldsymbol{R}$ is *continuous* at $a \in \boldsymbol{R}$ if and only if for every $\varepsilon > 0$ there is $\delta > 0$ such that $|f(x) - f(a)| < \varepsilon$ for all x such that $|x - a| < \delta$.

Continuity is also characterized by the following property:

(b) $f : \boldsymbol{R} \to \boldsymbol{R}$ is *continuous* at $a \in \boldsymbol{R}$ if and only if for every sequence $\langle x_n \mid n \in \boldsymbol{N}\rangle$ which converges to a, the sequence $\langle f(x_n) \mid n \in \boldsymbol{N}\rangle$ converges to $f(a)$.

It is easy to see that (a) implies (b): If $\langle x_n \mid n \in \boldsymbol{N}\rangle$ converges to a and if $\varepsilon > 0$ is given, then first we find $\delta > 0$ as in (a), and because $\langle x_n \mid n \in \boldsymbol{N}\rangle$ converges, there exists n_δ such that $|x_n - a| < \delta$ whenever $n \geq n_\delta$. Clearly, $|f(x_n) - f(a)| < \varepsilon$ for all such n.

If we assume the Axiom of Choice, then (b) also implies (a), and hence (a) and (b) are two equivalent definitions of continuity. Suppose that (a) fails, then there exists $\varepsilon > 0$ such that for each $\delta > 0$ there exists an x such that $|x - a| < \delta$

but $|f(x) - f(a)| \geq \varepsilon$. In particular, for each $k = 1, 2, 3, \ldots$, we can *choose* some x_k such that $|x_k - a| < 1/k$ and $|f(x_k) - f(a)| \geq \varepsilon$. The sequence $\langle x_k \mid k \in \boldsymbol{N} \rangle$ converges to a, but the sequence $\langle f(x_k) \mid k \in \boldsymbol{N} \rangle$ does not converge to $f(a)$, so (b) fails as well. □

As in Example 2.1, it can be shown that the equivalence of (a) and (b) cannot be proved from the axioms of Zermelo-Fraenkel set theory alone.

2.3 Example. Basis of a Vector Space. We assume that the reader is familiar with the notion of a vector space over a field (e.g., over the field of real numbers). Definitions and basic algebraic properties of vector spaces can be found in any text on linear algebra.

A set A of vectors is *linearly independent* if no finite linear combination $\alpha_1 v_1 + \cdots + \alpha_n v_n$ of elements $v_1, \ldots, v_n$ of A with nonzero coefficients $\alpha_1, \ldots, \alpha_n$ from the field is equal to the zero vector. A *basis* of a vector space V is a maximal (in the ordering by inclusion) linearly independent subset of V.

One of the fundamental facts about vector spaces is

2.4 Theorem* *Every vector space has a basis.*

Proof. The theorem is a straightforward application of Zorn's Lemma: If C is a $\subset$-chain of independent subsets of the given vector space, then the union of C is also an independent set. Consequently, a maximal independent set exists. [For more details, see the special case in the next example.] □

This theorem cannot be proved in Zermelo-Fraenkel set theory alone, without using the Axiom of Choice.

2.5 Example. Hamel Basis. Consider the set of all real numbers as a vector space over the field of rational numbers. By Theorem 2.4 this vector space has a basis, called a *Hamel basis* for $\boldsymbol{R}$.

In other words, a set $X \subseteq \boldsymbol{R}$ is a Hamel basis for $\boldsymbol{R}$ if every $x \in \boldsymbol{R}$ can be expressed in a unique way as

$$x = r_1 \cdot x_1 + \cdots + r_n \cdot x_n$$

for some mutually distinct $x_1, \ldots, x_n \in X$ and some nonzero rational numbers $r_1, \ldots, r_n$. Below we give a detailed argument that a set X with the last mentioned property exists.

A set of real numbers X is called *dependent* if there are mutually distinct $x_1, \ldots, x_n \in X$ and $r_1, \ldots, r_n \in \boldsymbol{Q}$ such that

$$r_1 \cdot x_1 + \cdots + r_n \cdot x_n = 0$$

and at least one of the coefficients $r_1, \ldots, r_n$ is not zero. A set which is not dependent is called *independent.* Let A be the system of all independent sets of real numbers. We use Zorn's Lemma to show that A has a maximal element in the ordering by $\subseteq$; the argument is then completed by showing that any $\subseteq$-maximal independent set is a Hamel basis.

To verify the assumptions of Zorn's Lemma, consider $A_0 \subseteq A$ linearly ordered by $\subseteq$. Let $X_0 = \bigcup A_0$; X_0 is an upper bound of A_0 in $(A, \subseteq)$ if $X_0 \in A$, i.e., if X_0 is independent. But this is true: Suppose there were $x_1, \ldots, x_n \in X_0$ and $r_1, \ldots, r_n \in \boldsymbol{Q}$, not all zero, such that $r_1 \cdot x_1 + \cdots + r_n \cdot x_n = 0$. Then there would be $X_1, \ldots, X_n \in A_0$ such that $x_1 \in X_1, \ldots, x_n \in X_n$. Since A_0 is linearly ordered by $\subseteq$, the finite subset $\{X_1, \ldots, X_n\}$ of A_0 would have a $\subseteq$-greatest element, say X_i. But then $x_1, \ldots, x_n \in X_i$, so X_i would not be independent.

By Zorn's Lemma, we conclude that $(A, \subseteq)$ has a maximal element X. It remains to be shown that X is a Hamel basis.

Suppose that $x \in \boldsymbol{R}$ cannot be expressed as $r_1 \cdot x_1 + \cdots + r_n \cdot x_n$ for any $r_1, \ldots, r_n \in \boldsymbol{Q}$ and $x_1, \ldots, x_n \in X$. Then $x \notin X$ (otherwise $x = 1 \cdot x$), so $X \cup \{x\} \supset X$, and $X \cup \{x\}$ is dependent (remember that X is a maximal independent set). Thus there are $x_1, \ldots, x_n \in X \cup \{x\}$ and $s_1, \ldots, s_n \in \boldsymbol{Q}$, not all zero, such that $s_1 \cdot x_1 + \cdots + s_n \cdot x_n = 0$. Since X is independent, $x \in \{x_1, \ldots, x_n\}$; say, $x = x_i$, and the corresponding coefficient $s_i \neq 0$. But then

$$\begin{aligned} x &= x_i \\ &= \left(-\frac{s_1}{s_i}\right) \cdot x_1 + \cdots + \left(-\frac{s_{i-1}}{s_i}\right) \cdot x_{i-1} + \left(-\frac{s_{i+1}}{s_i}\right) \cdot x_{i+1} + \cdots + \left(-\frac{s_n}{s_i}\right) \cdot x_n, \end{aligned}$$

where $x_1, \ldots, x_{i-1}, x_{i+1}, \ldots, x_n \in X$, and the coefficients are rational numbers. This contradicts the assumption on x.

Suppose now that some $x \in \boldsymbol{R}$ can be expressed in two ways: $x = r_1 \cdot x_1 + \cdots + r_n \cdot x_n = s_1 \cdot y_1 + \cdots + s_k \cdot y_k$, where $x_1, \ldots, x_n, y_1, \ldots, y_k \in X$, $r_1, \ldots, r_n, s_1, \ldots, s_k \in \boldsymbol{Q} - \{0\}$. Then

$$r_1 \cdot x_1 + \cdots + r_n \cdot x_n - s_1 \cdot y_1 - \cdots - s_k \cdot y_k = 0. \tag{2.6}$$

If $\{x_1, \ldots, x_n\} \neq \{y_1, \ldots, y_k\}$ (say, $x_1 \notin \{y_1, \ldots, y_k\}$), (2.6) can be written as a combination of distinct elements from X with at least one nonzero coefficient (namely, r_1), contradicting independence of X. We can conclude that $n = k$ and $x_1 = y_{i_1}, \ldots, x_n = y_{i_n}$ for some one-to-one mapping $\langle i_1, \ldots, i_n \rangle$ between indices $1, 2, \ldots, n$. We can thus write (2.6) in the form $(r_1 - s_{i_1}) \cdot x_1 + \cdots + (r_n - s_{i_n}) \cdot x_n = 0$. Since $x_1, \ldots, x_n$ are mutually distinct elements of X, we conclude that $r_1 - s_{i_1} = 0, \ldots, r_n - s_{i_n} = 0$, i.e., that $r_1 = s_{i_1}, \ldots, r_n = s_{i_n}$.

These arguments show that each $x \in \boldsymbol{R}$ has a unique expression in the desired form and so X is a Hamel basis.

The existence of a Hamel basis cannot be proved in Zermelo-Fraenkel set theory alone; it is necessary to use the Axiom of Choice.

2.7 Example. Additive Functions. A function $f : \boldsymbol{R} \to \boldsymbol{R}$ is called *additive* if $f(x + y) = f(x) + f(y)$ for all $x, y \in \boldsymbol{R}$. An example of an additive function if provided by f_a where $f_a(x) = a \cdot x$ for all $x \in \boldsymbol{R}$, and $a \in \boldsymbol{R}$ is fixed.

Easy computations indicate that any additive function looks much like f_a for some $a \in \boldsymbol{R}$. More precisely, let f be additive, and let us set $f(1) = a$. We then have

$$f(2) = f(1) + f(1) = a \cdot 2, \quad f(3) = f(2) + f(1) = a \cdot 3,$$

and, by induction, $f(b) = a \cdot b$ for all $b \in \boldsymbol{N} - \{0\}$. Since $f(0) + f(0) = f(0+0) = f(0)$, we get $f(0) = 0$. Next, $f(-b) + f(b) = f(0) = 0$, so $f(-b) = -f(b) = a \cdot (-b)$ for $b \in \boldsymbol{N}$. To compute $f(1/n)$, notice that

$$a = f(1) = \underbrace{f(1/n) + \cdots + f(1/n)}_{n \text{ times}} = n \cdot f(1/n);$$

consequently, $f(1/n) = a \cdot 1/n$. Continuing along these lines, we can easily prove that $f(x) = a \cdot x$ for all rational numbers x. It is now natural to conjecture that $f(x) = a \cdot x$ holds for all real numbers x; in other words, that every additive function is of the form f_a for some $a \in \boldsymbol{R}$. It turns out that this conjecture cannot be disproved in Zermelo-Fraenkel set theory, but that it is false if we assume the Axiom of Choice. We prove the following theorem.

2.8 Theorem* *There exists an additive function $f : \boldsymbol{R} \to \boldsymbol{R}$ such that $f \neq f_a$ for all $a \in \boldsymbol{R}$.*

Proof. Let X be a Hamel basis for $\boldsymbol{R}$. Choose a fixed $\overline{x} \in X$. Define

$$f(x) = \begin{cases} r_i & \text{if } x = r_1 \cdot x_1 + \cdots + r_i \cdot x_i + \cdots + r_n \cdot x_n \text{ and } x_i = \overline{x}, \\ 0 & \text{otherwise.} \end{cases}$$

It is easy to check that f is additive. Notice also that $0 \notin X$ and X is infinite (actually, $|X| = 2^{\aleph_0}$). We have $f(\overline{x}) = 1$ (because $\overline{x} = 1 \cdot \overline{x}$ is the basis representation of $\overline{x}$), while $f(\overline{\overline{x}}) = 0$ for any $\overline{\overline{x}} \in X$, $\overline{\overline{x}} \neq \overline{x}$ (because $\overline{x}$ does not occur in the basis representation of $\overline{\overline{x}} = 1 \cdot \overline{\overline{x}}$ of $\overline{\overline{x}}$). If $f = f_a$ were to hold for some $a \in \boldsymbol{R}$, we would have $f(\overline{x}) = 1 = a \cdot \overline{x}$, showing $a \neq 0$, and, on the other hand, $f(\overline{\overline{x}}) = 0 = a \cdot \overline{\overline{x}}$, showing $a = 0$. □

2.9 Example. Hahn-Banach Theorem. A function f defined on a vector space V over the field $\boldsymbol{R}$ of real numbers and with values in $\boldsymbol{R}$ is called a *linear functional* on V if

$$f(\alpha \cdot u + \beta \cdot v) = \alpha \cdot f(u) + \beta \cdot f(v)$$

holds for all $u, v \in V$ and $\alpha, \beta \in \boldsymbol{R}$.

A function p defined on V and with values in $\boldsymbol{R}$ is called a *sublinear functional* on V if

$$p(u + v) \leq p(u) + p(v) \quad \text{for all } u, v \in V$$

and

$$p(\alpha \cdot u) = \alpha \cdot p(u) \quad \text{for all } u \in V \text{ and } \alpha \geq 0.$$

The following theorem, due to Hans Hahn and Stefan Banach, is one of the cornerstones of functional analysis.

2.10 Theorem* *Let p be a sublinear functional on the vector space V, and let f_0 be a linear functional defined on a subspace V_0 of V such that $f_0(v) \le p(v)$ for all $v \in V_0$. Then there is a linear functional f defined on V such that $f \supseteq f_0$ and $f(v) \le p(v)$ for all $v \in V$.*

Proof. Let F be the set of all linear functionals g defined on some subspace W of V and such that

$$f_0 \subseteq g \quad \text{and} \quad g(v) \le p(v) \text{ for all } v \in W.$$

We obtain the desired linear functional f as a maximal element of $(F, \subseteq)$. To verify the assumptions of Zorn's Lemma, consider a nonempty $F_0 \subseteq F$ linearly ordered by $\subseteq$. If $g_0 = \bigcup F_0$, g_0 is an $\subseteq$-upper bound on F_0 provided $g_0 \in F$. Clearly, g_0 is a function with values in $\boldsymbol{R}$ and $g_0 \supseteq f_0$. Since the union of a set of subspaces of V linearly ordered by $\subseteq$ is a subspace of V, $\operatorname{dom} g_0 = \bigcup_{g \in F_0} \operatorname{dom} g$ is a subspace of V. To show that g_0 is linear consider $u, v \in \operatorname{dom} g_0$ and $\alpha, \beta \in \boldsymbol{R}$. Then there are $g, g' \in F_0$ such that $u \in \operatorname{dom} g$ and $v \in \operatorname{dom} g'$. Since F_0 is linearly ordered by $\subseteq$, we have either $g \subseteq g'$ or $g' \subseteq g$. In the first case, $u, v, \alpha \cdot u + \beta \cdot v \in \operatorname{dom} g'$ and $g_0(\alpha \cdot u + \beta \cdot v) = g'(\alpha \cdot u + \beta \cdot v) = \alpha \cdot g'(u) + \beta \cdot g'(v) = \alpha \cdot g_0(u) + \beta \cdot g_0(v)$; the second case is analogous. Finally, $g_0(u) = g(u) \le p(u)$ for any $u \in \operatorname{dom} g_0$ and $g \in F_0$ such that $u \in \operatorname{dom} g$. This completes the check of $g_0 \in F$.

By Zorn's Lemma, $(F, \subseteq)$ has a maximal element f. It remains to be shown that $\operatorname{dom} f = V$. We prove that $\operatorname{dom} f \subset V$ implies that f is not maximal. Fix $u \in V - \operatorname{dom} f$; let W be the subspace of V spanned by $\operatorname{dom} f$ and u. Since every $w \in W$ can be uniquely expressed as $w = x + \alpha \cdot u$ for some $x \in \operatorname{dom} f$ and $\alpha \in \boldsymbol{R}$, the function f_c defined by

$$f_c(w) = f(x) + \alpha \cdot c$$

is a linear functional on W and $f_c \supset f$. The proof is complete if we show that $c \in \boldsymbol{R}$ can be chosen so that

$$f_c(x + \alpha \cdot u) = f(x) + \alpha \cdot c \le p(x + \alpha \cdot u) \tag{2.11}$$

for all $x \in \operatorname{dom} f$ and $\alpha \in \boldsymbol{R}$.

The properties of f immediately guarantee (2.11) for $\alpha = 0$. So we have to choose c so as to satisfy two requirements:

(a) For all $\alpha > 0$ and $x \in \operatorname{dom} f$, $f(x) + \alpha \cdot c \le p(x + \alpha \cdot u)$.

(b) For all $\alpha > 0$ and $y \in \operatorname{dom} f$, $f(y) + (-\alpha) \cdot c \le p(y + (-\alpha) \cdot u)$.

Equivalently,

$$f(y) - p(y - \alpha \cdot u) \le \alpha \cdot c \le p(x + \alpha \cdot u) - f(x)$$

and then

$$f\left(\frac{1}{\alpha}\cdot y\right)-p\left(\frac{1}{\alpha}\cdot y-u\right)\le c\le p\left(\frac{1}{\alpha}\cdot x+u\right)-f\left(\frac{1}{\alpha}\cdot x\right) \tag{2.12}$$

should hold for all $x, y \in \operatorname{dom} f$ and $\alpha > 0$. But, for all $v, t \in \operatorname{dom} f$,

$$f(v)+f(t)=f(v+t)\le p(v+t)\le p(v-u)+p(t+u)$$

and thus

$$f(v)-p(v-u)\le p(t+u)-f(t).$$

If $A=\sup\{f(v)-p(v-u)\mid v\in\operatorname{dom} f\}$ and $B=\inf\{p(t+u)-f(t)\mid t\in\operatorname{dom} f\}$, we have $A\le B$. By choosing c such that $A\le c\le B$, we can make the identity (2.12) hold. □

2.13 Example. The Measure Problem. An important problem in analysis is to extend the notion of length of an interval to more complicated sets of real numbers. Ideally, one would like to have a function μ defined on $\mathcal{P}(\boldsymbol{R})$, with values in $[0,\infty)\cup\{\infty\}$, and having the following properties:

0) $\mu([a,b])=b-a$ for any $a,b\in\boldsymbol{R}$, $a<b$.

i) $\mu(\emptyset)=0$, $\mu(\boldsymbol{R})=\infty$.

ii) If $\{A_n\}_{n=0}^{\infty}$ is a collection of mutually disjoint subsets of $\boldsymbol{R}$, then

$$\mu\left(\bigcup_{n=0}^{\infty}A_n\right)=\sum_{n=0}^{\infty}\mu(A_n).$$

(This property is called *countable additivity* or *σ-additivity* of μ.)

iii) If $a\in\boldsymbol{R}$, $A\subseteq\boldsymbol{R}$, and $A+a=\{x+a\mid x\in A\}$, then $\mu(A+a)=\mu(A)$ (*translation invariance* of μ).

Several additional properties of μ follow immediately from 0)–iii) (Exercise 2.3):

iv) If $A\cap B=\emptyset$ then $\mu(A\cup B)=\mu(A)+\mu(B)$ (*finite additivity*).

v) If $A\subseteq B$ then $\mu(A)\le\mu(B)$ (*monotonicity*).

However, the Axiom of Choice implies that no function μ with the above-mentioned properties exists:

2.14 Theorem* *There is no function* $\mu : \mathcal{P}(\boldsymbol{R}) \to [0, \infty) \cup \{\infty\}$ *with the properties 0)–v).*

Proof. We define an equivalence relation $\approx$ on $\boldsymbol{R}$ by:

$$x \approx y \quad \text{if and only if } x - y \text{ is a rational number,}$$

and use the Axiom of Choice to obtain a set of representatives X for $\approx$. It is easy to see that

$$\boldsymbol{R} = \bigcup \{X + r \mid r \text{ is rational}\}; \tag{2.15}$$

moreover, if q and r are two distinct rationals, then $X + q$ and $X + r$ are disjoint. We note that $\mu(X) > 0$: if $\mu(X) = 0$, then $\mu(X + q) = 0$ for every $q \in \boldsymbol{Q}$, and

$$\mu(\boldsymbol{R}) = \sum_{q \in \boldsymbol{Q}} \mu(X + q) = 0,$$

a contradiction. By countable additivity, there is a closed interval $[a, b]$ such that $\mu(X \cap [a, b]) > 0$. Let $Y = X \cap [a, b]$. Then

$$\bigcup_{q \in \boldsymbol{Q} \cap [0,1]} (Y + q) \subseteq [a, b + 1] \tag{2.16}$$

and the left-hand side is the union of infinitely many mutually disjoint sets $Y + q$, each of measure $\mu(Y + q) = \mu(Y) > 0$. Thus the left-hand side of (2.16) has measure ∞, contrary to the fact that $\mu([a, b + 1]) = b + 1 - a$. □

The theorem we just established demonstrates that some of the requirements on μ have to be relaxed. For the purposes of mathematical analysis it is most fruitful to give up the condition that μ is defined for all subsets of $\boldsymbol{R}$, and require only that the domain of μ is closed under suitable set operations.

2.17 Definition Let S be a nonempty set. A collection $\mathfrak{S} \subseteq \mathcal{P}(S)$ is a σ-*algebra* of subsets of S if
(a) $\emptyset \in \mathfrak{S}$ and $S \in \mathfrak{S}$.
(b) If $X \in \mathfrak{S}$ then $S - X \in \mathfrak{S}$.
(c) If $X_n \in \mathfrak{S}$ for all n, then $\bigcup_{n=0}^{\infty} X_n \in \mathfrak{S}$ and $\bigcap_{n=0}^{\infty} X_n \in \mathfrak{S}$.

2.18 Definition A σ-*additive measure* on a σ-algebra $\mathfrak{S}$ of subsets of S is a function $\mu : \mathfrak{S} \to [0, \infty) \cup \{\infty\}$ such that

i) $\mu(\emptyset) = 0$, $\mu(S) > 0$.

ii) If $\{X_n\}_{n=0}^{\infty}$ is a collection of mutually disjoint sets from $\mathfrak{S}$, then

$$\mu\left(\bigcup_{n=0}^{\infty} X_n\right) = \sum_{n=0}^{\infty} \mu(X_n).$$

The elements of $\mathfrak{S}$ are called *μ-measurable sets.*

The reader can find some simple examples of σ-algebras and σ-additive measures in the exercises. In particular, $\mathcal{P}(S)$ is the largest σ-algebra of subsets of S; we refer to a measure defined on $\mathcal{P}(S)$ as a *measure on* S.

The theorem we proved in this example can now be reformulated as follows.

2.19 Corollary *Let μ be any σ-additive measure on a σ-algebra $\mathfrak{S}$ of subsets of $\boldsymbol{R}$ such that*

(0) $[a, b] \in \mathfrak{S}$ *and* $\mu([a, b]) = b - a$, *for all* $a, b \in \boldsymbol{R}$, $a < b$.

(iii) If $A \in \mathfrak{S}$ *then* $A + a \in \mathfrak{S}$ *and* $\mu(A + a) = \mu(A)$, *for all* $a \in \boldsymbol{R}$.

Then there exist sets of real numbers which are not μ-measurable. □

In real analysis, one constructs a particular σ-algebra $\mathfrak{M}$ of *Lebesgue measurable sets*, and a σ-additive measure μ on $\mathfrak{M}$, *the Lebesgue measure*, satisfying properties (0) and (iii) of the Corollary. So existence of Lebesgue nonmeasurable sets is a consequence of the Axiom of Choice. Robert Solovay showed that the Axiom of Choice is necessary to prove this result.

Other ways of weakening the properties 0)–iv) of μ have been considered, and led to very interesting questions in set theory. For example, it is possible to give up the requirement iii) (translation-invariance) and ask simply whether there exist any σ-additive measures μ on $\boldsymbol{R}$ such that $\mu([a, b]) = b - a$ for all $a, b \in \boldsymbol{R}$, $a < b$. This question has deep connections with the theory of large cardinals, and we return to it in Chapter 13. Another interesting possibility is to give up ii) (countable additivity) and require only iv) (finite additivity). Nontrivial finitely additive measures do exist (assuming the Axiom of Choice) and we study them further in Chapter 11.

Let us now resume our discussion of various aspects of the Axiom of Choice. First of all, there are many fundamental and intuitively very acceptable results concerning countable sets and topological and measure-theoretic properties of the real line, whose proofs depend on the Axiom of Choice. We have seen two such results in Examples 2.1 and 2.2. It is hard to image how one could study even advanced calculus without being able to prove them, yet it is known that they cannot be proved in Zermelo-Fraenkel set theory. This surely constitutes some justification for the Axiom of Choice. However, closer investigation of the proofs in Examples 2.1 and 2.2 reveals that only a very limited form of the Axiom is needed; indeed, all of the results can still be proved if one assumes only the Axiom of Countable Choice.

Axiom of Countable Choice There exists a choice function for every countable system of sets.

It might well be that the Axiom of Countable Choice is intuitively justified, but the full Axiom of Choice is not. Such a feeling might be strengthened by realizing that the full Axiom of Choice has some counterintuitive consequences, such as the existence of nonlinear additive functions from Example 2.7, or the existence of Lebesgue nonmeasurable sets from Example 2.13. Incidentally, none of these consequences follows from the Axiom of Countable Choice.

In our opinion, it is applications such as Example 2.9 which mostly account for the universal acceptance of the Axiom of Choice. The Hahn-Banach Theorem, Tichonov's Theorem (A topological product of any system of compact topological spaces is compact.) and Maximal Ideal Theorem (Every ideal in a ring can be extended to a maximal ideal.) are just a few examples of theorems of sweeping generality whose proofs require the Axiom of Choice in almost full strength; some of them are even equivalent to it. Even though it is true that we do not need such general results for applications to objects of more immediate mathematical concern, such as real and complex numbers and functions, the irreplaceable role of the Axiom of Choice is to simplify general topological and algebraic considerations which otherwise would be bogged down in irrelevant set-theoretic detail. For this pragmatic reason, we expect that the Axiom of Choice will always keep its place in set theory.

Exercises

2.1 Without using the Axiom of Choice, prove that the two definitions of closure points are equivalent if A is an open set. [*Hint*: X_n is open, so $X_n \cap \boldsymbol{Q} \neq \emptyset$, and $\boldsymbol{Q}$ can be well-ordered.]

2.2 Prove that every continuous additive function f is equal to f_a for some $a \in \boldsymbol{R}$.

2.3 Assume that μ has properties 0)–ii). Prove properties iv) and v). Also prove:

vi) $\mu(A \cup B) = \mu(A) + \mu(B) - \mu(A \cap B)$.

vii) $\mu(\bigcup_{n=0}^{\infty} A_n) \leq \sum_{n=0}^{\infty} \mu(A_n)$.

2.4* Let $\mathfrak{S} = \{X \subseteq S \mid |X| \leq \aleph_0 \text{ or } |S - X| \leq \aleph_0\}$. Prove that $\mathfrak{S}$ is a σ-algebra.

2.5 Let $\mathfrak{C}$ be any collection of subsets of S. Let $\mathfrak{S} = \bigcap\{\mathfrak{T} \mid \mathfrak{C} \subseteq \mathfrak{T} \text{ and } \mathfrak{T} \text{ is a } \sigma\text{-algebra of subsets of } S\}$. Prove that $\mathfrak{S}$ is a σ-algebra (it is called the *σ-algebra generated* by $\mathfrak{C}$).

2.6 Fix $a \in S$ and define μ on $\mathcal{P}(S)$ by: $\mu(A) = 1$ if $a \in A$, $\mu(A) = 0$ if $a \notin A$. Show that μ is a σ-additive measure on S.

2.7 For $A \subseteq S$ let $\mu(A) = 0$ if $A = \emptyset$, $\mu(A) = \infty$ otherwise. Show that μ is a σ-additive measure on S.

2.8 For $A \subseteq S$ let $\mu(A) = |A|$ if A is finite, $\mu(A) = \infty$ if A is infinite. μ is a σ-additive measure on S; it is called the *counting measure* on S.

Chapter 9

Arithmetic of Cardinal Numbers

1. Infinite Sums and Products of Cardinal Numbers

In Chapter 5 we introduced arithmetic operations on cardinal numbers. It is reasonable to generalize these operations and define sums and products of infinitely many cardinal numbers. For instance, it is natural to expect that

$$\underbrace{1 + 1 + \cdots}_{\aleph_0 \text{ times}} = \aleph_0$$

or, more generally,

$$\underbrace{\kappa + \kappa + \cdots}_{\lambda \text{ times}} = \kappa \cdot \lambda.$$

The sum of two cardinal numbers κ_1 and κ_2 was defined as the cardinality of $A_1 \cup A_2$, where A_1 and A_2 are disjoint sets such that $|A_1| = \kappa_1$ and $|A_2| = \kappa_2$. Thus we generalize the notion of sum as follows.

1.1 Definition Let $\langle A_i \mid i \in I \rangle$ be a system of mutually disjoint sets, and let $|A_i| = \kappa_i$ for all $i \in I$. We define the *sum* of $\langle \kappa_i \mid i \in I \rangle$ by

$$\sum_{i \in I} \kappa_i = \left| \bigcup_{i \in I} A_i \right|.$$

The definition of $\sum_{i \in I} \kappa_i$ uses particular sets A_i $(i \in I)$. In the finite case, when $I = \{1, 2\}$ and $\kappa_1 + \kappa_2 = |A_1 \cup A_2|$, we have shown that the choice of A_1 and A_2 is irrelevant. We have proved that if A'_1, A'_2 is another pair of disjoint sets such that $|A'_1| = \kappa_1$, $|A'_2| = \kappa_2$, then $|A'_1 \cup A'_2| = |A_1 \cup A_2|$.

In general, one needs the Axiom of Choice in order to prove the corresponding lemma for infinite sums. Without the Axiom of Choice, we cannot exclude the following possibility: There may exist two systems $\langle A_n \mid n \in \boldsymbol{N}\rangle$, $\langle A'_n \mid n \in \boldsymbol{N}\rangle$ of mutually disjoint sets such that each A_n and each A'_n has two elements, but $\bigcup_{n=0}^{\infty} A_n$ is not equipotent to $\bigcup_{n=0}^{\infty} A'_n$!

For this reason, and because many subsequent considerations depend heavily on the Axiom of Choice, we use the Axiom of Choice from now on without explicitly saying so each time.

1.2 Lemma *If $\langle A_i \mid i \in I\rangle$ and $\langle A'_i \mid i \in I\rangle$ are systems of mutually disjoint sets such that $|A_i| = |A'_i|$ for all $i \in I$, then $|\bigcup_{i\in I} A_i| = |\bigcup_{i \in I} A'_i|$.*

Proof. For each $i \in I$, choose a one-to-one mapping f_i of A_i onto A'_i. Then $f = \bigcup_{i\in I} f_i$ is a one-to-one mapping of $\bigcup_{i\in I} A_i$ onto $\bigcup_{i\in I} A'_i$. □

This lemma makes the definition of $\sum_{i\in I} \kappa_i$ legitimate. Since infinite unions of sets satisfy the associative law (Exercise 3.10 in Chapter 2), it follows that the infinite sums of cardinals are also associative (see Exercise 1.1). The operation $\sum$ has other reasonable properties, like: If $\kappa_i \leq \lambda_i$ for all $i \in I$, then $\sum_{i\in I} \kappa_i \leq \sum_{i\in I} \lambda_i$ (Exercise 1.2). However, if $\kappa_i < \lambda_i$ for all $i \in I$, it does not necessarily follow that $\sum_i \kappa_i < \sum_i \lambda_i$ (Exercise 1.3).

If the summands are all equal, then the following holds, as in the finite case: If $\kappa_i = \kappa$ for all $i \in \lambda$, then

$$\sum_{i\in\lambda} \kappa_i = \underbrace{\kappa + \kappa + \cdots}_{\lambda \text{ times}} = \kappa \cdot \lambda.$$

(Verify! See Exercise 1.4.)

It is not very difficult to evaluate infinite sums. For example, consider

$$\sum_{n\in\boldsymbol{N}} n = 1 + 2 + 3 + \cdots + n + \cdots \quad (n \in \boldsymbol{N}).$$

It is easy to see that this sum is equal to $\aleph_0$. In fact, this follows from a general theorem.

1.3 Theorem *Let λ be an infinite cardinal, let κ_α ($\alpha < \lambda$) be nonzero cardinal numbers, and let $\kappa = \sup\{\kappa_\alpha \mid \alpha < \lambda\}$. Then*

$$\sum_{\alpha<\lambda} \kappa_\alpha = \lambda \cdot \kappa = \lambda \cdot \sup\{\kappa_\alpha \mid \alpha < \lambda\}.$$

Proof. On the one hand, $\kappa_\alpha \leq \kappa$ for each $\alpha < \lambda$, and so $\sum_{\alpha<\lambda} \kappa_\alpha \leq \sum_{\alpha<\lambda} \kappa = \kappa \cdot \lambda$. On the other hand, we notice that $\lambda = \sum_{\alpha<\lambda} 1 \leq \sum_{\alpha<\lambda} \kappa_\alpha$. We also have $\kappa \leq \sum_{\alpha<\lambda} \kappa_\alpha$: the sum $\sum_{\alpha<\lambda} \kappa_\alpha$ is an upper bound of the κ_α's and κ is the least upper bound. Now since both κ and λ are $\leq \sum_{\alpha<\lambda} \kappa_\alpha$, it follows that $\kappa \cdot \lambda$, which is the greater of the two, is also $\leq \sum_{\alpha<\lambda} \kappa_\alpha$. The conclusion of Theorem 1.3 is now a consequence of the Cantor-Bernstein Theorem. □

1.4 Corollary *If κ_i $(i \in I)$ are cardinal numbers, and if $|I| \leq \sup\{\kappa_i \mid i \in I\}$, then*

$$\sum_{i \in I} \kappa_i = \sup_{i \in I} \kappa_i.$$

(In particular, the assumption is satisfied if all the κ_i's are mutually distinct.) □

The product of two cardinals κ_1 and κ_2 has been defined as the cardinality of the cartesian product $A_1 \times A_2$, where A_1 and A_2 are arbitrary sets such that $|A_1| = \kappa_1$ and $|A_2| = \kappa_2$. This is generalized as follows.

1.5 Definition Let $\langle A_i \mid i \in I \rangle$ be a family of sets such that $|A_i| = \kappa_i$ for all $i \in I$. We define the *product* of $\langle \kappa_i \mid i \in I \rangle$ by

$$\prod_{i \in I} \kappa_i = \left| \prod_{i \in I} A_i \right|.$$

We use the same symbol for the product of cardinals (the left-hand side) as for the cartesian product of the indexed family $\langle \kappa_i \mid i \in I \rangle$. It is always clear from the context which meaning the symbol $\prod$ has.

Again, the definition of $\prod_{i \in I} \kappa_i$ does not actually depend on the particular sets A_i.

1.6 Lemma *If $\langle A_i \mid i \in I \rangle$ and $\langle A'_i \mid i \in I \rangle$ are such that $|A_i| = |A'_i|$ for all $i \in I$, then $|\prod_{i \in I} A_i| = |\prod_{i \in I} A'_i|$.*

Proof. For each $i \in I$, choose a one-to-one mapping f_i of A_i onto A'_i. Let f be the function on $\prod_{i \in I} A_i$ defined as follows: If $x = \langle x_i \mid i \in I \rangle \in \prod_{i \in I} A_i$, let $f(x) = \langle f_i(x_i) \mid i \in I \rangle$. Then f is a one-to-one mapping of $\prod_{i \in I} A_i$ onto $\prod_{i \in I} A'_i$. □

The infinite products have many properties of finite products of natural numbers. For instance, if at least one κ_i is 0, then $\prod_{i \in I} \kappa_i = 0$. The products also satisfy the associative law (Exercise 1.7); another simple property is that if $\kappa_i \leq \lambda_i$ for all $i \in I$, then $\prod_{i \in I} \kappa_i \leq \prod_{i \in I} \lambda_i$ (Exercise 1.8). If all the factors κ_i are equal to κ, then we have, as in the finite case,

$$\prod_{i \in \lambda} \kappa_i = \underbrace{\kappa \cdot \kappa \cdot \cdots}_{\lambda \text{ times}} = \kappa^\lambda.$$

(Verify! See Exercise 1.10.) The following rules, involving exponentiation, also generalize from the finite to the infinite case (see Exercises 1.11 and 1.12):

$$\Big(\prod_{i \in I} \kappa_i\Big)^\lambda = \prod_{i \in I} (\kappa_i^\lambda),$$

$$\prod_{i \in I} (\kappa^{\lambda_i}) = \kappa^{\sum_{i \in I} \lambda_i}.$$

Infinite products are more difficult to evaluate than infinite sums. In some special cases, for instance when evaluating the product $\prod_{\alpha<\lambda} \kappa_\alpha$ of an increasing sequence $\langle \kappa_\alpha \mid \alpha < \lambda \rangle$ of cardinals, some simple rules can be proved. We consider only the following very special case:

$$\prod_{n=1}^{\infty} n = 1 \cdot 2 \cdot 3 \cdot \cdots \cdot n \cdot \cdots \quad (n \in \boldsymbol{N}).$$

First, we note that

$$\prod_{n=1}^{\infty} n \leq \prod_{n=1}^{\infty} \aleph_0 = \aleph_0^{\aleph_0} = 2^{\aleph_0}.$$

Conversely, we have

$$2^{\aleph_0} = \prod_{i=1}^{\infty} 2 \leq \prod_{n=2}^{\infty} n = \prod_{n=1}^{\infty} n,$$

and so we conclude that

$$1 \cdot 2 \cdot 3 \cdot \cdots \cdot n \cdot \cdots = 2^{\aleph_0}.$$

We now prove an important theorem, which can be used to derive various inequalities in cardinal arithmetic.

1.7 König's Theorem *If κ_i and λ_i $(i \in I)$ are cardinal numbers, and if $\kappa_i < \lambda_i$ for all $i \in I$, then*

$$\sum_{i \in I} \kappa_i < \prod_{i \in I} \lambda_i.$$

Proof. First, let us show that $\sum_{i\in I} \kappa_i \leq \prod_{i \in I} \lambda_i$. Let $\langle A_i \mid i \in I \rangle$ and $\langle B_i \mid i \in I \rangle$ be such that $|A_i| = \kappa_i$ and $|B_i| = \lambda_i$ for all $i \in I$ and the A_i's are mutually disjoint. We may further assume that $A_i \subset B_i$ for all $i \in I$. We find a one-to-one mapping f of $\bigcup_{i \in I} A_i$ into $\prod_{i\in I} B_i$.

We choose $d_i \in B_i - A_i$ for each $i \in I$, and define a function f as follows: For each $x \in \bigcup_{i\in I} A_i$, let i_x be the unique $i \in I$ such that $x \in A_i$. Let $f(x) = \langle a_i \mid i \in I \rangle$, where

$$a_i = \begin{cases} x & \text{if } i = i_x, \\ d_i & \text{if } i \neq i_x. \end{cases}$$

If $x \neq y$, let $f(x) = a$ and $f(y) = b$ and let us show that $a \neq b$. If $i_x = i_y = i$, then $a_i = x$ while $b_i = y$. If $i_x \neq i_y = i$, then $a_i = d_i \notin A$ while $b_i = y \in A$. In either case, $f(x) \neq f(y)$ and hence f is one-to-one.

Now let us show that $\sum_i \kappa_i < \prod_i \lambda_i$. Let B_i $(i \in I)$ be such that $|B_i| = \lambda_i$, for all $i \in I$. If the product $\prod_i \lambda_i$ were equal to the sum $\sum_i \kappa_i$, we could find mutually disjoint subsets X_i of the cartesian product $\prod_{i\in I} B_i$ such that $|X_i| = \kappa_i$ for all i and

$$\bigcup_{i\in I} X_i = \prod_{i \in I} B_i.$$

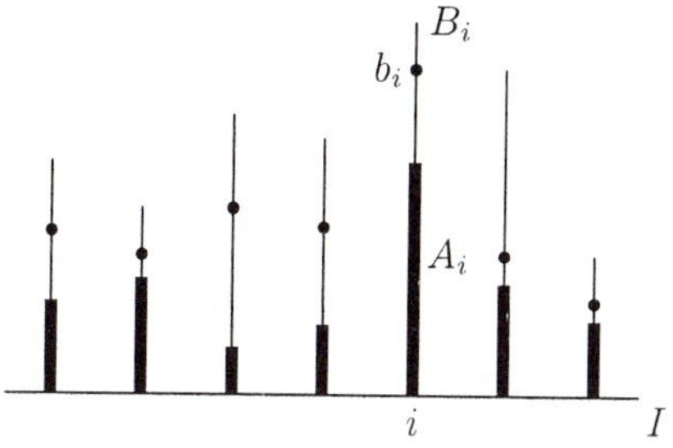

Figure 1

We show that this is impossible.

For each $i \in I$, let

$$A_i = \{a_i \mid a \in X_i\} \quad \text{(see Figure 1).} \tag{1.8}$$

For every $i \in I$, we have $A_i \subset B_i$, since $|A_i| \leq |X_i| = \kappa_i < \lambda_i = |B_i|$. Hence there exists $b_i \in B_i$ such that $b_i \notin A_i$. Let $b = \langle b_i \mid i \in I \rangle$. Now we can easily show that b is not a member of any X_i $(i \in I)$: For any $i \in I$, $b_i \notin A_i$, and so by (1.8), $b \notin X_i$. Hence $\bigcup_{i \in I} X_i$ is not the whole set $\prod_{i \in I} B_i$, a contradiction. □

We use König's Theorem in Section 3; at present, let us just mention that the theorem (and its proof) are generalizations of Cantor's Theorem which states that $2^\kappa > \kappa$ for all κ. If we express κ as the infinite sum

$$\kappa = 1 + 1 + \cdots \quad (\kappa \text{ times})$$

and 2^κ as the infinite product

$$2^\kappa = 2 \cdot 2 \cdot 2 \cdot \cdots \quad (\kappa \text{ times}),$$

we can apply König's Theorem (since $1 < 2$) and obtain

$$\kappa = \sum_{i \in \kappa} 1 < \prod_{i \in \kappa} 2 = 2^\kappa.$$

Exercises

1.1 If J_i $(i \in I)$ are mutually disjoint sets and $J = \bigcup_{i \in I} J_i$, and if κ_j $(j \in J)$ are cardinals, then

$$\sum_{i \in I} \Big(\sum_{j \in J_i} \kappa_j \Big) = \sum_{j \in J} \kappa_j$$

(*associativity* of $\sum$).

1.2 If $\kappa_i \leq \lambda_i$ for all $i \in I$, then $\sum_{i \in I} \kappa_i \leq \sum_{i \in I} \lambda_i$.

1.3 Find some cardinals κ_n, λ_n ($n \in \boldsymbol{N}$) such that $\kappa_n < \lambda_n$ for all n, but $\sum_{n=0}^{\infty} \kappa_n = \sum_{n=0}^{\infty} \lambda_n$.

1.4 Prove that $\kappa + \kappa + \cdots$ (λ times) $= \lambda \cdot \kappa$.

1.5 Prove the *distributive* law:

$$\lambda \cdot (\sum_{i \in I} \kappa_i) = \sum_{i \in I} (\lambda \cdot \kappa_i).$$

1.6 $|\bigcup_{i \in I} A_i| \leq \sum_{i \in I} |A_i|$.

1.7 If J_i ($i \in I$) are mutually disjoint sets and $J = \bigcup_{i \in I} J_i$, and if κ_j ($j \in J$) are cardinals, then

$$\prod_{i \in I} (\prod_{j \in J_i} \kappa_j) = \prod_{j \in J} \kappa_j$$

(*associativity* of $\prod$).

1.8 If $\kappa_i \leq \lambda_i$ for all $i \in I$, then

$$\prod_{i \in I} \kappa_i \leq \prod_{i \in I} \lambda_i.$$

1.9 Find some cardinals κ_n, λ_n ($n \in \boldsymbol{N}$) such that $\kappa_n < \lambda_n$ for all n, but $\prod_{n=0}^{\infty} \kappa_n = \prod_{n=0}^{\infty} \lambda_n$.

1.10 Prove that $\kappa \cdot \kappa \cdot \cdots$ (λ times) $= \kappa^{\lambda}$.

1.11 Prove the formula $(\prod_{i \in I} \kappa_i)^{\lambda} = \prod_{i \in I} (\kappa_i^{\lambda})$. [*Hint*: Generalize the proof of the special case $(\kappa^{\mu})^{\lambda} = (\kappa^{\lambda})^{\mu}$, given in Theorem 1.7 of Chapter 5.]

1.12 Prove the formula

$$\prod_{i \in I} (\kappa^{\lambda_i}) = \kappa^{\sum_{i \in I} \lambda_i}.$$

[*Hint*: Generalize the proof of the special case $\kappa^{\lambda} \cdot \kappa^{\mu} = \kappa^{\lambda + \mu}$ given in Theorem 1.7(a) of Chapter 5.]

1.13 Prove that if $1 < \kappa_i \leq \lambda_i$ for all $i \in I$, then $\sum_{i \in I} \kappa_i \leq \prod_{i \in I} \lambda_i$.

1.14 Evaluate the cardinality of $\prod_{0 < \alpha < \omega_1} \alpha$. [Answer: $2^{\aleph_1}$.]

1.15 Justify existence of the function f in the proof of Lemma 1.2 in detail by the axioms of set theory.

2. Regular and Singular Cardinals

Let $\langle \alpha_\nu \mid \nu < \vartheta \rangle$ be a transfinite sequence of ordinal numbers of length ϑ. We say that the sequence is *increasing* if $\alpha_\nu < \alpha_\mu$ whenever $\nu < \mu < \vartheta$. If ϑ is a limit ordinal number and if $\langle \alpha_\nu \mid \nu < \vartheta \rangle$ is an increasing sequence of ordinals, we define

$$\alpha = \lim_{\nu \to \vartheta} \alpha_\nu = \sup\{\alpha_\nu \mid \nu < \vartheta\}$$

and call α the *limit* of the increasing sequence.

2.1 Definition An infinite cardinal κ is called *singular* if there exists an increasing transfinite sequence $\langle \alpha_\nu \mid \nu < \vartheta \rangle$ of ordinals $\alpha_\nu < \kappa$ whose length ϑ is a limit ordinal less than κ, and $\kappa = \lim_{\nu\to\vartheta} \alpha_\nu$. An infinite cardinal that is not singular is called *regular.*

A subset $X \subseteq \kappa$ is *bounded* if $\sup X < \kappa$, and *unbounded* if $\sup X = \kappa$.

2.2 Theorem *Let κ be a regular cardinal.*
(a) If $X \subseteq \kappa$ is such that $|X| < \kappa$ then X is bounded. Hence every unbounded subset of κ has cardinality κ.
(b) If $\lambda < \kappa$ and $f : \lambda \to \kappa$, then $f[\lambda]$ is bounded.

Proof.
(a) This is clear if X has a greatest element. Thus assume that the order type of X is a limit ordinal, and let $\{\alpha_\nu \mid \nu < \vartheta\}$ be an increasing enumeration of X. As $|\vartheta| = |X| < \kappa$, we have $\vartheta < \kappa$, and because κ is a regular cardinal, it follows that $\sup X = \lim_{\nu\to\vartheta} \alpha_\nu < \kappa$.
(b) As $|f[\lambda]| \le \lambda < \kappa$, this follows from (a).

□

An example of a singular cardinal is the cardinal $\aleph_\omega$; we have

$$\aleph_\omega = \lim_{n\to\omega} \aleph_n,$$

where $\omega < \aleph_\omega$ and $\aleph_n < \aleph_\omega$ for each n.

Similarly, the cardinals $\aleph_{\omega+\omega}$, $\aleph_{\omega\cdot\omega}$, $\aleph_{\omega_1}$ are singular:

$$\aleph_{\omega+\omega} = \lim_{n\to\omega} \aleph_{\omega+n},$$
$$\aleph_{\omega\cdot\omega} = \lim_{n\to\omega} \aleph_{\omega\cdot n},$$
$$\aleph_{\omega_1} = \lim_{\alpha\to\omega_1} \aleph_\alpha.$$

On the other hand, $\aleph_0$ is a regular cardinal.

The following lemma gives a different characterization of singular cardinals.

2.3 Lemma *An infinite cardinal κ is singular if and only if it is the sum of less than κ smaller cardinals: $\kappa = \sum_{i\in I} \kappa_i$, where $|I| < \kappa$ and $\kappa_i < \kappa$ for all $i \in I$.*

Proof. If κ is singular, then there exists an increasing transfinite sequence such that $\kappa = \lim_{\nu\to\vartheta} \alpha_\nu$, where $\vartheta < \kappa$ and $\alpha_\nu < \kappa$ for all $\nu < \vartheta$. Since every ordinal is the set of all smaller ordinals, we can reformulate this as follows:

$$\kappa = \bigcup_{\nu\in\vartheta} \alpha_\nu = \bigcup_{\nu\in\vartheta} (\alpha_\nu - \bigcup_{\xi<\nu} \alpha_\xi).$$

If we let $A_\nu = \alpha_\nu - \bigcup_{\xi<\nu} \alpha_\xi$, then $\langle A_\nu \mid \nu < \vartheta\rangle$ is a sequence of fewer than κ sets of cardinality $\kappa_\nu = |A_\nu| = |\alpha_\nu - \bigcup_{\xi<\nu} \alpha_\xi| \le |\alpha_\nu| < \kappa$, and since the

A_ν's are mutually disjoint, this shows that $\kappa = \sum_{\nu<\vartheta} \kappa_\nu$ as required. Thus the condition in Lemma 2.3 is necessary.

To show that the condition is sufficient, let us assume that $\kappa = \sum_{\alpha<\lambda} \kappa_\alpha$, where λ is a cardinal less than κ, and for all $\alpha < \lambda$, κ_α are cardinals smaller than κ. By Theorem 1.3, $\kappa = \lambda \cdot \sup_{\alpha<\lambda} \kappa_\alpha$, and since $\lambda < \kappa$, we necessarily have $\kappa = \sup_{\alpha<\lambda} \kappa_\alpha$. Thus the range of the transfinite sequence $\langle \kappa_\alpha \mid \alpha < \lambda \rangle$ has supremum κ, and since $\kappa_\alpha < \kappa$ for all $\alpha < \lambda$, we can find (by transfinite recursion) a subsequence which is increasing and has limit κ. Clearly, the length of the subsequence is a limit ordinal $\vartheta \leq \lambda$, and it follows that κ is singular.

□

An infinite cardinal $\aleph_\alpha$ is called a *successor cardinal* if its index α is a successor ordinal, i.e., if $\aleph_\alpha = \aleph_{\beta+1}$ for some β. If $\kappa = \aleph_\beta$ then we call $\aleph_{\beta+1}$ the *successor* of κ and denote it κ^+. If α is a limit ordinal, then $\aleph_\alpha$ is called a *limit cardinal.* If $\alpha > 0$ is a limit ordinal, then $\aleph_\alpha$ is the limit of the sequence $\langle \aleph_\beta \mid \beta < \alpha \rangle$.

2.4 Theorem *Every successor cardinal $\aleph_{\alpha+1}$ is a regular cardinal.*

Proof. Otherwise, $\aleph_{\alpha+1}$ would be the sum of a smaller number of smaller cardinals:

$$\aleph_{\alpha+1} = \sum_{i\in I} \kappa_i,$$

where $|I| < \aleph_{\alpha+1}$, and $\kappa_i < \aleph_{\alpha+1}$ for all $i \in I$. Then $|I| \leq \aleph_\alpha$ and $\kappa_i \leq \aleph_\alpha$ for all $i \in I$, and we have

$$\aleph_{\alpha+1} = \sum_{i\in I} \kappa_i \leq \sum_{i\in I} \aleph_\alpha = \aleph_\alpha \cdot |I| \leq \aleph_\alpha \cdot \aleph_\alpha = \aleph_\alpha.$$

This is a contradiction, and hence $\aleph_{\alpha+1}$ is regular. □

By Theorem 2.4, every singular cardinal is a limit cardinal. Let us look now at limit cardinals.

2.5 Lemma *There are arbitrarily large singular cardinals.*

Proof. Let $\aleph_\alpha$ be an arbitrary cardinal. Consider the sequence

$$\aleph_\alpha,\ \aleph_{\alpha+1},\ \aleph_{\alpha+2},\ \ldots,\ \aleph_{\alpha+n},\ \ldots \quad (n \in \boldsymbol{N}).$$

Then

$$\aleph_{\alpha+\omega} = \lim_{n\to\omega} \aleph_{\alpha+n},$$

and hence $\aleph_{\alpha+\omega}$ is a singular cardinal greater than $\aleph_\alpha$. □

All uncountable limit cardinals we have seen so far were singular. A question naturally arises whether there are any uncountable regular limit cardinals. Suppose that $\aleph_\alpha$ is such a cardinal. Since α is a limit ordinal, we have

$$\aleph_\alpha = \lim_{\beta\to\alpha} \aleph_\beta,$$

i.e., $\aleph_\alpha$ is the limit of an increasing sequence of length α. Since $\aleph_\alpha$ is regular, we necessarily have $\alpha \geq \aleph_\alpha$, which together with $\alpha \leq \aleph_\alpha$ gives

$$(*) \qquad \alpha = \aleph_\alpha.$$

Already this property suggests that $\aleph_\alpha$ has to be very large.

Although the condition (*) seems to be strong, it is not as strong as it looks.

2.6 Lemma *There are arbitrarily large singular cardinals $\aleph_\alpha$ such that $\aleph_\alpha = \alpha$.*

Proof. Let $\aleph_\gamma$ be an arbitrary cardinal. Let us consider the following sequence:

$$\begin{aligned} \alpha_0 &= \omega_\gamma \\ \alpha_1 &= \omega_{\alpha_0} = \omega_{\omega_\gamma} \\ \alpha_2 &= \omega_{\alpha_1} = \omega_{\omega_{\omega_\gamma}} \\ &\dots \\ \alpha_{n+1} &= \omega_{\alpha_n} \\ &\dots \end{aligned}$$

for all $n \in \boldsymbol{N}$, and let $\alpha = \lim_{n\to\omega} \alpha_n$. It is clear that the sequence $\langle \aleph_{\alpha_n} \mid n \in \boldsymbol{N} \rangle$ has limit $\aleph_\alpha$. But then we have

$$\aleph_\alpha = \lim_{n\to\omega} \aleph_{\alpha_n} = \lim_{n\to\omega} \alpha_{n+1} = \alpha.$$

Since $\aleph_\alpha$ is the limit of a sequence of smaller cardinals of length ω, it is singular. □

An uncountable cardinal number $\aleph_\alpha$ that is both a limit cardinal and regular is called *inaccessible* (it is often called *weakly inaccessible* to distinguish this kind from cardinals that are defined by a stronger property — see Section 3). It is impossible to prove that inaccessible cardinals exist using only the axioms of Zermelo-Fraenkel set theory with Choice.

2.7 Definition If α is a limit ordinal, then the *cofinality* of α, $\mathrm{cf}(\alpha)$, is the least ordinal number ϑ such that α is the limit of an increasing sequence of ordinals of length ϑ.

(Note that $\mathrm{cf}(\alpha)$ is a limit ordinal and $\mathrm{cf}(\alpha) \leq \alpha$.) Thus $\aleph_\alpha$ is singular if $\mathrm{cf}(\omega_\alpha) < \omega_\alpha$ and is regular if $\mathrm{cf}(\omega_\alpha) = \omega_\alpha$.

Let α be a limit ordinal which is not a cardinal number. If we let $\kappa = |\alpha|$, there exists a one-to-one mapping of κ onto α, or, in other words, a one-to-one sequence $\langle \alpha_\nu \mid \nu < \kappa \rangle$ of length κ such that $\{\alpha_\nu \mid \nu < \kappa\} = \alpha$. Now we can find (by transfinite recursion) a subsequence which is increasing and has limit α. Since the length of the subsequence is at most κ, and since $\kappa = |\alpha|$ is less than α (because α is not a cardinal), we conclude that $\mathrm{cf}(\alpha) < \alpha$.

Thus we have proved the following:

2.8 Lemma *If a limit ordinal α is not a cardinal, then* $\mathrm{cf}(\alpha) < \alpha$. □

As a corollary, we have, for all limit ordinals α,

$$\mathrm{cf}(\alpha) = \alpha \text{ if and only if } \alpha \text{ is a regular cardinal.}$$

2.9 Lemma *For every limit ordinal α,* $\mathrm{cf}(\mathrm{cf}(\alpha)) = \mathrm{cf}(\alpha)$.

Proof. Let $\vartheta = \mathrm{cf}(\alpha)$. Clearly, ϑ is a limit ordinal, and $\mathrm{cf}(\vartheta) \leq \vartheta$. We have to show that $\mathrm{cf}(\vartheta)$ is not smaller than ϑ. If $\gamma = \mathrm{cf}(\vartheta) < \vartheta$, then there exists an increasing sequence of ordinals $\langle \nu_\xi \mid \xi < \gamma \rangle$ such that $\lim_{\xi \to \gamma} \nu_\xi = \vartheta$. Since $\vartheta = \mathrm{cf}(\alpha)$, there exists an increasing sequence of ordinals $\langle \alpha_\nu \mid \nu < \vartheta \rangle$ such that $\lim_{\nu \to \vartheta} \alpha_\nu = \alpha$. Then the sequence $\langle \alpha_{\nu_\xi} \mid \xi < \gamma \rangle$ has length γ and $\lim_{\xi \to \gamma} \alpha_{\nu_\xi} = \alpha$. But $\gamma < \vartheta$, and we reached a contradiction, since ϑ is supposed to be the least length of an increasing sequence with limit α. □

2.10 Corollary *For every limit ordinal α,* $\mathrm{cf}(\alpha)$ *is a regular cardinal.* □

Exercises

2.1 $\mathrm{cf}(\aleph_\omega) = \mathrm{cf}(\aleph_{\omega+\omega}) = \omega$.

2.2 $\mathrm{cf}(\aleph_{\omega_1}) = \omega_1$, $\mathrm{cf}(\aleph_{\omega_2}) = \omega_2$.

2.3 Let α be the cardinal number defined in the proof of Lemma 2.6. Show that $\mathrm{cf}(\alpha) = \omega$.

2.4 Show that $\mathrm{cf}(\alpha)$ is the least γ such that α is the union of γ sets of cardinality less than $|\alpha|$.

2.5 Let $\aleph_\alpha$ be a limit cardinal, $\alpha > 0$. Show that there is an increasing sequence of *alephs* of length $\mathrm{cf}(\aleph_\alpha)$ with limit $\aleph_\alpha$.

2.6 Let κ be a limit cardinal, and let $\lambda < \kappa$ be a regular infinite cardinal. Show that there is an increasing sequence $\langle \alpha_\nu \mid \nu < \mathrm{cf}(\kappa) \rangle$ of cardinals such that $\lim_{\nu \to \mathrm{cf}(\kappa)} \alpha_\nu = \kappa$ and $\mathrm{cf}(\alpha_\nu) = \lambda$ for all ν.

3. Exponentiation of Cardinals

While addition and multiplication of cardinals are simple (due to the fact that $\aleph_\alpha + \aleph_\beta = \aleph_\alpha \cdot \aleph_\beta =$ the greater of the two), the evaluation of cardinal exponentiation is rather complicated. Here, we do not give a complete set of rules (in fact, in a sense, the general problem of evaluation of κ^λ is still open), but prove only the basic properties of the operation κ^λ. It turns out that there is a difference between regular and singular cardinals.

First, we investigate the operation $2^{\aleph_\alpha}$. By Cantor's Theorem, $2^{\aleph_\alpha} > \aleph_\alpha$; in other words,

$$2^{\aleph_\alpha} \geq \aleph_{\alpha+1}. \tag{3.1}$$

Let us recall that Cantor's *Continuum Hypothesis* is the conjecture that $2^{\aleph_0} = \aleph_1$. A generalization of this conjecture is the *Generalized Continuum Hypothesis*:

$$2^{\aleph_\alpha} = \aleph_{\alpha+1} \quad \text{for all } \alpha.$$

As we show, the Generalized Continuum Hypothesis greatly simplifies the cardinal exponentiation; in fact, the operation κ^λ can then be evaluated by very simple rules.

The Generalized Continuum Hypothesis can be neither proved nor refuted from the axioms of set theory. (See the discussion of this subject in Chapter 15.)

Without assuming the Generalized Continuum Hypothesis, there is not much one can prove about $2^{\aleph_\alpha}$ except (3.1) and the trivial property:

$$2^{\aleph_\alpha} \leq 2^{\aleph_\beta} \quad \text{whenever} \quad \alpha \leq \beta. \tag{3.2}$$

The following fact is a consequence of König's Theorem.

3.3 Lemma *For every* α,

$$\operatorname{cf}(2^{\aleph_\alpha}) > \aleph_\alpha. \tag{3.4}$$

Thus $2^{\aleph_0}$ cannot be $\aleph_\omega$, since $\operatorname{cf}(2^{\aleph_\omega}) = \aleph_0$, but the lemma does not prevent $2^{\aleph_0}$ from being $\aleph_{\omega_1}$. Similarly, $2^{\aleph_1}$ cannot be either $\aleph_{\omega_1}$ or $\aleph_\omega$ or $\aleph_{\omega+\omega}$, etc.

Proof. Let $\vartheta = \operatorname{cf}(2^{\aleph_\alpha})$; ϑ is a cardinal. Thus $2^{\aleph_\alpha}$ is the limit of an increasing sequence of length ϑ, and it follows (see the proof of Lemma 2.3 for details) that

$$2^{\aleph_\alpha} = \sum_{\nu<\vartheta} \kappa_\nu,$$

where each κ_ν is a cardinal smaller than $2^{\aleph_\alpha}$. By König's Theorem (where we let $\lambda_\nu = 2^{\aleph_\alpha}$ for all $\nu < \vartheta$), we have

$$\sum_{\nu<\vartheta} \kappa_\nu < \prod_{\nu<\vartheta} 2^{\aleph_\alpha}$$

and hence $2^{\aleph_\alpha} < (2^{\aleph_\alpha})^\vartheta$. Now if ϑ were less than or equal to $\aleph_\alpha$, we would get

$$2^{\aleph_\alpha} < (2^{\aleph_\alpha})^\vartheta \leq (2^{\aleph_\alpha})^{\aleph_\alpha} = 2^{\aleph_\alpha \cdot \aleph_\alpha} = 2^{\aleph_\alpha},$$

a contradiction. □

The inequalities (3.1), (3.2), and (3.4) are the only properties that can be proved for the operation $2^{\aleph_\alpha}$ if the cardinal $\aleph_\alpha$ is regular. If $\aleph_\alpha$ is singular, then various additional rules restraining the behavior of $2^{\aleph_\alpha}$ are known. We prove one such theorem here (Theorem 3.5); in Chapter 11 we prove Silver's Theorem (Theorem 4.1): If $\aleph_\alpha$ is a singular cardinal of cofinality $\operatorname{cf}(\aleph_\alpha) \geq \aleph_1$, and if $2^{\aleph_\xi} = \aleph_{\xi+1}$ for all $\xi < \alpha$, then $2^{\aleph_\alpha} = \aleph_{\alpha+1}$.

3.5 Theorem *Let $\aleph_\alpha$ be a singular cardinal. Let us assume that the value of $2^{\aleph_\xi}$ is the same for all $\xi < \alpha$, say $2^{\aleph_\xi} = \aleph_\beta$. Then $2^{\aleph_\alpha} = \aleph_\beta$.*

Note that it is implicit in the theorem that $\aleph_\beta$ is greater than $\aleph_\alpha$. For instance, if we know that $2^{\aleph_n} = \aleph_{\omega+5}$ for all $n < \omega$, then $2^{\aleph_\omega} = \aleph_{\omega+5}$.

Proof. Since $\aleph_\alpha$ is singular, there exists, by Lemma 2.3, a collection $\langle \kappa_i \mid i \in I \rangle$ of cardinals such that $\kappa_i < \aleph_\alpha$ for all $i \in I$, and $|I| = \aleph_\gamma$ is a cardinal less than $\aleph_\alpha$, and $\aleph_\alpha = \sum_{i \in I} \kappa_i$. By the assumption, we have $2^{\kappa_i} = \aleph_\beta$ for all $i \in I$, and also $2^{\aleph_\gamma} = \aleph_\beta$, so

$$2^{\aleph_\alpha} = 2^{\sum_{i \in I} \kappa_i} = \prod_{i \in I} 2^{\kappa_i} = \prod_{i \in I} \aleph_\beta = \aleph_\beta^{\aleph_\gamma} = (2^{\aleph_\gamma})^{\aleph_\gamma} = 2^{\aleph_\gamma} = \aleph_\beta.$$

□

We now approach the problem of evaluating $\aleph_\alpha^{\aleph_\beta}$, where $\aleph_\alpha$ and $\aleph_\beta$ are arbitrary infinite cardinals. First, we make the following observation.

3.6 Lemma *If $\alpha \le \beta$, then $\aleph_\alpha^{\aleph_\beta} = 2^{\aleph_\beta}$.*

Proof. Clearly, $2^{\aleph_\beta} \le \aleph_\alpha^{\aleph_\beta}$. Since $\aleph_\alpha \le 2^{\aleph_\alpha}$, we also have

$$\aleph_\alpha^{\aleph_\beta} \le (2^{\aleph_\alpha})^{\aleph_\beta} = 2^{\aleph_\alpha \cdot \aleph_\beta} = 2^{\aleph_\beta}$$

because $\aleph_\beta = \max\{\aleph_\alpha, \aleph_\beta\}$. □

When trying to evaluate $\aleph_\alpha^{\aleph_\beta}$ for $\alpha > \beta$, we find the following useful.

3.7 Lemma *Let $\alpha \ge \beta$ and let S be the set of all subsets $X \subseteq \omega_\alpha$ such that $|X| = \aleph_\beta$. Then $|S| = \aleph_\alpha^{\aleph_\beta}$.*

Proof. We first show that $\aleph_\alpha^{\aleph_\beta} \le |S|$. Let S' be the set of all subsets $X \subseteq \omega_\beta \times \omega_\alpha$ such that $|X| = \aleph_\beta$. Since $\aleph_\beta \cdot \aleph_\alpha = \aleph_\alpha$, we have $|S'| = |S|$. Now every function $f : \omega_\beta \to \omega_\alpha$ is a member of the set S' and hence $\omega_\alpha^{\omega_\beta} \subseteq S'$. Therefore, $\aleph_\alpha^{\aleph_\beta} \le |S|$.

Conversely, if $X \in S$, then there exists a function f on ω_β such that X is the range of f. We pick one f for each $X \in S$ and let $f = F(X)$. Clearly, if $X \ne Y$ and $f = F(X)$ and $g = F(Y)$, we have $X = \operatorname{ran} f$ and $Y = \operatorname{ran} g$, and so $f \ne g$. Thus F is a one-to-one mapping of S into $\omega_\alpha^{\omega_\beta}$, and therefore $|S| \le \aleph_\alpha^{\aleph_\beta}$. □

We are now in a position to evaluate $\aleph_\alpha^{\aleph_\beta}$ for regular cardinals $\aleph_\alpha$, under the assumption of the Generalized Continuum Hypothesis.

3.8 Theorem *Let us assume the Generalized Continuum Hypothesis. If $\aleph_\alpha$ is a regular cardinal, then*

$$\aleph_\alpha^{\aleph_\beta} = \begin{cases} \aleph_\alpha & \text{if } \beta < \alpha, \\ \aleph_{\beta+1} & \text{if } \beta \ge \alpha. \end{cases}$$

Proof. If $\beta \geq \alpha$, then $\aleph_\alpha^{\aleph_\beta} = 2^{\aleph_\beta} = \aleph_{\beta+1}$ by Lemma 3.6. So let $\beta < \alpha$ and let $S = \{X \subseteq \omega_\alpha \mid |X| = \aleph_\beta\}$. By Lemma 3.7, $|S| = \aleph_\alpha^{\aleph_\beta}$. By Theorem 2.2(a), every $X \in S$ is a bounded subset of ω_α. Thus, let $B = \bigcup_{\delta<\omega_\alpha} \mathcal{P}(\delta)$ be the collection of all bounded subsets of ω_α. We will show that $|B| \leq \aleph_\alpha$; as $S \subset B$, it then follows that $\aleph_\alpha^{\aleph_\beta} = \aleph_\alpha$.

Since $B = \bigcup_{\delta<\omega_\alpha} \mathcal{P}(\delta)$, we have

$$|B| \leq \sum_{\delta<\omega_\alpha} 2^{|\delta|}.$$

However, for every cardinal $\aleph_\gamma < \aleph_\alpha$, we have $2^{\aleph_\gamma} = \aleph_{\gamma+1} \leq \aleph_\alpha$ and so $2^{|\delta|} \leq \aleph_\alpha$ for every $\delta < \omega_\alpha$, and we get

$$|B| \leq \sum_{\delta<\omega_\alpha} 2^{|\delta|} \leq \sum_{\delta<\omega_\alpha} \aleph_\alpha = \aleph_\alpha \cdot \aleph_\alpha = \aleph_\alpha.$$

□

We prove a similar (but a little more complicated) formula for singular $\aleph_\alpha$, but first we need a generalization of Lemma 3.3.

3.9 Lemma *For every cardinal $\kappa > 1$ and every α, $\operatorname{cf}(\kappa^{\aleph_\alpha}) > \aleph_\alpha$.*

Proof. Exactly like the proof of Lemma 3.3, except that $2^{\aleph_\alpha}$ is replaced by $\kappa^{\aleph_\alpha}$. □

3.10 Theorem *Let us assume the Generalized Continuum Hypothesis. If $\aleph_\alpha$ is a singular cardinal, then*

$$\aleph_\alpha^{\aleph_\beta} = \begin{cases} \aleph_\alpha & \text{if } \aleph_\beta < \operatorname{cf}(\aleph_\alpha), \\ \aleph_{\alpha+1} & \text{if } \operatorname{cf}(\aleph_\alpha) \leq \aleph_\beta \leq \aleph_\alpha, \\ \aleph_{\beta+1} & \text{if } \aleph_\beta \geq \aleph_\alpha. \end{cases}$$

Proof. If $\beta \geq \alpha$, then $\aleph_\alpha^{\aleph_\beta} = 2^{\aleph_\beta} = \aleph_{\beta+1}$. If $\aleph_\beta < \operatorname{cf}(\aleph_\alpha)$, then every subset $X \subseteq \omega_\alpha$ such that $|X| = \aleph_\beta$ is a bounded subset, and we get $\aleph_\alpha^{\aleph_\beta} = \aleph_\alpha$ by exactly the same argument as in the case of regular $\aleph_\alpha$.

Thus let us assume that $\operatorname{cf}(\aleph_\alpha) \leq \aleph_\beta \leq \aleph_\alpha$. On the one hand, we have

$$\aleph_\alpha \leq \aleph_\alpha^{\aleph_\beta} \leq \aleph_\alpha^{\aleph_\alpha} = 2^{\aleph_\alpha} = \aleph_{\alpha+1}.$$

On the other hand, $\operatorname{cf}(\aleph_\alpha^{\aleph_\beta}) > \aleph_\beta$ by Lemma 3.9, and since $\aleph_\beta \geq \operatorname{cf}(\aleph_\alpha)$, we have $\operatorname{cf}(\aleph_\alpha^{\aleph_\beta}) \neq \operatorname{cf}(\aleph_\alpha)$, and therefore $\aleph_\alpha^{\aleph_\beta} \neq \aleph_\alpha$. Thus necessarily $\aleph_\alpha^{\aleph_\beta} = \aleph_{\alpha+1}$. □

If we do not assume the Generalized Continuum Hypothesis, the situation becomes much more complicated. We only prove the following theorem.

3.11 Hausdorff's Formula *For every α and every β,*

$$\aleph_{\alpha+1}^{\aleph_\beta} = \aleph_\alpha^{\aleph_\beta} \cdot \aleph_{\alpha+1}.$$

Proof. If $\beta \geq \alpha + 1$, then $\aleph_{\alpha+1}^{\aleph_\beta} = 2^{\aleph_\beta}$, $\aleph_\alpha^{\aleph_\beta} = 2^{\aleph_\beta}$, and $\aleph_{\alpha+1} \leq \aleph_\beta \leq 2^{\aleph_\beta}$; hence the formula holds. Thus let us assume that $\beta \leq \alpha$. Since $\aleph_\alpha^{\aleph_\beta} \leq \aleph_{\alpha+1}^{\aleph_\beta}$ and $\aleph_{\alpha+1} \leq \aleph_{\alpha+1}^{\aleph_\beta}$, it suffices to show that $\aleph_{\alpha+1}^{\aleph_\beta} \leq \aleph_\alpha^{\aleph_\beta} \cdot \aleph_{\alpha+1}$.

Each function $f : \omega_\beta \to \omega_{\alpha+1}$ is bounded; i.e., there is $\gamma < \omega_{\alpha+1}$ such that $f(\xi) < \gamma$ for all $\xi < \omega_\beta$ (this is because $\omega_{\alpha+1}$ is regular and $\omega_\beta < \omega_{\alpha+1}$). Hence,

$$\omega_{\alpha+1}^{\omega_\beta} = \bigcup_{\gamma<\omega_{\alpha+1}} \gamma^{\omega_\beta}.$$

Now every $\gamma < \omega_{\alpha+1}$ has cardinality $|\gamma| \leq \aleph_\alpha$, and we have (by Exercise 1.6) $|\bigcup_{\gamma<\omega_{\alpha+1}} \gamma^{\omega_\beta}| \leq \sum_{\gamma<\omega_{\alpha+1}} |\gamma|^{\aleph_\beta}$. Thus

$$\aleph_{\alpha+1}^{\aleph_\beta} \leq \sum_{\gamma<\omega_{\alpha+1}} |\gamma|^{\aleph_\beta} \leq \sum_{\gamma<\omega_{\alpha+1}} \aleph_\alpha^{\aleph_\beta} = \aleph_\alpha^{\aleph_\beta} \cdot \aleph_{\alpha+1}.$$

□

This theorem enables us to evaluate some simple cases of $\aleph_\alpha^{\aleph_\beta}$ (see Exercise 3.5).

An infinite cardinal $\aleph_\alpha$ is a *strong limit cardinal* if $2^{\aleph_\beta} < \aleph_\alpha$ for all $\beta < \alpha$.

Clearly, a strong limit cardinal is a limit cardinal, since if $\aleph_\alpha = \aleph_{\gamma+1}$, then $2^{\aleph_\gamma} \geq \aleph_\alpha$. Not every limit cardinal is necessarily a strong limit cardinal: If $2^{\aleph_0}$ is greater than $\aleph_\omega$, then $\aleph_\omega$ is a counterexample. However, if we assume the Generalized Continuum Hypothesis, then every limit cardinal is a strong limit cardinal.

3.12 Theorem *If $\aleph_\alpha$ is a strong limit cardinal and if κ and λ are infinite cardinals such that $\kappa < \aleph_\alpha$ and $\lambda < \aleph_\alpha$, then $\kappa^\lambda < \aleph_\alpha$.*

Proof. $\kappa^\lambda \leq (\kappa \cdot \lambda)^{\kappa\cdot\lambda} = 2^{\kappa\cdot\lambda} < \aleph_\alpha$. □

An uncountable cardinal number κ is *strongly inaccessible* if it is regular and a strong limit cardinal. (Thus every strongly inaccessible cardinal is weakly inaccessible, and, if we assume the Generalized Continuum Hypothesis, every weakly inaccessible cardinal is strongly inaccessible.) The reason why such cardinal numbers are called inaccessible is that they cannot be obtained by the usual set-theoretic operations from smaller cardinals:

3.13 Theorem *Let κ be a strongly inaccessible cardinal.*
(a) *If X has cardinality $< \kappa$, then $\mathcal{P}(X)$ has cardinality $< \kappa$.*
(b) *If each $X \in S$ has cardinality $< \kappa$ and $|S| < \kappa$, then $\bigcup S$ has cardinality $< \kappa$.*
(c) *If $|X| < \kappa$ and $f : X \to \kappa$, then $\sup f[X] < \kappa$.*

Proof.
(a) κ is a strong limit cardinal.
(b) Let $\lambda = |S|$ and $\mu = \sup\{|X| \mid X \in S\}$. Then (by Theorem 2.2(a)) $\mu < \kappa$ because κ is regular, and $|\bigcup S| \leq \lambda \cdot \mu < \kappa$.
(c) By Theorem 2.2(b).

□

Exercises

3.1 If $2^{\aleph_\beta} \geq \aleph_\alpha$, then $\aleph_\alpha^{\aleph_\beta} = 2^{\aleph_\beta}$.

3.2 Verify this generalization of Exercise 3.1: If there is $\gamma < \alpha$ such that $\aleph_\gamma^{\aleph_\beta} \geq \aleph_\alpha$, say $\aleph_\gamma^{\aleph_\beta} = \aleph_\delta$, then $\aleph_\alpha^{\aleph_\beta} = \aleph_\delta$.

3.3 Let α be a limit ordinal and let $\aleph_\beta < \mathrm{cf}(\aleph_\alpha)$. Show that if $\aleph_\xi^{\aleph_\beta} \leq \aleph_\alpha$ for all $\xi < \alpha$, then $\aleph_\alpha^{\aleph_\beta} = \aleph_\alpha$. [*Hint*: If $X \subseteq \omega_\alpha$ is such that $|X| = \aleph_\beta$, then $X \subseteq \omega_\xi$ for some $\xi < \alpha$.]

3.4 If $\aleph_\alpha$ is strongly inaccessible and $\beta < \alpha$, then $\aleph_\alpha^{\aleph_\beta} = \aleph_\alpha$. [*Hint*: Use Exercise 3.3.]

3.5 If $n < \omega$, then $\aleph_n^{\aleph_\beta} = \aleph_n \cdot 2^{\aleph_\beta}$. [*Hint*: Apply Hausdorff's formula n times.]

3.6 Prove that $\prod_{n<\omega} \aleph_n = \aleph_\omega^{\aleph_0}$. [*Hint*: Let A_i $(i < \omega)$ be mutually disjoint infinite subsets of ω. Then

$$\prod_{n<\omega} \aleph_n \geq \prod_{i<\omega}\Big(\prod_{n\in A_i} \aleph_n\Big) \geq \prod_{i<\omega}\Big(\sum_{n\in A_i} \aleph_n\Big) \geq \prod_{i<\omega} \aleph_\omega = \aleph_\omega^{\aleph_0}.$$

The other direction is easy.]

3.7 Prove that

$$\aleph_\omega^{\aleph_1} = \aleph_\omega^{\aleph_0} \cdot 2^{\aleph_1}.$$

[*Hint*: $\aleph_\omega^{\aleph_1} = (\sum_{n<\omega} \aleph_n)^{\aleph_1} \leq (\prod_{n<\omega} \aleph_n)^{\aleph_1} = \prod_{n<\omega} \aleph_n^{\aleph_1} = \prod_{n<\omega}(\aleph_n \cdot 2^{\aleph_1}) = (\prod_{n<\omega} \aleph_n) \cdot (2^{\aleph_1})^{\aleph_0} = \aleph_\omega^{\aleph_0} \cdot 2^{\aleph_1}$.]

Chapter 10

Sets of Real Numbers

1. Integers and Rational Numbers

In Chapter 3 we defined natural numbers and their ordering and indicated how arithmetic operations on natural numbers can be defined. The next logical step in the development of foundations for mathematics is to define integers and rational numbers. The guiding idea in both cases is to make an arithmetic operation that is only partially defined on natural numbers (subtraction in the case of integers, division in the case of rationals) into a total operation, and belongs more properly in the realm of algebra than set theory. We thus limit ourselves to outlining the main ideas, and leave out almost all proofs. Those can be found in most textbooks on abstract algebra. Better still, the reader may work out some or all of them as exercises.

In Exercise 4.3 of Chapter 3 we defined *subtraction* for those pairs (n, m) of natural numbers where $n \geq m$. In this case, $n - m$ is the unique natural number k for which $n = m + k$. If $n < m$, no such natural number k exists, and $n - m$ is undefined. If $n - m$ is to be defined, it has to be a "new" object; for the time being, we represent it simply by the ordered pair (n, m). However, intuitively familiar properties of integer arithmetic suggest that different ordered pairs may have to represent the same integer: for example, $(2, 5)$ and $(6, 9)$ both represent -3 $[2 - 5 = 6 - 9 = -3]$. In general, (n_1, m_1) and (n_2, m_2) represent the same integer if and only if $n_1 - m_1 = n_2 - m_2$. At this point, this makes sense only intuitively, but it can be rewritten in the form

$$n_1 + m_2 = n_2 + m_1,$$

which involves only addition of natural numbers (which has been previously defined). These remarks motivate the following definitions and results.

Let $\boldsymbol{Z}' = \boldsymbol{N} \times \boldsymbol{N}$. Define a relation $\approx$ on $\boldsymbol{Z}'$ by $(a, b) \approx (c, d)$ if and only if $a + d = b + c$. The relation $\approx$ is an equivalence relation on $\boldsymbol{Z}'$. (This has to be checked, of course.) Let $\boldsymbol{Z} = \boldsymbol{Z}'/\approx$ be the set of all equivalence classes of $\boldsymbol{Z}'$ modulo $\approx$. We call $\boldsymbol{Z}$ the *set of all integers*; its elements are *integers*.

One immediate consequence of the definition is of set-theoretic interest.

1.1 Theorem *The set of all integers $\boldsymbol{Z}$ is countable.*

Proof. Theorem 3.13 in Chapter 4. Theorem 3.12 in Chapter 4 gives another proof. □

Next, define a relation $<$ on $\boldsymbol{Z}$ as follows: $[(a,b)] < [(c,d)]$ if and only if $a+d < b+c$. [Recall that intuitively (a,b) represents $a-b$ and (c,d) represents $c-d$, so $a-b < c-d$ should mean $a+d < b+c$.]

One can prove that $<$ is well defined (i.e., truth or falsity of $[(a,b)] < [(c,d)]$ does not depend on the choice of representatives (a,b) and (c,d) but only on their respective equivalence classes) and that it is a linear ordering.

Finally, we observe that for each integer $[(a,b)]$ either $a \geq b$, in which case $(a,b) \approx (a-b,0)$ (here $-$ stands for subtraction of natural numbers, which is defined in this case), or $a < b$, in which case $(a,b) \approx (0,b-a)$. It follows that each integer contains a unique pair of the form $(n,0)$, $n \in \boldsymbol{N}$ or $(0,n)$, $n \in \boldsymbol{N} - \{0\}$. So $[(n,0)]$ are the positive integers and $[(0,n)]$ are the negative ones. The mapping $F : \boldsymbol{N} \to \boldsymbol{Z}$ defined by $F(n) = [(n,0)]$ is one-to-one and order-preserving [i.e., $m < n$ implies that $F(m) < F(n)$]. We identify each integer of the form $[(n,0)]$ with the corresponding natural number n, and denote each integer of the form $[(0,n)]$ by $-n$. So, e.g., $-3 = [(0,3)] = [(2,5)] = [(6,9)]$, as expected.

The rest of the theory is now straightforward. One can prove that $(\boldsymbol{Z}, <)$ has no endpoints and that $\{x \in \boldsymbol{Z} \mid a < x < b\}$, $a, b \in \boldsymbol{Z}$, $a < b$, has a finite number of elements. Also, every nonempty set of integers bounded from above has a greatest element, and every nonempty set of integers bounded from below has a least element. One can define *addition* and *multiplication* of integers by

$$[(a,b)] + [(c,d)] = [(a+c, b+d)],$$
$$[(a,b)] \cdot [(c,d)] = [(ac+bd, ad+bc)]$$

and prove that these operations satisfy the usual laws of algebra (commutativity, associativity, and distributivity of multiplication over addition), and that for those integers that are natural numbers, addition and multiplication of integers agree with addition and multiplication of natural numbers.

We define *subtraction* by

$$[(a,b)] - [(c,d)] = [(a,b)] + (-[(c,d)]),$$

where $-[(c,d)] = [(d,c)]$ is the *opposite* of $[(c,d)]$. Notice that $-[(n,0)] = [(0,n)] = -n$ and $-[(0,n)] = [(n,0)] = n$, in agreement with our previous notation.

Absolute value of an integer a, $|a|$, is defined by

$$|a| = \begin{cases} a & \text{if } a \geq 0; \\ -a & \text{if } a < 0. \end{cases}$$

Additional properties of these concepts can be proved as needed. Addition, subtraction, and multiplication are defined for all pairs of integers. There remains the inconvenience that division cannot always be performed.

We say that an integer a is *divisible* by an integer b if there is a unique integer x such that $a = b \cdot x$; this unique x is then called the *quotient* of a and b. We would like to extend the system of integers so that any a is divisible by any b and all useful arithmetic laws remain valid in the extended system. Now, if $0 \cdot x = 0$ is to be valid in the extended system for all x, we see immediately that no number a can be divisible by 0; the equation $a = 0 \cdot x$ has either none or many solutions, depending on whether $a \neq 0$ or $a = 0$. The best we can hope for is an extension in which for all a and all $b \neq 0$ there is a unique x such that $a = b \cdot x$.

Let $\boldsymbol{Q}' = \boldsymbol{Z} \times (\boldsymbol{Z} - \{0\}) = \{(a, b) \in \boldsymbol{Z}^2 \mid b \neq 0\}$. We call $\boldsymbol{Q}'$ the set of *fractions over* $\boldsymbol{Z}$ and write a/b in place of (a, b) for $(a, b) \in \boldsymbol{Q}'$. We define an equivalence $\approx$ on the set $\boldsymbol{Q}'$ by

$$\frac{a}{b} \approx \frac{c}{d} \quad \text{if and only if} \quad a \cdot d = b \cdot c.$$

Let $\boldsymbol{Q} = \boldsymbol{Q}'/\approx$ be the set of equivalence classes of $\boldsymbol{Q}$ modulo $\approx$. Elements of $\boldsymbol{Q}$ are called *rational numbers*; the rational number represented by a/b is denoted $[a/b]$. [Later, the brackets are habitually dropped and we do not distinguish between a rational number and the (many) fractions that represent it.]

There is an obvious one-to-one mapping i of the set $\boldsymbol{Z}$ of integers into the rationals:

$$i(a) = \left[\frac{a}{1}\right].$$

(Later, we identify integers with the corresponding rationals.) We now define addition and multiplication of rationals:

$$\left[\frac{a}{b}\right] + \left[\frac{c}{d}\right] = \left[\frac{a \cdot d + b \cdot c}{b \cdot d}\right],$$
$$\left[\frac{a}{b}\right] \cdot \left[\frac{c}{d}\right] = \left[\frac{a \cdot c}{b \cdot d}\right].$$

In order to lay the foundations satisfactorily, one should prove the following:

(a) Addition and multiplication of the rationals are well defined (i.e., independent of the choice of representatives).

(b) For integers, the new definitions agree with the old ones; i.e., $i(a + b) = i(a) + i(b)$ and $i(a \cdot b) = i(a) \cdot i(b)$ for all $a, b \in \boldsymbol{Z}$.

(c) Addition and multiplication of rationals satisfy the usual laws of algebra.

(d) If $A \in \boldsymbol{Q}$, $B \in \boldsymbol{Q}$, and $B \neq [0/1]$, then the equation $A = B \cdot X$ has a unique solution $X \in \boldsymbol{Q}$. Thus *division* of rational numbers is defined, as long as the divisor is not zero; we denote this operation $\div$: $X = A \div B$.

Finally, we extend the ordering of integers to the rationals.

First, notice that each rational can be represented by a fraction a/b where the denominator b is greater than 0:

$$\left[\frac{a}{b}\right] = \left[\frac{-a}{-b}\right] \quad \text{and either} \quad b > 0 \text{ or } -b > 0.$$

We now define the natural ordering of rationals:

$$\text{If } b > 0 \text{ and } d > 0, \text{ let } \left[\frac{a}{b}\right] < \left[\frac{c}{d}\right] \text{ if and only if } a \cdot d < b \cdot c.$$

We again leave to the reader the proof that the definition does not depend on the choice of representatives as long as $b > 0$ and $d > 0$, that $<$ is really a linear ordering, that for $a, b \in \boldsymbol{Z}$, $a < b$ if and only if $[a/1] < [b/1]$, and that the usual algebraic laws (such as: if $a < b$ then $a + c < b + c$, etc.) hold. We have again

1.2 Theorem *The set of rationals $\boldsymbol{Q}$ is countable.*

Proof. Theorem 3.13 in Chapter 4. Theorem 3.12 in Chapter 4 gives another proof. □

The next result has been referred to in Chapter 4.

1.3 Theorem *$(\boldsymbol{Q}, <)$ is a dense linearly ordered set and has no endpoints. In fact, for every $r \in \boldsymbol{Q}$ there exists $n \in \boldsymbol{N}$ such that $r < n$.*

Proof. $\boldsymbol{Q}$ is infinite; since $a/b - 1 < a/b < a/b + 1$, $\boldsymbol{Q}$ has no endpoints. If $r \leq 0$ we can take $n = 1$. If $r > 0$, we write $r = a/b$ where $a > 0$, $b > 0$, $a, b \in \boldsymbol{N}$, and take $n = a + 1$.

It remains to show that $(\boldsymbol{Q}, <)$ is dense. Let r, s be rationals such that $r < s$; assume that $r = a/b$ and $s = c/d$, where $b > 0$ and $d > 0$. Now we let

$$x = \frac{a \cdot d + b \cdot c}{2 \cdot b \cdot d}$$

[i.e., $x = (r + s)/2$]. Then $r < x < s$. □

We conclude this section with a few remarks on decimal (or, in general, base p) expansions of rationals. Every integer $p > 1$ can serve as a *base* of a number system. The one generally used has $p = 10$. Another useful case is $p = 2$.

1.4 Lemma *Given a rational number r, there is a unique integer e such that $e \leq r < e + 1$. We call e the* integer part *of r, $e = [\![r]\!]$.*

Proof. Let $r = a/b$, $b > 0$. Assume that $a \geq 0$, $1 \leq b$, so $a \leq a \cdot b$ and $r = a/b \leq a \in \boldsymbol{Z}$. It now follows that $S = \{x \in \boldsymbol{Z} \mid x \leq r\} \subseteq \boldsymbol{Z}$ has an upper bound a in $\boldsymbol{Z}$. If $a < 0$, then 0 is an upper bound on S. Therefore, S has a greatest element e. Then $e \leq r < e + 1$, and clearly e is the unique integer with this property. □

The expansions of integers in base p are sufficiently well known. By Lemma 1.4, $r = [\![r]\!] + q$ where $[\![r]\!]$ is an integer and $q \in \boldsymbol{Q}$, $0 \leq q < 1$. We concentrate on the expansion of q.

Construct a sequence of *digits* $0, 1, \ldots, p - 1$ by recursion, as follows:

Find $a_1 \in \{0, \ldots, p-1\}$ such that $a_1/p \le q < (a_1+1)/p$ (let $a_1 = [\![q \cdot p]\!]$).

Then find $a_2 \in \{0, \ldots, p-1\}$ such that $a_1/p + a_2/p^2 \le q < a_1/p + (a_2+1)/p^2$ (let $a_2 = [\![(q - a_1/p) \cdot p^2]\!]$).

In general, find $a_k \in \{0, \ldots, p-1\}$ such that

$$\frac{a_1}{p} + \ldots + \frac{a_k}{p^k} \le q < \frac{a_1}{p} + \ldots + \frac{a_k + 1}{p^k}$$

$$\left(\text{take } a_k = \left[\!\!\left[\left(q - \frac{a_1}{p} - \ldots - \frac{a_{k-1}}{p^{k-1}}\right) \cdot p^k\right]\!\!\right]\right).$$

We call the sequence $\langle a_i \mid i \in \boldsymbol{N} \rangle$ the *expansion of q in base p*. When $p = 10$, it is customary to write $q = 0.a_1a_2a_3 \ldots$.

One can show

(a) There is no i such that $a_j = p - 1$ for all $j \ge i$.

(b) There exist $n \in \boldsymbol{N}$ and $l > 0$ such that $a_{n+l} = a_n$ for all $n \ge n_0$ (the expansion is *eventually periodic*, with *period* l).

Moreover, if $q = a/b$, then we can find a period l such that $l \le |b|$. Conversely, each sequence $\langle a_i \mid i \in \boldsymbol{N} \rangle$ with the properties (a) and (b) is an expansion of some rational number q ($0 \le q < 1$).

Exercises

1.1 Prove some of the claims made in the text of this section.

1.2 If $r \in \boldsymbol{Q}$, $r > 0$, then there exists $n \in \boldsymbol{N} - \{0\}$ such that $1/n < r$. [*Hint:* Use the fact mentioned in Theorem 1.3.]

2. Real Numbers

The set $\boldsymbol{R}$ of all real numbers and its natural linear ordering $<$ have been defined in Chapter 4, Section 5, as the completion of the rationals. In particular, every nonempty set of real numbers bounded from above has a supremum, and every nonempty set of real numbers bounded from below has an infimum. Here we consider the algebraic operations on real numbers. We first prove a useful lemma.

2.1 Lemma *For every $x \in \boldsymbol{R}$ and $n \in \boldsymbol{N} - \{0\}$ there exist $r, s \in \boldsymbol{Q}$ such that $r < x \le s$ and $s - r \le 1/n$.*

Proof. Fix some $r_0, s_0 \in \boldsymbol{Q}$ such that $r_0 < x < s_0$, and some $k \in \boldsymbol{N} - \{0\}$ such that $k > n(s_0 - r_0)$. Consider the increasing finite sequence of rational numbers $\langle r_i \rangle_{i=0}^k$ where $r_i = r_0 + i/n$. Let j be the greatest i for which $r_i < x$; notice that $j < k$. Now we have $r_j < x \le r_{j+1}$ and $r_{j+1} - r_j = 1/n$. It suffices to take $r = r_j$, $s = r_{j+1}$. □

We now define the operation of addition on real numbers.

2.2 Definition Let $x, y \in \boldsymbol{R}$. We let $x+y = \inf\{r+s \mid r, s \in \boldsymbol{Q},\ x \leq r,\ y \leq s\}$. (The symbol $+$ on the right-hand side refers to the addition of rational numbers.)

We note that the infimum exists, because the nonempty set in question is bounded below (by $p+q$, for any $p, q \in \boldsymbol{Q}$, $p < x$, $q < y$). It is also clear that if both x and y are rational numbers, then $x+y$, under the new definition, is equal to $x+y$, where $+$ is the previously defined addition of rationals.

2.3 Lemma *Let $x, y, z \in \boldsymbol{R}$.*

(i) $x + y = y + x$.

(ii) $(x+y)+z = x+(y+z)$.

(iii) $x + 0 = x$.

(iv) There exists a unique $w \in \boldsymbol{R}$ such that $x + w = 0$. We denote $w = -x$, the opposite *of x.*

(v) If $x < y$ then $x + z < y + z$.

Proof. We use some simple properties of suprema and infima (see Exercise 2.1). (i), (ii), and (iii) follow immediately from the corresponding properties of rational numbers.

(iv) We recall that $x = \inf\{s \in \boldsymbol{Q} \mid x \leq s\} = \sup\{r \in \boldsymbol{Q} \mid r < x\}$. This suggests letting $w = \inf\{-r \mid r \in \boldsymbol{Q},\ r < x\}$. We have $x + w = \inf\{s - r \mid r, s \in \boldsymbol{Q},\ x \leq s,\ r < x\}$. As $r < x$, $x \leq s$ imply $s - r > 0$, it follows $x+w \geq 0$. Assume $x+w > 0$; by density of $\boldsymbol{Q}$ in $\boldsymbol{R}$ and Exercise 1.2, there exists $n \in \boldsymbol{N} - \{0\}$ such that $1/n < x + w$. But, Lemma 2.1 guarantees existence of some $r, s \in \boldsymbol{Q}$, $r < x \leq s$, such that $s - r \leq 1/n$. Therefore $x + w \leq 1/n$, a contradiction. This shows the existence of a w with the property $x + w = 0$. If also $x + v = 0$, we have (using (i), (ii), and (iii)) $w = w + 0 = 0 + w = (v + x) + w = v + (x + w) = v + 0 = v$, so w is uniquely determined.

(v) If $x < y$ then $x + z \leq y + z$ is immediate from the definition of addition (and Exercise 2.1). If $x + z = y + z$, we have $x = x + 0 = x + (z + (-z)) = (x+z)+(-z) = (y+z)+(-z) = y+(z+(-z)) = y+0 = y$, a contradiction. □

The operation of multiplication on positive real numbers can be defined in an analogous way. We let $\boldsymbol{R}^+ = \{x \in \boldsymbol{R} \mid x > 0\}$.

2.4 Definition Let $x, y \in \boldsymbol{R}^+$. We let $x \cdot y = \inf\{r \cdot s \mid r, s \in \boldsymbol{Q},\ x \leq r,\ y \leq s\}$.

2.5 Lemma *Let $x, y, z \in \boldsymbol{R}^+$.*

(vi′) $x \cdot (y + z) = x \cdot y + x \cdot z$.

(vii′) $x \cdot y = y \cdot x$.

(viii′) $(x \cdot y) \cdot z = x \cdot (y \cdot z)$.

(ix′) $x \cdot 1 = x$.

(x′) *There exists a unique $w \in \boldsymbol{R}^+$ such that $x \cdot w = 1$. We denote $w = 1/x$, the* reciprocal *of x.*

(xi′) *If $x < y$ then $x \cdot z < y \cdot z$.*

Proof. Entirely analogous to the proof of Lemma 2.3. For (x′) we let $w = \inf\{1/r \mid r \in \boldsymbol{Q},\ 0 < r < x\}$. □

It is now a straightforward matter to extend multiplication to all real numbers. We first define the *absolute value* of $x \in \boldsymbol{R}$:

$$|x| = \begin{cases} x & \text{if } x \geq 0; \\ -x & \text{if } x < 0 \end{cases}$$

and notice that for $x \neq 0$, $|x| \in \boldsymbol{R}^+$ (if $x < 0$ then $0 = x + (-x) < 0 + (-x) = -x$).

2.6 Definition For $x, y \in \boldsymbol{R}$ we let

$$x \cdot y = \begin{cases} |x| \cdot |y| & \text{if } x > 0,\ y > 0 \text{ or } x < 0,\ y < 0; \\ -(|x| \cdot |y|) & \text{if } x > 0,\ y < 0 \text{ or } x < 0,\ y > 0; \\ 0 & \text{if } x = 0 \text{ or } y = 0. \end{cases}$$

We leave to the reader the straightforward but tedious exercise of showing that this definition agrees with the definition of multiplication of rationals from Section 1, and proving the following lemma.

2.7 Lemma *Let $x, y, z \in \boldsymbol{R}$.*

(vi) $x \cdot (y + z) = x \cdot y + x \cdot z$.

(vii) $x \cdot y = y \cdot x$.

(viii) $(x \cdot y) \cdot z = x \cdot (y \cdot z)$.

(ix) $x \cdot 1 = x$.

(x) *For each $x \in \boldsymbol{R}$, if $x \neq 0$, then there exists a unique $w \in \boldsymbol{R}$ such that $x \cdot w = 1$.*

(xi) If $x < y$ and $z > 0$ then $x \cdot z < y \cdot z$.

As before, we denote the unique w in (x) by $1/x$. We also define *division* by a nonzero real number x: $y \div x = y \cdot (1/x)$.

A structure $\mathfrak{A} = (A, <, +, \cdot, 0, 1)$ where $<$ is a linear ordering, $+$ and $\cdot$ are binary operations, and 0, 1 are constants, such that all the properties (i)–(v) and (vi)–(xi) are satisfied is called an *ordered field* in algebra. The contents of Lemmas 2.3 and 2.5 can thus be summarized by saying that the real numbers (with the usual ordering and arithmetic operations as defined above) are an ordered field. As the ordering of the real numbers is complete, they are a *complete ordered field.*

2.8 Theorem *The structure $\mathfrak{R} = \langle \boldsymbol{R}, <, +, \cdot, 0, 1\rangle$ is a complete ordered field.*

It is possible to prove that the complete ordered field is unique, i.e., if $\mathfrak{A} = \langle A, <, +, \cdot, 0, 1\rangle$ is also a complete ordered field, then $\mathfrak{A}$ and $\mathfrak{R}$ are isomorphic. As the proof requires a fairly heavy dose of algebra, we do not present it here (but see Exercise 2.5).

We conclude this section by mentioning a well-known fact about real numbers that we use in Section 6 of Chapter 4. We leave the proof as an exercise.

2.9 Theorem (Expansion of real numbers in base p) *Let $p \geq 2$ be a natural number. For every real number $0 \leq a < 1$ there is a unique sequence of real numbers $\langle a_n\rangle_{n=1}^{\infty}$ such that*
(a) $0 \leq a_n < p$, for each $n = 1, 2, \ldots$
(b) There is no n_0 such that $a_n = p - 1$ for all $n > n_0$.
(c) $a_1/p + \cdots + a_n/p^n \leq a < a_1/p + \cdots + (a_n + 1)/p^n$, for each $n \geq 1$.
The real number a is rational if and only if $\langle a_n\rangle_{n=1}^{\infty}$ is eventually periodic.

Exercises

2.1 Let $A \subseteq B \subseteq \boldsymbol{R}$, $A \neq \emptyset$. If B is bounded from below then $\inf B \leq \inf A$. If B is bounded from above then $\sup A \leq \sup B$. If, in addition, for every $b \in B$ there exists $a \in A$ such that $a \leq b$, then $\inf B = \inf A$. Similarly for sup.

2.2 For every $x \in \boldsymbol{R}^+$ and $n \in \boldsymbol{N} - \{0\}$ there exist $r, s \in \boldsymbol{Q}$ such that $0 < r < x \leq s$ and $1 < s/r \leq 1 + 1/n$. [*Hint*: Fix $r_0 \in \boldsymbol{Q}$ such that $0 < r_0 < x$, and $k \in \boldsymbol{N} - \{0\}$ such that $kr_0 > n$. Use Lemma 2.1 to find $r, s \in \boldsymbol{Q}$, $r_0 < r < x \leq s$, so that $s - r < 1/k$, and estimate s/r.]

2.3 Prove that if $a, b \in \boldsymbol{R}$, $b \neq 0$, then the equation $a = b \cdot x$ has a unique solution.

2.4 Prove that for every $a \in \boldsymbol{R}^+$ there exists a unique $x \in \boldsymbol{R}^+$ such that $x \cdot x = a$.

2.5 Prove that every complete ordered field is isomorphic to $\boldsymbol{R}$.
[*Hints:*

1. Given a complete ordered field $\mathfrak{A} = \langle A, <, +, \cdot, 0, 1\rangle$ consider the closure C of $\{0, 1\}$ under $+$ and $\cdot$. Let $\mathfrak{C}$ be the structure obtained by restricting the ordering and the operations of $\mathfrak{A}$ to C. Show that $\mathfrak{C}$ is (uniquely) isomorphic to $\mathfrak{Q} = \langle \boldsymbol{Q}, <, +, \cdot, 0, 1\rangle$. This is the "algebraic" part of the proof.
2. Show that for every $a \in A$ there is $c \in C$ such that $a \leq c$. (If not, then $S = \{a \in A \mid a > c \text{ for all } c \in C\}$ is nonempty and bounded below, but does not have an infimum, because $a \in S$ implies $a - 1 \in S$.)
3. Use 2. to show that C is dense in $(A, <)$.
4. Use 3. to show that the isomorphism between $\mathfrak{Q}$ and $\mathfrak{C}$ extends uniquely to an isomorphism between $\mathfrak{R}$ and $\mathfrak{A}$.]

2.6 (Complex Numbers) Let $\boldsymbol{C} = \boldsymbol{R} \times \boldsymbol{R}$, and let us define addition and multiplication on $\boldsymbol{C}$ as follows:

$$(a_1, a_2) + (b_1, b_2) = (a_1 + b_1, a_2 + b_2);$$
$$(a_1, a_2) \cdot (b_1, b_2) = (a_1 \cdot b_1 - a_2 \cdot b_2, a_1 \cdot b_2 + a_2 \cdot b_1).$$

Show that $+$ and $\cdot$ satisfy (i)–(iv) of Lemma 2.3 and (vi)–(x) of Lemma 2.7.

3. Topology of the Real Line

The main result of the preceding section is a characterization of the real number system as a complete ordered field. This is the usual departure point for the study of topological properties of the real line. We give some basic definitions and theorems of the subject in this section. Our objectives are to justify the claim that set theory, as we developed it so far, provides a satisfactory foundation for analysis, to show some consequences of the previously proved results, and to provide a convenient reference for some concepts and theorems used elsewhere. Readers who studied advanced calculus should be familiar with most of this material. In any case, one can skip this section and refer to it only as needed.

We begin with a definition of a familiar concept.

3.1 Definition Let $(P, <)$ be a linear ordering, $a, b \in P$, $a < b$. A (bounded) *open interval* with endpoints a and b is the set $(a, b) = \{x \in P \mid a < x < b\}$. A (bounded) *closed interval* with endpoints a and b is the set $[a, b] = \{x \in P \mid a \leq x \leq b\}$. We also define *unbounded open* and *closed intervals* (a, ∞), $(-\infty, a)$, $[a, \infty)$, $(-\infty, a]$, and $(-\infty, \infty) = P$ in the usual way: for example, $(a, \infty) = \{x \in P \mid a < x\}$. We employ the standard notation (a, b) for open intervals, even though it clashes with our notation for ordered pairs; the meaning should always be clear from the context.

The next theorem is a consequence of the existence of a countable dense subset in $\boldsymbol{R}$.

3.2 Theorem *Every system of mutually disjoint open intervals in* $\boldsymbol{R}$ *is at most countable.*

Proof. Let $P = \boldsymbol{Q} \cap \bigcup S$ where S is a system of mutually disjoint open intervals in $\boldsymbol{R}$ and $\boldsymbol{Q}$ is the set of all rational numbers. As $\boldsymbol{Q}$ is dense in $\boldsymbol{R}$, each open interval in S contains at least one element of P. As elements of S are mutually disjoint, each element of P is contained in a unique open interval from S. The function assigning to each element of P the unique open interval from S to which it belongs is a mapping of an at most countable set P onto S. Hence S is at most countable, by Theorem 3.4 in Chapter 4. □

This is a convenient place to mention a famous problem in set theory dating from the beginning of the century: Let $(P, <)$ be a complete linearly ordered set without endpoints where every system of mutually disjoint open intervals is at most countable. Is $(P, <)$ isomorphic to the real line? This is the *Suslin's Problem.* Like the Continuum Hypothesis, it remained unsolved for decades. With the help of models of set theory, it has been established that it, like the Continuum Hypothesis, can be neither proved nor refuted from the axioms of Zermelo-Fraenkel set theory. The reader can find out more about these matters in Chapter 15.

3.3 Definition A system of sets S has the *finite intersection property* if every nonempty finite subsystem of S has a nonempty intersection.

An example of a system of sets with the finite intersection property is the range of a nonincreasing sequence of nonempty sets, i.e., $S = \{A_n \mid n \in \boldsymbol{N}\}$ where $A_n \neq \emptyset$, $A_n \supseteq A_{n'}$ for all $n, n' \in \boldsymbol{N}$, $n \leq n'$.

The next theorem is a consequence of the completeness of the real line.

3.4 Theorem *Any nonempty system of closed and bounded intervals in* $\boldsymbol{R}$ *with the finite intersection property has nonempty intersection.*

Proof. Let S be such a system. Let $A = \{x \in \boldsymbol{R} \mid [x, y] \in S \text{ for some } y \in \boldsymbol{R}\}$ be the set of all left endpoints of intervals in S. We show that A is bounded from above. Fix $[a, b] \in S$, then certainly $x \leq b$, because in the opposite case we would have $a < b < x < y$ and the intervals $[x, y]$, $[a, b]$ would be disjoint, violating the finite intersection property for the 2-element subset $\{[x, y], [a, b]\}$ of S. So b is an upper bound on A. By the completeness property, A has a supremum $\overline{a}$. If $[a, b]$ is an interval in S, we have $a \leq \overline{a}$ because $a \in A$, and $\overline{a} \leq b$ because b is an upper bound on A. So $\overline{a} \in [a, b]$ for any $[a, b] \in S$ and $\bigcap S$ is not empty. □

Completeness of the real line is also the reason why the notion of limit works as well as it does. We recall some very familiar definitions from calculus.

3.5 Definition Let $\langle a_n \mid n \in \boldsymbol{N} \rangle$ be an infinite sequence of real numbers.

(a) $\langle a_n \rangle$ is *nondecreasing* if $a_n \le a_{n'}$ holds for all $n, n' \in \boldsymbol{N}$ such that $n < n'$. It is *increasing* if $a_n < a_{n'}$ holds whenever $n < n'$.

(b) $\langle a_n \rangle$ is *bounded from above* if its range $\{a_n \mid n \in \boldsymbol{N}\}$ is bounded from above. Similarly for *bounded from below*. $\langle a_n \rangle$ is *bounded* if it is bounded from above and below.

(c) $\langle a_n \rangle$ has a *limit a* [*converges to a*] if for every $\varepsilon > 0$, $\varepsilon \in \boldsymbol{R}$, there is $n_0 \in \boldsymbol{N}$ such that, for all $n \ge n_0$, $|a_n - a| < \varepsilon$.

(d) $\langle a_n \rangle$ is a *Cauchy sequence* if for every $\varepsilon > 0$, $\varepsilon \in \boldsymbol{R}$, there is $n_0 \in \boldsymbol{N}$ such that, for all $m, n \ge n_0$, $|a_m - a_n| < \varepsilon$.

Here are a few fundamental theorems about convergence. Note the crucial role the completeness of $\boldsymbol{R}$ plays in the proofs.

3.6 Theorem *Every nondecreasing sequence of real numbers bounded from above has a limit.*

Proof. If $\{a_n \mid n \in \boldsymbol{N}\}$ is bounded from above, then it has a supremum $\overline{a}$. We prove that $\langle a_n \mid n \in \boldsymbol{N} \rangle$ converges to $\overline{a}$. Let $\varepsilon > 0$ be given. Since $\overline{a} - \varepsilon < \overline{a}$, and $\overline{a}$ is the least upper bound on $\{a_n \mid n \in \boldsymbol{N}\}$, there is n_0 such that $\overline{a} - \varepsilon < a_{n_0}$. We now have, for all $n \ge n_0$, $\overline{a} - \varepsilon < a_{n_0} \le a_n \le \overline{a} < \overline{a} + \varepsilon$, i.e., $|a_n - \overline{a}| < \varepsilon$. This shows that $\langle a_n \rangle$ converges to $\overline{a}$. □

3.7 Definition Let $\langle k_n \mid n \in \boldsymbol{N} \rangle$ be an increasing sequence of natural numbers. The sequence $\langle a_{k_n} \mid n \in \boldsymbol{N} \rangle$ is called a *subsequence* of $\langle a_n \mid n \in \boldsymbol{N} \rangle$. (Note that it is just the composition of $\langle a_n \rangle \circ \langle k_n \rangle$.)

3.8 Theorem *Every bounded sequence of real numbers has a convergent subsequence.*

Proof. Assume that $\langle a_n \mid n \in \boldsymbol{N} \rangle$ is bounded. Let $b_n = \inf\{a_k \mid k \ge n\}$ and observe that the sequence $\langle b_n \mid n \in \boldsymbol{N} \rangle$ is nondecreasing and bounded from above (every upper bound on $\langle a_n \rangle$ is an upper bound on $\langle b_n \rangle$). By Theorem 3.6, $\langle b_n \rangle$ has a limit $\overline{a}$. We construct a subsequence of $\langle a_n \rangle$ with limit $\overline{a}$ by recursion:

$k_0 = 0$, i.e., $a_{k_0} = a_0$;

given k_n, let k_{n+1} be the least $k \in \boldsymbol{N}$ such that $k > k_0$ and

$$\overline{a} - \frac{1}{n+1} < a_k < \overline{a} + \frac{1}{n+1}.$$

We have to show that such k's exist. But $\overline{a} = \sup\{b_n \mid n \in \boldsymbol{N}\}$ (see the proof of Theorem 3.6) and so there is $i > k_n$ such that

$$\overline{a} - \frac{1}{n+1} < b_i \le \overline{a}.$$

Also, $b_i = \inf\{a_k \mid k \geq i\}$ and so there is $k \geq i$ such that

$$b_i \leq a_k < \overline{a} + \frac{1}{n+1}.$$

We now have $k > k_n$ with

$$\overline{a} - \frac{1}{n+1} < a_k < \overline{a} + \frac{1}{n+1},$$

as required.

The subsequence $\langle a_{k_n} \mid n \in \boldsymbol{N}\rangle$ of $\langle a_n\rangle$ is such that for every $n \geq 1$, $|a_{k_n} - \overline{a}| < 1/n$. From this it follows easily that $\langle a_{k_n}\rangle$ converges to $\overline{a}$ (see Exercise 3.4). □

3.9 Theorem *Every Cauchy sequence of real numbers converges.*

Proof. Every Cauchy sequence is bounded: If $\langle a_n\rangle$ is a Cauchy sequence, let $\varepsilon = 1$ in Definition 3.5(d); the result is $n_0 \in \boldsymbol{N}$ such that for all $n \geq n_0$, $|a_{n_0} - a_n| < 1$ (we also let $m = n_0$). So $a_{n_0} - 1 < a_n < a_{n_0} + 1$ for all $n \geq n_0$, and letting

$$M_1 = \max\{a_0, \dots, a_{n_0-1}, a_{n_0} + 1\}$$
$$M_2 = \min\{a_0, \dots, a_{n_0-1}, a_{n_0} - 1\}$$

produces the desired upper and lower bounds on $\{a_n \mid n \in \boldsymbol{N}\}$.

By Theorem 3.8, $\langle a_n\rangle$ has a subsequence $\langle a_{k_n}\rangle$ converging to $\overline{a} \in \boldsymbol{R}$; we prove that $\langle a_n\rangle$ itself converges to $\overline{a}$. So let $\varepsilon > 0$ be given; since $\langle a_{k_n}\rangle$ converges to $\overline{a}$, there is $n_0 \in \boldsymbol{N}$ such that for all $n \geq n_0$, $|a_{k_n} - \overline{a}| < \varepsilon/2$. Since $\langle a_n\rangle$ is a Cauchy sequence, there is $n_1 \in \boldsymbol{N}$ such that for all $m, n \geq n_1$, $|a_m - a_n| < \varepsilon/2$. Let $n_2 = \max\{n_0, n_1\}$. If $n \geq n_2$, then $|a_n - \overline{a}| \leq |a_n - a_{k_{n_2}}| + |a_{k_{n_2}} - \overline{a}| < \varepsilon/2 + \varepsilon/2 = \varepsilon$ and we have established that $\langle a_n\rangle$ has a limit $\overline{a}$. □

We now turn our attention to continuous functions and open and closed sets.

3.10 Definition A function $f : \boldsymbol{R} \to \boldsymbol{R}$ is *continuous at* $a \in \boldsymbol{R}$ if for every $\varepsilon > 0$ there is $\delta > 0$ such that for all $x \in \boldsymbol{R}$, if $|x - a| < \delta$, then $|f(x) - f(a)| < \varepsilon$. f is *continuous* if it is continuous at every $a \in \boldsymbol{R}$. A set $A \subseteq \boldsymbol{R}$ is *open* if for every $a \in A$ there is $\delta > 0$ such that $|x - a| < \delta$ implies that $x \in A$. [Or, in other words, the open interval $(a - \delta, a + \delta) \subseteq A$.] A set B is *closed* if its relative complement $\boldsymbol{R} - B$ is open.

Continuous functions are particularly simple and are a main object of study in mathematical analysis. For our purposes, we need only a result showing that a continuous function is uniquely determined by its values on a dense subset of $\boldsymbol{R}$ (such as the rationals).

3.11 Theorem *Let $D \subseteq \boldsymbol{R}$ be dense in $\boldsymbol{R}$. If $f \restriction D = g \restriction D$ where f and g are continuous, then $f = g$.*

Proof. Suppose that $f \neq g$; then $f(a) \neq g(a)$ for some $a \in \boldsymbol{R}$. Let $\varepsilon = |f(a) - g(a)|/2$. As both f and g are assumed continuous at a, there exist $\delta_1, \delta_2 > 0$ such that if $|x - a| < \delta_1$, then $|f(x) - f(a)| < \varepsilon$, and if $|x - a| < \delta_2$, then $|g(x) - g(a)| < \varepsilon$. The density of D guarantees existence of $x \in D$ such that $|x - a| < \min\{\delta_1, \delta_2\}$. But now $|f(a) - g(a)| \leq |f(a) - f(x)| + |f(x) - g(x)| + |g(x) - g(a)| < \varepsilon + 0 + \varepsilon = |f(a) - g(a)|$, a contradiction. [We have used the fact that $f(x) = g(x)$ for $x \in D$ in replacing the middle term by 0.] □

Similarly, open and closed sets are relatively simple and well-behaved. It is obvious from the definition that a set is open if and only if it is a union of a system of open intervals. Consequently, the union of any system of open sets is open and the intersection of any system of closed sets is closed. Open intervals are the simplest examples of open sets; closed intervals, finite sets, and $\{1/n \mid n \in \boldsymbol{N} - \{0\}\} \cup \{0\}$ are typical examples of closed sets. Other examples can be obtained by using the next lemma.

3.12 Lemma *The intersection of a finite system of open sets is open. The union of a finite system of closed sets is closed.*

Proof. Let A_1 and A_2 be open sets. If $a \in A_1 \cap A_2$, there are $\delta_1, \delta_2 > 0$ such that $|x - a| < \delta_1$ implies $x \in A_1$ and $|x - a| < \delta_2$ implies $x \in A_2$. Put $\delta = \min\{\delta_1, \delta_2\}$; then $\delta > 0$ and $|x - a| < \delta$ implies $x \in A_1 \cap A_2$. This argument proves that the intersection of two open sets is open. The claim for any finite system of open sets is proved by induction, and the claim for closed sets by taking relative complements in $\boldsymbol{R}$ and using the De Morgan Laws (Section 4 in Chapter 1). □

3.13 Theorem *The following statements about $f : \boldsymbol{R} \to \boldsymbol{R}$ are equivalent:*
(a) f is continuous.
(b) $f^{-1}[A]$ is open whenever $A \subseteq \boldsymbol{R}$ is open.
(c) $f^{-1}[B]$ is closed whenever $B \subseteq \boldsymbol{R}$ is closed.

Proof. (a) implies (b). Assume that f is continuous. If $a \in f^{-1}[A]$, $f(a) \in A$ and so there is $\varepsilon > 0$ such that $|y - f(a)| < \varepsilon$ implies $y \in A$ (because A is open). By definition of continuity there is $\delta > 0$ such that $|x - a| < \delta$ implies $|f(x) - f(a)| < \varepsilon$. So we found $\delta > 0$ such that $|x - a| < \delta$ implies $f(x) \in A$, i.e., $x \in f^{-1}[A]$, showing $f^{-1}[A]$ open.

(b) implies (a). Let $a \in \boldsymbol{R}$ and $\varepsilon > 0$. The assumption (b) implies that $f^{-1}[(f(a) - \varepsilon, f(a) + \varepsilon)]$ is open (and contains a). Thus there is $\delta > 0$ such that $|x - a| < \delta$ implies $x \in f^{-1}[(f(a) - \varepsilon, f(a) + \varepsilon)]$, i.e., $f(x) \in (f(a) - \varepsilon, f(a) + \varepsilon)$, establishing the continuity of f at a.

The equivalence of (b) and (c) follows immediately from Exercise 3.6(b) in Chapter 2. □

In the rest of this section we prove some results about open and closed sets. The following lemma is used in Chapter 5 to determine the cardinality of the system of all open sets.

3.14 Lemma *Every open set is a union of a system of open intervals with rational endpoints.*

Proof. Let A be open and let S be the system of all open intervals with rational endpoints included in A. Clearly, $\bigcup S \subseteq A$; if $a \in A$, then $(a-\delta, a+\delta) \subseteq A$ for some $\delta > 0$ and we use the density of $\boldsymbol{Q}$ in $\boldsymbol{R}$ to find $r_1, r_2 \in \boldsymbol{Q}$ such that $a - \delta < r_1 < a < r_2 < a + \delta$. Now $a \in (r_1, r_2) \subseteq A$, so $a \in \bigcup S$, showing that $\bigcup S = A$. □

We have defined closed sets as complements of open sets. It is useful to have a characterization of them in terms of the behavior of their points.

3.15 Definition $a \in \boldsymbol{R}$ is an *accumulation point* of $A \subseteq \boldsymbol{R}$ if for every $\delta > 0$ there is $x \in A$, $x \neq a$, such that $|x - a| < \delta$. $a \in \boldsymbol{R}$ is an *isolated point* of $A \subseteq \boldsymbol{R}$ if $a \in A$ and there is $\delta > 0$ such that $|x - a| < \delta$, $x \neq a$, implies $x \notin A$.

It is easy to see that every element of A is either an isolated point or an accumulation point, and that there may be accumulation points of A that do not belong to A.

3.16 Lemma *$A \subseteq \boldsymbol{R}$ is closed if and only if all accumulation points of A belong to A.*

Proof. Assume that A is closed, i.e., $\boldsymbol{R} - A$ is open. If $a \in \boldsymbol{R} - A$, then there is $\delta > 0$ such that $|x-a| < \delta$ implies $x \in \boldsymbol{R}-A$, so a is not an accumulation point of A. Thus all accumulation points of A belong to A.

Conversely, assume that all accumulation points of A belong to A. If $a \in \boldsymbol{R} - A$, then it is not an accumulation point of A, so there is $\delta > 0$ such that there is no $x \in A$, $x \neq a$, with $|x - a| < \delta$. But then $|x - a| < \delta$ implies $x \in \boldsymbol{R} - A$, so $\boldsymbol{R} - A$ is open and A is closed. □

As an application, we generalize Theorem 3.4.

3.17 Theorem *Any nonempty system of closed and bounded sets with the finite intersection property has a nonempty intersection.*

Proof. Let S be such a system. We note that the intersection of any nonempty finite subsystem T of S is a nonempty closed and bounded set. So $\overline{S} = \{\bigcap T \mid T \text{ is a nonempty finite subsystem of } S\}$ is again a nonempty system of closed and bounded sets with the finite intersection property, and the additional property that the intersection of any nonempty finite subsystem of $\overline{S}$ actually belongs to $\overline{S}$. Since obviously $\bigcap S = \bigcap \overline{S}$, it suffices to prove that $\bigcap \overline{S} \neq \emptyset$.

The proof closely follows that of Theorem 3.4. Let $A = \{\inf F \mid F \in \overline{S}\}$. We show that A is bounded from above. Fix $G \in \overline{S}$; then for any $F \in \overline{S}$ we have $F \cap G \in \overline{S}$ and $\inf F \leq \inf(F \cap G) \leq \sup(F \cap G) \leq \sup G$, so A is bounded by $\sup G$. By the completeness property, A has a supremum $\overline{a}$. The proof is concluded by showing $\overline{a} \in F$ for any $F \in \overline{S}$.

Suppose that $\bar{a} \notin F$. For any $\delta > 0$ there is $H \in \overline{S}$ such that $\bar{a} - \delta < \inf H \leq \bar{a}$, by definition of $\bar{a}$. Since $H \cap F \in \overline{S}$ and $\inf H \leq \inf(H \cap F)$, we have also $\bar{a} - \delta < \inf(H \cap F) \leq \bar{a} < \bar{a} + \delta$. It now follows from the definition of infimum that there is $x \in H \cap F$ for which $\inf(H \cap F) \leq x < \bar{a} + \delta$. We conclude that for each $\delta > 0$, there exists $x \in F$ such that $x \neq \bar{a}$ (we assumed that $\bar{a} \notin F$) and $|x - \bar{a}| < \delta$. But this means that $\bar{a}$ is an accumulation point of F, and as F is closed, $\bar{a} \in F$, a contradiction. □

Closed sets contain all of their accumulation points; in addition, they may or may not contain isolated points. An example of a closed set without isolated points is any closed interval. $[0, 1] \cup \boldsymbol{N}$ is an example of a closed set with (infinitely many) isolated points. Nonempty closed sets without isolated points are particularly well-behaved; they are called *perfect sets.* We use them in the next section as a tool for analyzing the structure of closed sets. We conclude this section with an example showing that, their good behavior notwithstanding, perfect sets may differ quite substantially from the familiar examples, the closed intervals.

3.18 Example Cantor Set. Let $S = \text{Seq}(\{0, 1\}) = \bigcup_{n \in \boldsymbol{N}} \{0, 1\}^{\boldsymbol{N}}$ be the set of all finite sequences of 0's and 1's. We construct a system of closed intervals $D = \langle D_s \mid s \in S \rangle$ as follows:

$$D_{\langle\rangle} = [0, 1];$$

$$D_{\langle 0 \rangle} = [0, \frac{1}{3}], \quad D_{\langle 1 \rangle} = [\frac{2}{3}, 1];$$

$$D_{\langle 0,0 \rangle} = [0, \frac{1}{9}], \quad D_{\langle 0,1 \rangle} = [\frac{2}{9}, \frac{1}{3}], \quad D_{\langle 1,0 \rangle} = [\frac{2}{3}, \frac{7}{9}], \quad D_{\langle 1,1 \rangle} = [\frac{8}{9}, 1], \quad \text{etc.}$$

In general, if $D_{\langle s_0, \ldots, s_{n-1} \rangle} = [a, b]$, we let $D_{\langle s_0, \ldots, s_{n-1}, 0 \rangle} = [a, a + \frac{1}{3}(b - a)]$ and $D_{\langle s_0, \ldots, s_{n-1}, 1 \rangle} = [a + \frac{2}{3}(b - a), b]$ be the first and last thirds of $[a, b]$. The existence of the system D is justified by the Recursion Theorem, although it takes some thought to see: we actually define recursively a sequence $\langle D^n \mid n \in \boldsymbol{N} \rangle$, where each D^n is a system of closed intervals indexed by $\{0, 1\}^n$: $D^0_{\langle\rangle} = [0, 1]$; having defined D^n, we define D^{n+1} so that for all $\langle s_0, \ldots, s_{n-1} \rangle \in \{0, 1\}^n$,

$D^{n+1}_{\langle s_0,\ldots,s_{n-1},0\rangle} = [a, a+\frac{1}{3}(b-a)]$ and $D^{n+1}_{\langle s_0,\ldots,s_{n-1},1\rangle} = [a+\frac{2}{3}(b-a), b]$ where $[a,b] = D^n_{s_0,\ldots,s_{n-1}}$, and then let $D = \bigcup_{n\in \boldsymbol{N}} D^n$.

Now let $F_n = \bigcup\{D_s \mid s \in \{0,1\}^n\}$, so that $F_0 = [0,1]$, $F_1 = [0,\frac{1}{3}] \cup [\frac{2}{3},1]$, etc. Notice that each F_n is a union of a finite system of closed intervals, and hence closed (see Lemma 3.12). The set $F = \bigcap_{n\in \boldsymbol{N}} F_n$ is thus also closed; it is called the *Cantor set*.

For any $f \in \{0,1\}^{\boldsymbol{N}}$ we let $D_f = \bigcap_{n\in \boldsymbol{N}} D_{f\restriction n}$. We note that $D_f \subseteq F$ and $D_f \neq \emptyset$ (Theorem 3.4). Moreover, the length of the interval $D_{f\restriction n}$ is $1/3^n$, so the infimum of the lengths of the intervals in the system $\langle D_{f\restriction n} \mid n \in \boldsymbol{N}\rangle$ is zero. It follows that D_f contains a unique element: $D_f = \{d_f\}$. Conversely, for any $a \in F$ there is a unique $f \in \{0,1\}^{\boldsymbol{N}}$ such that $a = d_f$: let $f = \bigcup\{s \in \mathrm{Seq}(\{0,1\}) \mid a \in D_s\}$. (Compare these arguments with Exercise 3.13 in Chapter 2.) $d = \langle d_f \mid f \in \{0,1\}^{\boldsymbol{N}}\rangle$ is thus a one-to-one mapping of $\{0,1\}^{\boldsymbol{N}}$ onto F, and we conclude that the Cantor set has the cardinality of the continuum $2^{\aleph_0}$.

We prove two out of many other interesting properties of the set F.

(1) *The Cantor set is a perfect set.*

Proof. We know that F is closed and nonempty already, so it remains to prove that it has no isolated points. Let $a \in F$ and $\delta > 0$; we have to find $x \in F$, $x \neq a$, such that $|x-a| < \delta$. Take $n \in \boldsymbol{N}$ for which $1/3^n < \delta$. As we know, $a = d_f$ for some $f \in \{0,1\}^{\boldsymbol{N}}$; let $x = d_g$ where g is any sequence of 0's and 1's such that $g \restriction n = f \restriction n$ and $f \neq g$. Then $a \neq x$, but $a, x \in D_{f\restriction n} = D_{g\restriction n}$ which has length $1/3^n < \delta$, so $|x-a| < \delta$, as required. □

(2) *The relative complement of the Cantor set in* $[0,1]$ *is dense in* $[0,1]$.

Proof. Let $0 \le a < b \le 1$; we show that (a,b) contains elements not in F. Take $n \in \boldsymbol{N}$ such that $\frac{1}{3^n} < \frac{1}{2}(b-a)$. Let k be the least natural number for which $\frac{k}{3^n} \ge a$; we then have $a \le \frac{k}{3^n} < \frac{(k+1)}{3^n} < b$. The open interval

$$\left(\frac{3k+1}{3^{n+1}}, \frac{3k+2}{3^{n+1}}\right) \quad \left(\text{the middle third of } \left[\frac{k}{3^n}, \frac{k+1}{3^n}\right]\right)$$

is certainly disjoint from F, and we see that (a,b) contains elements not in F. □

Exercises

3.1 Every system of mutually disjoint open sets is at most countable. The statement is false for closed sets.

3.2 Let S be a nonempty system of closed and bounded intervals in $\boldsymbol{R}$. If $\inf\{b-a \mid [a,b] \in S\} = 0$, then $\bigcap S$ contains at most one element.

3.3 Let $(P, <)$ be a dense linearly ordered set where every nonempty system of closed and bounded intervals with the finite intersection property has a nonempty intersection. Then P is complete. (This is a converse to Theorem 3.4.)

3.4 $\langle a_n \rangle_{n=0}^{\infty}$ converges to a if and only if for every $k \in \boldsymbol{N}$, $k > 0$, there is $n_0 \in \boldsymbol{N}$ such that, for all $n \geq n_0$, $|a_n - a| < 1/k$.

3.5 If $\langle a_n \rangle_{n=0}^{\infty}$ converges to a and to b, then $a = b$ (the limit is unique). This justifies the usual notation $\lim_{n\to\infty} a_n$ for the limit, if it exists.

3.6 Every convergent sequence of reals is a Cauchy sequence.

3.7 The number $\overline{a} = \sup\{\inf\{a_k \mid k \geq n\} \mid n \in \boldsymbol{N}\}$ used in the proof of Theorem 3.8 is called the *lower limit* of $\langle a_n \rangle$ and is denoted $\liminf_{n\to\infty} a_n$. Similarly, $\overline{b} = \inf\{\sup\{a_k \mid k \geq n\} \mid n \in \boldsymbol{N}\}$ is called the *upper limit* of $\langle a_n \rangle$ and denoted $\limsup_{n\to\infty} a_n$. Prove that
(a) $\overline{b}$ exists for every bounded sequence and $\overline{a} \leq \overline{b}$.
(b) If c is a limit of some subsequence of $\langle a_n \rangle$, then $\overline{a} \leq c \leq \overline{b}$.
(c) $\langle a_n \rangle$ converges if and only if $\overline{a} = \overline{b}$ (and if it does, $\overline{a}$ is its limit).

3.8 There is a sequence $\langle a_n \rangle$ with the property that for any $a \in \boldsymbol{R}$, $\langle a_n \rangle$ has a subsequence converging to a. [*Hint*: Consider an enumeration of all rational numbers.]

3.9 Show that $f : \boldsymbol{R} \to \boldsymbol{R}$ is continuous at $a \in \boldsymbol{R}$ if and only if for every $n \in \boldsymbol{N}$ there is $k \in \boldsymbol{N}$ such that for all $x \in \boldsymbol{R}$, $|x - a| < 1/k$ implies $|f(x) - f(a)| < 1/n$.

3.10 Let $f : \boldsymbol{R} \to \boldsymbol{R}$ be continuous, $a, b, y \in \boldsymbol{R}$, $a < b$ and $f(a) \leq y \leq f(b)$. Prove that there exists $x \in \boldsymbol{R}$, $a \leq x \leq b$, such that $f(x) = y$ (the Intermediate Value Theorem). [*Hint*: Consider $x = \sup\{z \in [a, b] \mid f(z) \leq y\}$.]

3.11 Let $f : \boldsymbol{R} \to \boldsymbol{R}$ be continuous, $a, b \in \boldsymbol{R}$, $a < b$. Prove that the image of $[a, b]$ under f is closed.

3.12 Let $f : \boldsymbol{R} \to \boldsymbol{R}$ be continuous, $a, b \in \boldsymbol{R}$, $a < b$. Prove that there is $x \in [a, b]$ such that $f(x) \geq f(z)$ for all $z \in [a, b]$. Therefore, the image of $[a, b]$ under f is bounded.

3.13 If A is open and $\langle x_n \rangle$ converges to $a \in A$, then there is $n_0 \in \boldsymbol{N}$ such that for all $n \geq n_0$, $a_n \in A$.

3.14 $a \in \boldsymbol{R}$ is a *closure point* of $A \subseteq \boldsymbol{R}$ if for every $\delta > 0$ there is $x \in A$ such that $|x - a| < \delta$. $a \in \boldsymbol{R}$ is a closure point if and only if it is either an isolated point of A or an accumulation point of A. Let $\overline{A}$ be the set of all closure points of A. A is closed if and only if $A = \overline{A}$.

3.15 Every open set is a union of a system of mutually disjoint open intervals. [*Hint*: If A is open and $a \in A$, $\bigcup\{(x, y) \mid a \in (x, y) \subseteq A\}$ is an open interval.]

3.16 Give an example of a decreasing sequence of closed sets with empty intersection.

3.17 Show that Exercises 3.11 and 3.12 remain true for any closed and bounded set C in place of the closed interval $[a, b]$.

3.18 We describe an alternative construction of the completion of $(\boldsymbol{Q}, <)$. A sequence $\langle a_n \rangle_{n=0}^{\infty}$ of rational numbers is a *Cauchy sequence* if for each rational $p > 0$ there is n_0 such that $|a_n - a_{n_0}| < p$ whenever $n \geq n_0$. Let C be the set of all Cauchy sequences of rational numbers. We define an equivalence relation on C by $\langle a_n \rangle_{n=0}^{\infty} \approx \langle b_n \rangle_{n=0}^{\infty}$ if and only if for each $p > 0$ there is n_0 such that $|a_n - b_n| < p$ whenever $n \geq n_0$, and a preordering of C by $\langle a_n \rangle_{n=0}^{\infty} \preccurlyeq \langle b_n \rangle_{n=0}^{\infty}$ if and only if for each $p > 0$ there is n_0 such that $a_n - b_n < p$ whenever $n \geq n_0$. Prove that

(a) $\approx$ is an equivalence relation on C.
(b) $\preccurlyeq$ defines a linear ordering of $C/\approx$.
(c) The ordered set $C/\approx$ is isomorphic to $\boldsymbol{R}$.

4. Sets of Real Numbers

The Continuum Hypothesis postulates that every set of real numbers is either at most countable or has the cardinality of the continuum. Although the Continuum Hypothesis cannot be proved in Zermelo-Fraenkel set theory, the situation is different when one restricts attention to sets that are *simple* in some sense, for example, topologically. In the present section we look at several results of this type.

4.1 Theorem *Every nonempty open set of reals has cardinality* $2^{\aleph_0}$.

Proof. The function $\tan x$, familiar from calculus, is a one-to-one mapping of the open interval $(-\pi/2, \pi/2)$ onto the real line $\boldsymbol{R}$; therefore, the interval $(-\pi/2, \pi/2)$ has cardinality $2^{\aleph_0}$. If (a, b) and (c, d) are two open intervals, the function

$$f(x) = c + \frac{d-c}{b-a}(x-a)$$

is a one-to-one mapping of (a, b) onto (c, d); this shows that all open intervals have cardinality $2^{\aleph_0}$. Finally, every nonempty open set is the union of a nonempty system of open intervals, and thus has cardinality at least $2^{\aleph_0}$. As a subset of $\boldsymbol{R}$, it has cardinality at most $2^{\aleph_0}$ as well. □

4.2 Theorem *Every perfect set has cardinality* $2^{\aleph_0}$.

We first prove a lemma showing that every perfect set contains two disjoint perfect subsets (the "splitting" lemma). Let $\mathcal{P}$ denote the system of all perfect subsets of $\boldsymbol{R}$.

4.3 Lemma *There are functions G_0 and G_1 from $\mathcal{P}$ into $\mathcal{P}$ such that, for each $F \in \mathcal{P}$, $G_0(F) \subseteq F$, $G_1(F) \subseteq F$ and $G_0(F) \cap G_1(F) = \emptyset$.*

Proof. We show that for every perfect set F there are rational numbers r, s, $r < s$, such that $F \cap (-\infty, r]$ and $F \cap [s, +\infty)$ are both perfect.

Let $\alpha = \inf F$ if F is bounded from below (notice that in this case $\alpha \in F$) and $\alpha = -\infty$ otherwise. Similarly, let $\beta = \sup F$ if F is bounded from above ($\beta \in F$) and $\beta = +\infty$ otherwise. There are two cases to consider:

(a) $(\alpha, \beta) \subseteq F$. In this case, any $r, s \in \boldsymbol{Q}$ such that $\alpha < r < s < \beta$ have the desired property: for example, if $\alpha \in \boldsymbol{R}$, then $F \cap (-\infty, r] = [\alpha, r]$; otherwise, $F \cap (-\infty, r] = (-\infty, r]$ (and we noted that all closed intervals are perfect sets).

(b) $(\alpha, \beta) \not\subseteq F$. Then there exists $a \in (\alpha, \beta)$, $a \notin F$. The set F is closed (i.e., its complement is open), and thus there is $\delta > 0$ such that $(a-\delta, a+\delta) \cap F = \emptyset$. We note that $\alpha < a - \delta$ (because $\alpha = \inf F$, there exists $x \in F$, $\alpha < x < a$) and similarly $a + \delta < \beta$. Any $r, s \in \boldsymbol{Q}$ such that $a - \delta < r < a < s < a + \delta$ have the desired property: for example, $F \cap (-\infty, r]$ is clearly closed and nonempty ($F \cap (-\infty, r] = F \cap (-\infty, a]$). If it had an isolated point b, we could find $\delta' \leq \delta$ such that $x \in (b - \delta', b + \delta')$, $x \neq b$, implies $x \notin F \cap (-\infty, r] = F \cap (-\infty, a+\delta)$. But for such x, $x < b+\delta' \leq r+\delta' < a+\delta' \leq a+\delta$, so $x \in (-\infty, a+\delta)$. We conclude $x \notin F$, showing that b is an isolated point of F. This contradicts the assumption that F is perfect.

To complete the proof of the lemma, we fix an enumeration $\langle (r_n, s_n) \mid n \in \boldsymbol{N} \rangle$ of the countable set of all ordered pairs of rationals, and define $G_0(F) = F \cap (-\infty, r_n]$, $G_1(F) = F \cap [s_n, \infty)$, where (r_n, s_n) is the first (in our enumeration) pair of rationals for which $r_n < s_n$ and both $F \cap (-\infty, r_n]$ and $F \cap [s_n, \infty)$ are perfect. □

One more result is needed to prove Theorem 4.2. For any bounded nonempty set $A \subseteq \boldsymbol{R}$, we define the *diameter* of A, $\operatorname{diam}(A)$, as $\sup A - \inf A$. The next lemma shows that each perfect set has a perfect subset of an arbitrarily small diameter.

4.4 Lemma *There is a function $H : \mathcal{P} \times (\boldsymbol{N} - \{0\}) \to \mathcal{P}$ such that, for each $F \in \mathcal{P}$ and each $n \in \boldsymbol{N}$, $n \neq 0$, $H(F, n) \subseteq F$ and* $\operatorname{diam}(H(F, n)) \leq 1/n$.

Proof. Let $F \in \mathcal{P}$, $n \in \boldsymbol{N}$, $n > 0$. There exists $m \in \boldsymbol{Z}$ such that $F \cap (m/n, (m+1)/n) \neq \emptyset$. (Otherwise, $F \subseteq \{m/n \mid m \in \boldsymbol{Z}\}$, so all points of F would be isolated.) Take the least $m \geq 0$ with this property, or, if none exists, the greatest $m < 0$ with this property. Let $a = \inf F \cap (m/n, (m+1)/n)$ and $b = \sup F \cap (m/n, (m+1)/n)$. Clearly, $m/n \leq a \leq b \leq (m+1)/n$, so $b - a \leq 1/n$. We define $H(F, n) = F \cap [a, b]$. Clearly, F is closed, $F \neq \emptyset$. The definition of infimum implies that a is not an isolated point of $H(F, n)$, and neither is b. An argument similar to the one used in case (b) in the proof of the previous lemma shows that no $x \in (a, b) \subseteq (m/n, (m+1)/n)$ is an isolated point of $H(F, n)$ either, so $H(F, n)$ is perfect. □

Proof of Theorem 4.2. This proof is now easy. It closely resembles the argument used in Section 3 to show that the Cantor set has the cardinality $2^{\aleph_0}$. Let F be a perfect set; we construct a system of its perfect subsets $\langle F_s \mid s \in S \rangle$, where $S = \operatorname{Seq}(\{0, 1\})$, as follows:

$$
\begin{aligned}
F_{\langle\rangle} &= F; \\
F_{\langle s_0, \ldots, s_{n-1}, 0 \rangle} &= H(G_0(F_{\langle s_0, \ldots, s_{n-1} \rangle}), n); \\
F_{\langle s_0, \ldots, s_{n-1}, 1 \rangle} &= H(G_1(F_{\langle s_0, \ldots, s_{n-1} \rangle}), n).
\end{aligned}
$$

For any $f \in \{0,1\}^{\boldsymbol{N}}$ we let $F_f = \bigcap_{n \in \boldsymbol{N}} F_{f \restriction n}$. $F_f \neq \emptyset$ by Theorem 3.17. Moreover, F_f contains a unique element: If $x, y \in F_f$, $|x - y| \leq \operatorname{diam}(F_f) \leq \operatorname{diam}(F_{f \restriction n}) \leq 1/n$ for all $n \in \boldsymbol{N}$, $n \neq 0$, so $x = y$. Let d_f be the unique element of F_f. The function $\langle d_f \mid f \in \{0,1\}^{\boldsymbol{N}} \rangle$ is a one-to-one mapping of $\{0,1\}^{\boldsymbol{N}}$ into (although not necessarily onto) F, showing that the cardinality of F is at least $2^{\aleph_0}$. Since $F \subseteq \boldsymbol{R}$, it is also at most $2^{\aleph_0}$. Theorem 4.2 now follows from the Cantor-Bernstein theorem. □

Our next goal is to show that closed sets also behave in accordance with the Continuum Hypothesis: They are either at most countable or have cardinality $2^{\aleph_0}$. This is an immediate consequence of Theorem 4.2 and the following:

4.5 Theorem *Every uncountable closed set contains a perfect subset.*

4.6 Corollary *Every closed set of reals is either at most countable or has the cardinality* $2^{\aleph_0}$.

We give two proofs of Theorem 4.5. The first proof is quite simple; however, it uses the fact that the union of any countable system of at most countable sets is at most countable. The proof of this requires the Axiom of Choice (see Chapter 8). Our second proof does not depend on the Axiom of Choice at all; it also provides a deeper analysis of the structure of closed sets, which is of independent interest. It uses transfinite recursion (see Chapter 6).

Proof of Theorem 3.5. Let A be a set of real numbers. We call $a \in \boldsymbol{R}$ a *condensation point* of A if for every $\delta > 0$ the set of all $x \in A$ such that $|x - a| < \delta$ is uncountable (i.e., neither finite nor countable). It is obvious from the definition that any condensation point of A is an accumulation point of A. We denote the set of all condensation points of A by A^c.

(1) A^c is a closed set.

Proof. By Lemma 3.16 it suffices to show that every accumulation point of A^c belongs to A^c. If a is an accumulation point of A^c and $\delta > 0$, then there is $x \in A^c$ such that $|x - a| < \delta$. Let $\varepsilon = \delta - |x - a| > 0$. There are uncountably many $y \in A$ such that $|y - x| < \varepsilon$; but our choice of ε guarantees that if $|y - x| < \varepsilon$, then $|y - a| \leq |y - x| + |x - a| < \varepsilon + |x - a| = \delta$, so there are uncountably many $y \in A$ such that $|y - a| < \delta$, and $a \in A^c$. □

Now let F be a closed set; then $F^c \subseteq F$ (see Lemma 3.16). Our next observation is

(2) $C = F - F^c$ is at most countable.

Proof. If $a \in C$, then a is not a condensation point of F, so there is $\delta > 0$ such that $F \cap (a - \delta, a + \delta)$ is at most countable. The density of the rationals in the real line guarantees the existence of rational numbers r, s such that $a - \delta < r < a < s < a + \delta$. This shows that for each $a \in C$ there is an open interval (r, s) with rational endpoints such that $a \in F \cap (r, s)$ and $F \cap (r, s)$ is at most countable; in other words, $C \subseteq \bigcup\{F \cap (r, s) \mid r, s \in \boldsymbol{Q},\ r < s$, and $F \cap (r, s)$ is at most countable$\}$. But the set on the right side of the last inclusion is a union of an at most countable system of at most countable sets, so it, and consequently C, too, is at most countable. (Here we use the fact dependent on the Axiom of Choice.) □

(3) If F is an uncountable set, then F^c is perfect.

Proof. We know F^c is closed [observation (1)] and nonempty [observation (2)]. It remains to show that F^c has no isolated points. So assume that $a \in F^c$ is an isolated point of F^c. Then there is $\delta > 0$ such that $|x - a| < \delta$, $x \neq a$, implies $x \notin F^c$. But then, for this δ, $|x - a| < \delta$ and $x \in F$ imply $x = a$ or $x \in F - F^c$, which is an at most countable set by observation (2). Thus there are at most countably many $x \in F$ for which $|x - a| < \delta$, contradicting the assumption $a \in F^c$. □

Observation (3) completes the proof of Theorem 4.5. □

For an alternative proof of Theorem 4.5, and for its own sake, we begin a somewhat deeper analysis of the structure of closed sets. In general, closed sets differ from perfect sets in that they may have isolated points. We first prove

4.7 Theorem *Every closed set has at most countably many isolated points.*

Proof. If a is an isolated point of a closed set F, then there is $\delta > 0$ such that the only element of F in the open interval $(a - \delta, a + \delta)$ is a. Using the density of the rationals in the reals, we can find $r, s \in \boldsymbol{Q}$ such that $a - \delta < r < a < s < a + \delta$, so that a is the only element of F in the open interval (r, s) with rational endpoints. The function that assigns to each open interval with rational endpoints containing a unique element of F that element as a value, is a mapping of an at most countable set onto the set of all isolated points of F. The conclusion follows from Theorem 3.4 in Chapter 4. □

4.8 Definition Let $A \subseteq \boldsymbol{R}$; the *derived set* of A, denoted by A', is the set of all accumulation points of A.

It is easy to see that a set A is closed if and only if $A' \subseteq A$ and perfect if and only if $A \neq \emptyset$ and $A' = A$. Also, the derived set is itself a closed set. (See Exercise 4.4 for these results.) Theorem 4.7 says that for any closed F, the set $F - F'$ of its isolated points is at most countable. These considerations suggest a strategy for an investigation of the cardinality of closed sets.

Let $F \neq \emptyset$ be a closed set. If $F = F'$, F is perfect and $|F| = 2^{\aleph_0}$. Otherwise, $F = (F - F') \cup F'$ and $|F - F'| \leq \aleph_0$, so it remains to determine the cardinality of the smaller closed set F'. If $F' = \emptyset$ then $F = F - F'$ has only isolated points and $|F| \leq \aleph_0$.

If F' is perfect, then $|F'| = 2^{\aleph_0}$, so $|F| = 2^{\aleph_0}$. It is possible, however, that F' again has isolated points; for an example, consider $F = \boldsymbol{N} \cup \{m - \frac{1}{n} \mid m, n \in \boldsymbol{N},\ m \geq 1,\ n > 1\}$. In that case we can repeat the procedure and examine $F'' = (F')'$. If $F'' = \emptyset$ or F'' is perfect, we get $|F| \leq \aleph_0$ or $|F| = 2^{\aleph_0}$. But F'' may again have isolated points, and the analysis must continue. We can define an infinite sequence $\langle F_n \mid n \in \boldsymbol{N} \rangle$ by recursion:

$$F_0 = F;$$
$$F_{n+1} = (F_n)'.$$

All F_n are closed sets; if for some $n \in \boldsymbol{N}$ $F_n = \emptyset$ or F_n is perfect, we get $|F| \leq \aleph_0$ or $|F| = 2^{\aleph_0}$. Unfortunately, it may happen that all of the F_n's have isolated points. We could then define $F_\omega = \bigcap_{n \in \boldsymbol{N}} F_n$; this is again a closed set, smaller than all F_n. If F_ω is perfect, we get $|F| \geq |F_\omega| = 2^{\aleph_0}$, and if $F_\omega = \emptyset$, $F = \bigcup_{n \in \boldsymbol{N}} (F_n - F_{n+1})$ is a countable union of countable sets, which can be proved to be countable (without use of the Axiom of Choice). The problem is that even F_ω may have isolated points! (Construction of a set F for which this happens is somewhat complicated, but see Exercises 4.5 and 4.6.) So we have to consider further $F_{\omega+1} = F'_\omega$, $F_{\omega+2} = F'_{\omega+1}$, ..., and, if necessary, even $F_{\omega+\omega} = \bigcap_{n \in \boldsymbol{N}} F_{\omega+n}$ and so on. As the sets are getting smaller, there is good hope that the process will stop eventually, by reaching either a perfect set or $\emptyset$. This was the original motivation that led Georg Cantor to the development of his theory of transfinite or *ordinal numbers*.

4.9 Definition Let A be a set of reals. For every ordinal α we define, by recursion on α,

$$A^{(0)} = A;$$
$$A^{(\alpha+1)} = (A^{(\alpha)})' = \text{the derived set of } A^{(\alpha)};$$
$$A^{(\alpha)} = \bigcap_{\xi<\alpha} A^{(\xi)} \quad (\alpha \text{ a limit ordinal}).$$

4.10 Theorem *Let F be a closed set of reals. There exists an at most countable ordinal Θ such that*

(a) For every $\alpha < \Theta$, the set $F^{(\alpha)} - F^{(\alpha+1)}$ is nonempty and at most countable.
(b) $F^{(\Theta+1)} = F^{(\Theta)}$.
(c) $P = F^{(\Theta)}$ is either empty or perfect.
(d) $C = F - P$ is at most countable.

In particular, the theorem gives a decomposition of each uncountable closed set F into a perfect set P and an at most countable set C.

Proof. For each α, the set $F^{(\alpha)}$ is closed. This is easily seen by induction, because the derived set of a closed set is closed, and the intersection of closed sets is a closed set. For each α, the set $F^{(\alpha)} - F^{(\alpha+1)}$ is the set of all isolated points of $F^{(\alpha)}$, and so is at most countable by Theorem 4.7.

As long as $F^{(\alpha+1)} \neq F^{(\alpha)}$, the transfinite sequence $\langle F^{(\alpha)} \rangle$ is decreasing (i.e., $F^{(\alpha)} \supset F^{(\beta)}$ when $\alpha < \beta$). By the Axiom Schema of Replacement there exists an ordinal number Θ such that $F^{(\Theta+1)} = F^{(\Theta)}$. [To see this, assume that $F^{(\alpha)} - F^{(\alpha+1)} \neq \emptyset$ for all α and let γ be the Hartogs number of $\mathcal{P}(F)$. The function $g(\alpha) = F^{(\alpha)} - F^{(\alpha+1)}$, $\alpha < \gamma$, is a one-to-one mapping of γ into $\mathcal{P}(F)$, a contradiction.] Let Θ be the least such ordinal. [As a matter of fact, a consequence of the argument given below is that Θ is an ordinal less than ω_1.]

Let $P = F^{(\Theta)}$. We have $(F^{(\Theta)})' = F^{(\Theta)}$, so the set P is either empty or perfect. Let $C = F - P$.

Let $\langle J_0, J_1, \dots, J_n, \dots \rangle$ be a sequence of all open intervals with rational endpoints. For each $a \in C$, let α_a be the unique $\alpha < \Theta$ such that $a \in F^{(\alpha)} - F^{(\alpha+1)}$, and let

$$f(a) = \text{the least } n \text{ such that } J_n \cap F^{(\alpha_a)} = \{a\}.$$

Since a is an isolated point of $F^{(\alpha_a)}$, there is an interval J_n that has only the point a in common with $F^{(\alpha_a)}$, so f is well defined.

The function f maps C into ω. We complete the proof by showing that f is one-to-one. Thus let $a, b \in C$, and assume that $f(a) = f(b) = n$. Let us say that $\alpha_a \leq \alpha_b$. Then $F^{(\alpha_b)} \subseteq F^{(\alpha_a)}$, and $b \in J_n \cap F^{(\alpha_b)} \subseteq J_n \cap F^{(\alpha_a)}$, so $b = a$. Hence f is one-to-one and so C is at most countable. □

Every uncountable closed set has a perfect subset. This suggests that there may not be a simple example of an uncountable set of reals that does *not* have a perfect subset. And indeed, it is necessary to use the Axiom of Choice to find such an example.

4.11 Example An uncountable set without a perfect subset.

Using the Axiom of Choice, we construct two disjoint sets X and Y, both of cardinality $2^{\aleph_0}$, such that neither X nor Y have a perfect subset. We construct X and Y as the ranges of two one-to-one sequences $\langle x_\alpha \mid \alpha < 2^{\aleph_0} \rangle$, $\langle y_\alpha \mid \alpha < 2^{\aleph_0} \rangle$ to be defined below.

As proved in Chapter 5, the number of all closed sets of reals is $2^{\aleph_0}$. Thus there are only $2^{\aleph_0}$ perfect sets of reals, and we let $\langle P_\alpha \mid \alpha < 2^{\aleph_0} \rangle$ be some one-to-one mapping of $2^{\aleph_0}$ onto the set of all perfect reals. To construct $\langle x_\alpha \rangle$ and $\langle y_\alpha \rangle$, we proceed by transfinite recursion (for $\alpha < 2^{\aleph_0}$).

We choose x_0 and y_0 to be two distinct elements of P_0. Having constructed $\langle x_\xi \mid \xi < \alpha \rangle$ and $\langle y_\xi \mid \xi < \alpha \rangle$ (where $\alpha < 2^{\aleph_0}$), we observe that the set

$$P_\alpha - (\{x_\xi \mid \xi < \alpha\} \cup \{y_\xi \mid \xi < \alpha\})$$

has cardinality $2^{\aleph_0}$ (because $|P_\alpha| = 2^{\aleph_0}$ and $|\alpha| < 2^{\aleph_0}$) and we choose distinct $x_\alpha, y_\alpha \in P_\alpha$ such that both x_α and y_α differ from all x_ξ, y_ξ $(\xi < \alpha)$.

The sequences $\langle x_\xi \mid \xi < 2^{\aleph_0} \rangle$ and $\langle y_\xi \mid \xi < 2^{\aleph_0} \rangle$ are one-to-one and their ranges X and Y are disjoint sets of cardinality $2^{\aleph_0}$. Neither X nor Y have a perfect subset: if P is perfect, then $P = P_\alpha$ for some $\alpha < 2^{\aleph_0}$, and $P \cap X \neq \emptyset \neq P \cap Y$ because $x_\alpha \in P \cap X$ and $y_\alpha \in P \cap Y$. □

Exercises

4.1 Every perfect set is either an interval of the form $[a, b]$, $(-\infty, a]$, $[b, +\infty)$, or $\boldsymbol{R} = (-\infty, +\infty)$, or it is the union of two disjoint perfect sets.

4.2 Assume that the union of any countable system of countable sets is countable. Prove that every uncountable closed set contains two disjoint uncountable closed sets (a "splitting" lemma for closed sets).

4.3 Use Exercise 4.2 to give yet another proof of the fact that every uncountable closed set has cardinality $2^{\aleph_0}$.

4.4 Let A' be the derived set of $A \subseteq \boldsymbol{R}$. The set A' is closed. A is closed if and only if $A' \subseteq A$. A is perfect if and only if $A' = A$ and $A \neq \emptyset$.

4.5 The set $F = \{1\} \cup \{1 - 1/2^{n_1} \mid n_1 \geq 1\} \cup \{1 - 1/2^{n_1} - 1/2^{n_1+n_2} \mid n_1 \geq 1,\ n_2 \geq 1\}$ is closed, $F' = \{1\} \cup \{1 - 1/2^{n_1} \mid n_1 \geq 1\}$, $F'' = \{1\}$, and $F''' = \emptyset$.

4.6 The set $F = \{1\} \cup \{1 - 1/2^{n_1} - 1/2^{n_1+n_2} - \cdots - 1/2^{n_1+n_2+\cdots+n_k} \mid k \leq n_1 \text{ and } n_i \geq 1 \text{ for all } i \leq k\}$ has $F_\omega = \bigcap_{n \in \boldsymbol{N}} F_n = \{1\}$.

4.7 The decomposition of a closed set F into $P \cup C$ where $P \cap C = \emptyset$, P is perfect, and C is at most countable, is unique; i.e., if $F = P_1 \cup C_1$ where P_1 is perfect and P_1 and C_1 are disjoint and $|C_1| \leq \aleph_0$, then $P_1 = P$ and $C_1 = C$. [*Hint*: Show that P is the set of all condensation points of F.]

5. Borel Sets

Theorem 4.1 and Corollary 4.6 show that open and closed sets are either at most countable or have cardinality $2^{\aleph_0}$. It is tempting to try to prove similar results for more complicated, but still relatively simple, sets. How do we obtain such sets? One idea is to use set-theoretic operations, such as unions and intersections. Now, a union or an intersection of a finite system of open sets is again open, so no new sets can be obtained in this way, and the same holds true for closed sets. The next logical step is to consider unions and intersections of countable systems. We define *Borel sets* as those sets of reals that can be obtained from open and closed sets by means of repeatedly taking countable unions and intersections.

5.1 Definition A set $B \subseteq \boldsymbol{R}$ is *Borel* if it belongs to every system of sets $\mathcal{S} \subseteq \mathcal{P}(\boldsymbol{R})$ with the following properties.
(a) All open sets and all closed sets belong to $\mathcal{S}$.
(b) If $B_n \in \mathcal{S}$ for each $n \in \boldsymbol{N}$, then $\bigcup_{n=0}^{\infty} B_n$ and $\bigcap_{n=0}^{\infty} B_n$ belong to $\mathcal{S}$.
$\mathcal{B}$ denotes the system of all Borel sets.

In other words, $\mathcal{B} = \bigcap\{\mathcal{S} \subseteq \mathcal{P}(\boldsymbol{R}) \mid \mathcal{S} \text{ has properties (a) and (b)}\}$ is the smallest collection of sets of reals that contains all open and closed sets and is "closed" under countable unions and intersections. [Note that $\mathcal{P}(\boldsymbol{R})$ has properties (a) and (b), so the intersection is defined.] It is possible to prove some properties of Borel sets using this definition. We give one example (see also Exercises 5.1, 5.2).

The notion of a σ-algebra of subsets of S is defined in 2.17, Chapter 8.

5.2 Theorem *The collection $\mathcal{B}$ of all Borel sets is a σ-algebra of subsets of $\boldsymbol{R}$.*

Proof. Let $\mathcal{C} = \{X \subseteq \boldsymbol{R} \mid \boldsymbol{R} - X \in \mathcal{B}\}$. We show that $\mathcal{C}$ has properties (a) and (b) from Definition 5.1.

If X is open, then $\boldsymbol{R} - X$ is closed, so $\boldsymbol{R} - X \in \mathcal{B}$ and $X \in \mathcal{C}$. Similarly, X closed implies $X \in \mathcal{C}$. This shows $\mathcal{C}$ has property (a).

If $B_n \in \mathcal{C}$ for each $n \in \boldsymbol{N}$, then $\boldsymbol{R} - B_n \in \mathcal{B}$ for each $n \in \boldsymbol{N}$, hence $\boldsymbol{R} - \bigcup_{n=0}^{\infty} B_n = \bigcap_{n=0}^{\infty}(\boldsymbol{R} - B_n) \in \mathcal{B}$, and $\bigcup_{n=0}^{\infty} B_n \in \mathcal{C}$. This, and a similar argument for intersections, shows that $\mathcal{C}$ has property (b).

We conclude that $\mathcal{B} \subseteq \mathcal{C}$. Hence $X \in \mathcal{B}$ implies $X \in \mathcal{C}$, which implies $\boldsymbol{R} - X \in \mathcal{B}$. □

Theorem 5.2 shows that the collection $\mathcal{B}$ of all Borel sets is the σ-algebra generated by all open sets (or, by all closed sets); see Exercise 2.5 in Chapter 8. However, for a deeper study of Borel sets it is desirable to have some more explicit characterization of them. Let us consider which sets are Borel once more.

First, all open sets and all closed sets are Borel by clause (a). We define

$$\boldsymbol{\Sigma}_1^0 = \{B \subseteq \boldsymbol{R} \mid B \text{ is open}\};$$
$$\boldsymbol{\Pi}_1^0 = \{B \subseteq \boldsymbol{R} \mid B \text{ is closed}\}.$$

Thus $\boldsymbol{\Sigma}_1^0 \subseteq \mathcal{B}$, $\boldsymbol{\Pi}_1^0 \subseteq \mathcal{B}$. Next, countable unions and intersections of open sets must be Borel by clause (b). Countable unions of open sets are open (see the remark following Theorem 3.11) and do not produce new sets; so we define

$$\boldsymbol{\Pi}_2^0 = \{B \subseteq \boldsymbol{R} \mid B = \bigcap_{n=0}^{\infty} B_n \text{ where each } B_n \in \boldsymbol{\Sigma}_1^0\}.$$

Clearly $\boldsymbol{\Sigma}_1^0 \subseteq \boldsymbol{\Pi}_2^0$ and it is possible to prove that the inclusion is proper (see Exercise 5.3), so we obtain some new Borel sets in this way. Similarly, one defines

$$\boldsymbol{\Sigma}_2^0 = \{B \subseteq \boldsymbol{R} \mid B = \bigcup_{n=0}^{\infty} B_n \text{ where each } B_n \in \boldsymbol{\Pi}_1^0\}$$

and proves that $\boldsymbol{\Pi}_1^0 \subset \boldsymbol{\Sigma}_2^0 \subseteq \mathcal{B}$.

It is also true that $\boldsymbol{\Pi}_1^0 \subset \boldsymbol{\Pi}_2^0$ and $\boldsymbol{\Sigma}_1^0 \subset \boldsymbol{\Sigma}_2^0$ (see Exercise 5.3). The process can be continued recursively. We define

$$\boldsymbol{\Sigma}_3^0 = \{B \subseteq \boldsymbol{R} \mid B = \bigcup_{n=0}^{\infty} B_n \text{ where each } B_n \in \boldsymbol{\Pi}_2^0\}$$

and similarly

$$\boldsymbol{\Pi}_3^0 = \{B \subseteq \boldsymbol{R} \mid B = \bigcap_{n=0}^{\infty} B_n \text{ where each } B_n \in \boldsymbol{\Sigma}_2^0\}.$$

In general, for any $k > 0$,

$$\boldsymbol{\Sigma}^0_{k+1} = \{B \subseteq \boldsymbol{R} \mid B = \bigcup_{n=0}^{\infty} B_n \text{ where each } B_n \in \boldsymbol{\Pi}^0_k\}$$

and

$$\boldsymbol{\Pi}^0_{k+1} = \{B \subseteq \boldsymbol{R} \mid B = \bigcap_{n=0}^{\infty} B_n \text{ where each } B_n \in \boldsymbol{\Sigma}^0_k\}.$$

By induction, $\boldsymbol{\Sigma}^0_k \subseteq \mathcal{B}$ and $\boldsymbol{\Pi}^0_k \subseteq \mathcal{B}$, for all $k \in \boldsymbol{N}$, $k > 0$. It can also be shown (with the help of the Axiom of Choice; we omit the proof) that

$$\boldsymbol{\Sigma}^0_k \cup \boldsymbol{\Pi}^0_k \subset \boldsymbol{\Sigma}^0_{k+1} \text{ and } \boldsymbol{\Sigma}^0_k \cup \boldsymbol{\Pi}^0_k \subset \boldsymbol{\Pi}^0_{k+1},$$

so the hierarchy produces new Borel sets at each level. One might hope that all Borel sets are obtained in this fashion, that is, that $\mathcal{B} = \bigcup_{k=1}^{\infty} \boldsymbol{\Sigma}^0_k = \bigcup_{k=1}^{\infty} \boldsymbol{\Pi}^0_k$, but it is not so. There are sequences $\langle B_n \mid n \in \boldsymbol{N}\rangle$ such that $B_n \in \boldsymbol{\Sigma}^0_{n+1} \subseteq \mathcal{B}$ for each $n \in \boldsymbol{N}$, but $\bigcup_{n=0}^{\infty} B_n$ or $\bigcap_{n=0}^{\infty} B_n$ does not belong to any $\boldsymbol{\Sigma}^0_k$ or $\boldsymbol{\Pi}^0_k$ (although it is of course Borel). We see again the need for a continuation of the construction of the hierarchy into "transfinite."

5.3 Definition For all ordinals $\alpha < \omega_1$ we define collections of sets of reals $\boldsymbol{\Sigma}^0_\alpha$ and $\boldsymbol{\Pi}^0_\alpha$ as follows:

(a) $\boldsymbol{\Sigma}^0_1 = \{B \subseteq \boldsymbol{R} \mid B \text{ is open}\}$,
$\boldsymbol{\Pi}^0_1 = \{B \subseteq \boldsymbol{R} \mid B \text{ is closed}\}$.

(b) $\boldsymbol{\Sigma}^0_{\alpha+1} = \{B \subseteq \boldsymbol{R} \mid B = \bigcup_{n=0}^{\infty} B_n \text{ where each } B_n \in \boldsymbol{\Pi}^0_\alpha\}$,
$\boldsymbol{\Pi}^0_{\alpha+1} = \{B \subseteq \boldsymbol{R} \mid B = \bigcap_{n=0}^{\infty} B_n \text{ where each } B_n \in \boldsymbol{\Sigma}^0_\alpha\}$.

(c) $\boldsymbol{\Sigma}^0_\alpha = \{B \subseteq \boldsymbol{R} \mid B = \bigcup_{n=0}^{\infty} B_n \text{ where each } B_n \in \boldsymbol{\Pi}^0_\beta \text{ for some } \beta < \alpha\}$,
$\boldsymbol{\Pi}^0_\alpha = \{B \subseteq \boldsymbol{R} \mid B = \bigcap_{n=0}^{\infty} B_n \text{ where each } B_n \in \boldsymbol{\Sigma}^0_\beta \text{ for some } \beta < \alpha\}$ (α a limit ordinal).

[In view of the next lemma, one could use clause (c) for successor ordinals as well, instead of (b).]

5.4 Lemma

(i) For every $\alpha < \omega_1$,
$\boldsymbol{\Sigma}^0_\alpha \subseteq \boldsymbol{\Pi}^0_{\alpha+1}$ *and* $\boldsymbol{\Pi}^0_\alpha \subseteq \boldsymbol{\Sigma}^0_{\alpha+1}$.

(ii) For every $\alpha < \omega_1$,
$\boldsymbol{\Sigma}^0_\alpha \subseteq \boldsymbol{\Sigma}^0_{\alpha+1}$ *and* $\boldsymbol{\Pi}^0_\alpha \subseteq \boldsymbol{\Pi}^0_{\alpha+1}$.

(iii) If $B \in \boldsymbol{\Sigma}^0_\alpha$, *then* $\boldsymbol{R} - B \in \boldsymbol{\Pi}^0_\alpha$; *if* $B \in \boldsymbol{\Pi}^0_\alpha$, *then* $\boldsymbol{R} - B \in \boldsymbol{\Sigma}^0_\alpha$.

(iv) If $B_n \in \boldsymbol{\Sigma}^0_\alpha$ *for each* $n \in \boldsymbol{N}$, *then* $\bigcup_{n=0}^{\infty} B_n \in \boldsymbol{\Sigma}^0_\alpha$; *if* $B_n \in \boldsymbol{\Pi}^0_\alpha$ *for each* $n \in \boldsymbol{N}$, *then* $\bigcap_{n=0}^{\infty} B_n \in \boldsymbol{\Pi}^0_\alpha$.

(v) If $\alpha < \beta$ *then* $\boldsymbol{\Sigma}^0_\alpha \subseteq \boldsymbol{\Sigma}^0_\beta$, $\boldsymbol{\Sigma}^0_\alpha \subseteq \boldsymbol{\Pi}^0_\beta$, *and* $\boldsymbol{\Pi}^0_\alpha \subseteq \boldsymbol{\Pi}^0_\beta$, $\boldsymbol{\Pi}^0_\alpha \subseteq \boldsymbol{\Sigma}^0_\beta$.

Proof.

(i) Trivial: in Definition 5.3(b), let $B_n = B$ for all n.

(v) Every open set is in $\mathbf{\Sigma}_2^0$ (see Exercise 5.3), and similarly, every closed set is in $\mathbf{\Pi}_2^0$. Hence $\mathbf{\Sigma}_1^0 \subseteq \mathbf{\Sigma}_2^0$ and $\mathbf{\Pi}_1^0 \subseteq \mathbf{\Pi}_2^0$. Also, $\mathbf{\Pi}_1^0 \subseteq \mathbf{\Sigma}_2^0$ and $\mathbf{\Sigma}_1^0 \subseteq \mathbf{\Pi}_2^0$, by (i). Then we proceed to prove that the assertion holds for every $\alpha < \beta$, by induction on β.

(ii) Follows from (v).

(iii) By induction on α.

(iv) This innocuous statement requires the Countable Axiom of Choice: For each n, if $B_n \in \mathbf{\Sigma}_\alpha^0$, then $B_n = \bigcup_{m=0}^\infty B_{mn}$ for some countable collection $\{B_{mn} \mid m \in \boldsymbol{N}\}$ of sets $B_{mn} \in \bigcup_{\gamma<\alpha} \mathbf{\Pi}_\gamma^0$. For each n, we choose one such collection (along with its enumeration $\langle B_{mn} \mid m \in \boldsymbol{N}\rangle$), and then

$$\bigcup_{n=0}^\infty B_n = \bigcup_{n=0}^\infty \bigcup_{m=0}^\infty B_{mn}$$

is the union of the countable collection $\{B_{mn} \mid (n,m) \in \boldsymbol{N} \times \boldsymbol{N}\}$ and thus in $\mathbf{\Sigma}_\alpha^0$.

The proof for $\mathbf{\Pi}_\alpha^0$ is similar.

□

It is clear (by induction on α) that each $\mathbf{\Sigma}_\alpha^0$ and each $\mathbf{\Pi}_\alpha^0$ consists of Borel sets. What is important, however, is that this hierarchy exhausts *all* Borel sets.

5.5 Theorem *A set of reals is a Borel set if and only if it belongs to $\mathbf{\Sigma}_\alpha^0$ for some $\alpha < \omega_1$ (if and only if it belongs to $\mathbf{\Pi}_\alpha^0$ for some $\alpha < \omega_1$).*

Proof. It suffices to show that the set

$$\mathcal{S} = \bigcup_{\alpha<\omega_1} \mathbf{\Sigma}_\alpha^0 = \bigcup_{\alpha<\omega_1} \mathbf{\Pi}_\alpha^0$$

is closed under countable unions and intersections. Thus let $\{B_n \mid n \in \boldsymbol{N}\}$ be a countable collection of sets in $\mathcal{S}$. We show that, e.g., $\bigcup_{n=0}^\infty B_n$ is in $\mathcal{S}$.

For each n let α_n be the least α such that $B_n \in \mathbf{\Pi}_\alpha^0$. The set $\{\alpha_n \mid n \in \boldsymbol{N}\}$ is an at most countable set of ordinals less than ω_1 and thus (by the Countable Axiom of Choice) its supremum $\alpha = \sup\{\alpha_n \mid n \in \boldsymbol{N}\} = \bigcup\{\alpha_n \mid n \in \boldsymbol{N}\}$ is an ordinal less than ω_1. By Lemma 5.4 we have $B_n \in \mathbf{\Pi}_\alpha^0$ for each n. Consequently, $\bigcup_{n=0}^\infty B_n$ belongs to $\mathbf{\Sigma}_{\alpha+1}^0$ and hence to $\mathcal{S}$. □

We note, without proof, that Theorem 4.5 can be extended to all Borel sets: Every uncountable Borel set contains a perfect subset. In particular, every uncountable Borel set has cardinality $2^{\aleph_0}$. In Chapter 8 we mentioned the σ-algebra $\mathfrak{M}$ of Lebesgue measurable sets of reals; of course, $\mathcal{B} \subseteq \mathfrak{M}$, but there exist also Lebesgue measurable sets which are not Borel. In fact, one cannot prove in Zermelo-Fraenkel set theory that all uncountable Lebesgue measurable sets have cardinality $2^{\aleph_0}$.

The construction of the Borel hierarchy is just a special case of a very general procedure. We observe that the infinitary unions and intersections used in it are, in some sense, generalizations of the usual operations $\cup$ and $\cap$. We define: F is a (countably) *infinitary operation* on A if F is a function on a subset of $A^{\boldsymbol{N}}$ (the set of all infinite sequences of elements of A) into A, and introduce *structures with infinitary operations.* This can be done formally by extending the definition of type (see Section 5 in Chapter 3) by allowing $f_j \in \boldsymbol{N} \cup \{\boldsymbol{N}\}$ and postulating that F_j is an infinitary operation whenever $f_j = \boldsymbol{N}$. The definition of a set $B \subseteq A$ being *closed* remains essentially unchanged; if $f_j = \boldsymbol{N}$ we require $F_j(\langle a_i \mid i \in \boldsymbol{N}\rangle) \in B$ for all $\langle a_i \mid i \in \boldsymbol{N}\rangle$ such that $a_i \in B$ for all $i \in \boldsymbol{N}$ and F_j is defined. The closure $\overline{C}$ of a set $C \subseteq A$ is still the least closed set containing all elements of C.

In this terminology, the system of Borel sets is just the closure of the system of all open and closed sets under the infinitary union and intersection. More precisely, we let $\mathfrak{A} = \langle A, F_1, F_2\rangle$ where $A = \mathcal{P}(\boldsymbol{R})$, $F_1(\langle B_i \mid i \in \boldsymbol{N}\rangle) = \bigcup_{i=0}^{\infty} B_i$ and $F_2(\langle B_i \mid i \in \boldsymbol{N}\rangle) = \bigcap_{i=0}^{\infty} B_i$, and $C = \{B \subseteq \boldsymbol{R} \mid B \text{ is open or } B \text{ is closed}\}$. Then $\mathcal{B} = \overline{C}$.

Exercise 5.9 provides another example of a structure with infinitary operations and a closed subset of some mathematical interest.

One important general question about the closure is that of its cardinality. We proved the simplest result providing an answer in Theorem 3.14 of Chapter 4. There it was assumed that C is at most countable, and all operations are finitary. We now generalize this, first, to arbitrary C, and then to structures with infinitary operations.

5.6 Theorem *Let $\mathfrak{A} = (A, \langle R_0, \ldots, R_{m-1}\rangle, \langle F_0, \ldots, F_{n-1}\rangle)$ be a structure and let $C \subseteq A$. Let $\overline{C}$ be the closure of C. If C is finite, then $\overline{C}$ is at most countable; if C is infinite, then $|\overline{C}| = |C|$.*

Proof. We proved in Theorem 5.10 in Chapter 3 that $\overline{C} = \bigcup_{i=0}^{\infty} C_i$ where $C_0 = C$ and $C_{i+1} = C_i \cup F_0[C_i^{f_0}] \cup \cdots \cup F_{n-1}[C_i^{f_{n-1}}]$. If C is finite, then each C_i is finite, and $|\overline{C}| \le \aleph_0$. If C is infinite and $|C| = \aleph_\alpha$; then we prove by induction that $|C_k| = \aleph_\alpha$ for all k: Assume that it is true for C_i; then $|C_i^{f_j}| \le \aleph_\alpha^{f_j} = \aleph_\alpha$, $|F_j[C_i^{f_j}]| \le \aleph_\alpha$, and $|C_{i+1}| = \aleph_\alpha$. Hence $|\overline{C}| = \aleph_0 \cdot \aleph_\alpha = \aleph_\alpha$. □

When $\mathfrak{A}$ is a structure with infinitary operations, Theorem 5.6 can be generalized but the cardinality estimate is different. Let $\mathfrak{A}$ be a structure, and for simplicity let us assume that $\mathfrak{A}$ has just one operation F and that F is an infinitary operation (the result can of course be generalized to structures with more than one operation). $\mathfrak{A}$ has universe A, and F is a function on a subset of A^ω into A, where A^ω is the set of all sequences $\langle a_n \mid n \in \boldsymbol{N}\rangle$ with values in A.

5.7 Theorem *Let $\mathfrak{A} = \langle A, F\rangle$ be a structure where F is a function on a subset of A^ω into A, and let $C \subseteq A$. Let $\overline{C}$ be the closure of C in $\mathfrak{A}$, i.e.,*

$$\overline{C} = \bigcap\{X \subseteq A \mid F[X^\omega] \subseteq X \text{ and } C \subseteq X\}.$$

If C has at least two elements, then $|\overline{C}| \leq |C|^{\aleph_0}$.

Proof. We construct an ω_1-sequence

$$C_0 \subseteq C_1 \subseteq \cdots \subseteq C_\alpha \subseteq \cdots \quad (\alpha < \omega_1)$$

of subsets of A as follows:

$$\begin{aligned} C_0 &= C; \\ C_{\alpha+1} &= C_\alpha \cup F[C_\alpha^\omega]; \\ C_\alpha &= \bigcup_{\xi<\alpha} C_\xi \quad \text{if } \alpha \text{ is a limit ordinal.} \end{aligned}$$

Let $\overline{C}$ be the closure of C in $\mathfrak{A}$ and let $D = \bigcup_{\alpha<\omega_1} C_\alpha$. By induction on α we have $C_\alpha \subseteq \overline{C}$, so $D \subseteq \overline{C}$. On the other hand, if $\langle a_n \mid n \in \boldsymbol{N}\rangle$ is a sequence in D, then $\langle a_n \mid n \in \boldsymbol{N}\rangle \in C_\alpha$ for some $\alpha < \omega_1$: for each n let ξ_n be the least ξ such that $a_n \in C_\xi$, and let $\alpha = \sup\{\xi_n \mid n \in \boldsymbol{N}\}$. And if $\langle a_n \mid n \in \boldsymbol{N}\rangle \in \operatorname{dom} F$ then $F(\langle a_n \mid n \in \boldsymbol{N}\rangle) \in C_{\alpha+1}$; hence $F[D^\omega] \subseteq D$, so $D \supseteq \overline{C}$.

Thus $\overline{C} = \bigcup_{\alpha<\omega_1} C_\alpha$. To estimate the size of $\overline{C}$, we first prove, by induction on α, that $|C_\alpha| \leq |C|^{\aleph_0} \cdot 2^{\aleph_0}$. This is certainly true for $\alpha = 0$. If the estimate is true for α, then

$$|C_{\alpha+1}| \leq |C_\alpha| + |C_\alpha|^{\aleph_0} \leq (|C|^{\aleph_0} \cdot 2^{\aleph_0})^{\aleph_0} = |C|^{\aleph_0} \cdot 2^{\aleph_0}.$$

If α is a limit ordinal, and if the estimate is true for all $\xi < \alpha$, then

$$|C_\alpha| \leq |\alpha| \cdot |C|^{\aleph_0} \cdot 2^{\aleph_0} = |C|^{\aleph_0} \cdot 2^{\aleph_0}$$

(because $|\alpha| \leq \aleph_0$). Finally, we have

$$|\overline{C}| \leq \aleph_1 \cdot |C|^{\aleph_0} \cdot 2^{\aleph_0} = |C|^{\aleph_0} \cdot 2^{\aleph_0},$$

because $\aleph_1 \cdot 2^{\aleph_0} = 2^{\aleph_0}$. And if $|C| \geq 2$, then $|C|^{\aleph_0} \geq 2^{\aleph_0}$, and we have $|\overline{C}| \leq |C|^{\aleph_0}$. □

Exercises

5.1 If X is Borel and $a \in \boldsymbol{R}$ then $X + a = \{x + a \mid x \in X\}$ is Borel. [*Hint*: Imitate the proof of Theorem 5.2.]

5.2 If X is Borel and $f : \boldsymbol{R} \to \boldsymbol{R}$ is a continuous function, then $f^{-1}[X]$ is Borel.

5.3 Prove that every open interval is a union of a countable system of closed intervals and conclude from this fact that $\mathbf{\Sigma}_1^0 \subseteq \mathbf{\Sigma}_2^0$. Since $\mathbf{\Pi}_1^0 \subseteq \mathbf{\Sigma}_2^0$, conclude that $\mathbf{\Sigma}_1^0 \subset \mathbf{\Sigma}_2^0$ and $\mathbf{\Pi}_1^0 \subset \mathbf{\Sigma}_2^0$. This yields immediately that $\mathbf{\Pi}_1^0 \subset \mathbf{\Pi}_2^0$ and $\mathbf{\Sigma}_1^0 \subset \mathbf{\Pi}_2^0$.

5.4 Prove by induction that $\mathbf{\Sigma}_n^0 \cup \mathbf{\Pi}_n^0 \subseteq \mathbf{\Sigma}_{n+1}^0 \cap \mathbf{\Pi}_{n+1}^0$ for all $n \in \boldsymbol{N}$.

5.5 Prove that for each α, both $\mathbf{\Sigma}_\alpha^0$ and $\mathbf{\Pi}_\alpha^0$ are closed under finite unions and intersections, i.e., if B_1 and $B_2 \in \mathbf{\Sigma}_\alpha^0$, then $B_1 \cup B_2 \in \mathbf{\Sigma}_\alpha^0$ and $B_1 \cap B_2 \in \mathbf{\Sigma}_\alpha^0$, and similarly for $\mathbf{\Pi}_\alpha^0$.

5.6 Let $f : \boldsymbol{R} \to \boldsymbol{R}$ be a continuous function. If $B \in \mathbf{\Sigma}_\alpha^0$, then $f^{-1}[B] \in \mathbf{\Sigma}_\alpha^0$. Similarly for $\mathbf{\Pi}_\alpha^0$.

5.7 Show that the cardinality of the set $\mathcal{B}$ of all Borel sets is $2^{\aleph_0}$.

5.8 Prove that the Borel sets are the closure of the set of all open intervals with rational endpoints under infinitary unions and intersections. This shows that in a structure with infinitary operations, the closure of a countable set may be uncountable.

5.9 Let Fn be the set of all functions from a subset of $\boldsymbol{R}$ into $\boldsymbol{R}$. The infinitary operation lim (limit) is defined on Fn as follows: $\lim(\langle f_i \mid i \in \boldsymbol{N}\rangle) = f$ where $f(x) = \lim_{i\to\infty} f_i(x)$ whenever the limit on the right side exists, and is undefined otherwise. The elements of the closure of the set of all continuous functions under lim are called *Baire functions*. Show that Baire functions need not be continuous. In particular, show that the characteristic functions of integers and rationals are Baire functions.

5.10 For $\alpha < \omega_1$, we define functions (on a subset of $\boldsymbol{R}$ into $\boldsymbol{R}$) of *class* α as follows: the functions of class 0 are all the continuous functions; f is of class α if $f = \lim_{n\to\infty} f_n$ where each f_n is of some class $< \alpha$. Show that Baire functions are exactly the functions of class α for some $\alpha < \omega_1$.

5.11 Show that the cardinality of the set of all Baire functions is $2^{\aleph_0}$.

Chapter 11

Filters and Ultrafilters

1. Filters and Ideals

This chapter studies deeper properties of sets in general. Unlike earlier chapters, we do not restrict ourselves to sets of natural or real numbers, and neither do we concern ourselves only with cardinals and ordinals. Here we deal with abstract collections of sets and investigate their properties. We introduce several concepts that have become of fundamental importance in applications of set theory. It is interesting that a deeper investigation of such concepts leads to the theory of large cardinals. Large cardinals are studied in Chapter 13.

We start by introducing a *filter* of sets. Filters play an important role in many mathematical disciplines.

1.1 Definition Let S be a nonempty set. A *filter* on S is a collection F of subsets of S that satisfies the following conditions:
(a) $S \in F$ and $\emptyset \notin F$.
(b) If $X \in F$ and $Y \in F$, then $X \cap Y \in F$.
(c) If $X \in F$ and $X \subseteq Y \subseteq S$, then $Y \in F$.

A trivial example of a filter on S is the collection $F = \{S\}$ that consists only of the set S itself. This *trivial filter* on S is the smallest filter on S; it is included in every filter on S.

Let A be a nonempty subset of S, and let us consider the collection

$$F = \{X \subseteq S \mid X \supseteq A\}.$$

The collection F so defined satisfies Definition 1.1 and so is a filter on S. It is called the *principal filter* on S *generated* by A.

If in this example, the set A has just one element a, i.e., $A = \{a\}$ where $a \in S$, then the principal filter

$$F = \{X \subseteq S \mid a \in X\}$$

is *maximal*: There is no filter F' on S such that $F' \supset F$ (because if $X \in F' - F$ then $a \notin X$, but $\{a\} \in F'$ as well, and so $\emptyset = X \cap \{a\} \in F'$, a contradiction).

We shall prove shortly that there are maximal filters that are not principal.

To give an example of a *nonprincipal* filter, let S be an infinite set, and let

$$F = \{X \subseteq S \mid S - X \text{ is finite}\}. \tag{1.2}$$

F is the filter of all *cofinite* subsets of S. (X is a *cofinite* subset of S if $S - X$ is finite.) F is a filter because the intersection of two cofinite subsets of S is a cofinite subset of S. F is not a principal filter because whenever $A \in F$ then there is a proper subset X of A such that $X \in F$ (let X be any cofinite subset of A, $X \neq A$).

1.3 Definition Let S be a nonempty set. An *ideal* on S is a collection I of subsets of S that satisfies the following conditions:
(a) $\emptyset \in I$ and $S \notin I$.
(b) If $X \in I$ and $Y \in I$, then $X \cup Y \in I$.
(c) If $Y \in I$ and $X \subseteq Y$, then $X \in I$.

The *trivial* ideal on S is the ideal $\{\emptyset\}$. A *principal* ideal is an ideal of the form

$$I = \{X \mid X \subseteq A\} \tag{1.4}$$

where $A \subseteq S$.

To see how filters and ideals are related, note that if F is a filter on S then

$$I = \{S - X \mid X \in F\} \tag{1.5}$$

is an ideal, and vice versa, if I is an ideal, then

$$F = \{S - X \mid X \in I\} \tag{1.6}$$

is a filter. The two objects related by (1.5) and (1.6) are called *dual* to each other.

The filter of cofinite subsets of S is the dual of the ideal of finite subsets of S. In Exercises 1.2 and 1.3 the reader can find other examples of nonprincipal ideals.

We recall (Definition 3.3 in Chapter 10) that a nonempty collection G has the *finite intersection property* if every nonempty finite subcollection $\{X_1, \ldots, X_n\}$ of G has nonempty intersection $X_1 \cap \cdots \cap X_n \neq \emptyset$.

It follows from (a) and (b) of Definition 1.1 that every filter has the finite intersection property. In fact, if G is any subcollection of a filter F, then G has the finite intersection property.

Conversely, every set that has the finite intersection property is a subcollection of some filter.

1.7 Lemma *Let $G \neq \emptyset$ be a collection of subsets of S and let G have the finite intersection property. Then there is a filter F on S such that $G \subseteq F$.*

Proof. Let F be the collection of all subsets X of S with the property that there is a finite subset $\{X_1, \dots, X_n\}$ of G such that

$$X_1 \cap \dots \cap X_n \subseteq X.$$

Clearly, S itself is in F, and $\emptyset$ is not in F because G has the finite intersection property. The condition (c) of Definition 1.1 is certainly satisfied by F. As for the condition (b), if $X \supseteq X_1 \cap \dots \cap X_n$ for some $X_1, \dots, X_n \in G$, and if $Y \supseteq Y_1 \cap \dots \cap Y_m$ for some $Y_1, \dots, Y_m \in G$, then $X \cap Y \supseteq X_1 \cap \dots \cap X_n \cap Y_1 \cap \dots \cap Y_m$, so $X \cap Y \in F$. Thus F is a filter. □

The filter F constructed in Lemma 1.7 is the smallest filter on S that extends the collection G. We say that G *generates* the filter F. See Exercise 1.6.

1.8 Example Let S be a Euclidean space and let a be a point in S. Consider the collection G of all open sets U in S such that $a \in U$. Then G has the finite intersection property and hence it generates a filter F on S. F is the *neighborhood filter* of a.

1.9 Example. Density. Let A be a set of natural numbers. For each $n \in \boldsymbol{N}$, let $A(n) = |A \cap \{0, ..., n-1\}|$ denote the number of elements of A that are smaller than n. The limit

$$d(A) = \lim_{n \to \infty} \frac{A(n)}{n},$$

if it exists, is called the *density* of A. For example, the set of all even numbers has density 1/2 (Exercise 1.7). Every finite set has density 0, and there exist infinite sets that have density 0. (For example, the set $\{2^n \mid n \in \boldsymbol{N}\}$ of all powers of 2 — Exercise 1.8.)

Let A and B be sets of natural numbers. If $A \subseteq B$ then $A(n) \leq B(n)$ for all n, and so if both A and B have density then $d(A) \leq d(B)$. In particular, if $d(B) = 0$ then $d(A) = 0$.

Also, for every n, $(A \cup B)(n) \leq A(n) + B(n)$, and if A and B are disjoint then $(A \cup B)(n) = A(n) + B(n)$. Hence $d(A \cup B) \leq d(A) + d(B)$ (provided that the densities exist), and $d(A \cup B) = d(A) + d(B)$ if A and B are disjoint. If both $d(A)$ and $d(B)$ are zero, then $d(A \cup B) = 0$.

This gives us an example of an ideal on $\boldsymbol{N}$, the ideal of sets of density 0:

$$I_d = \{A \mid d(A) = 0\}.$$

We have $\emptyset \in I_d$ and $\boldsymbol{N} \notin I_d$ (because $d(\boldsymbol{N}) = 1$). If $A \subseteq B$ and $B \in I_d$ then $A \in I_d$, and if both $A \in I_d$ and $B \in I_d$ then $A \cup B \in I_d$. Thus I_d is an ideal. As noted above, I_d contains all finite sets, but also some infinite sets.

As a final remark, not every set $A \subset \boldsymbol{N}$ has density. It is not difficult to construct a set A such that the limit $\lim_{n \to \infty} A(n)/n$ does not exist (Exercise 1.9).

We conclude this section with the introduction of an important concept that has many applications in analysis:

1.10 Definition A *measure* on a set S is a real-valued function m defined on $\mathcal{P}(S)$ that satisfies the following conditions:
(a) $m(\emptyset) = 0$, $m(S) > 0$.
(b) If $A \subseteq B$ then $m(A) \leq m(B)$.
(c) If A and B are disjoint then $m(A \cup B) = m(A) + m(B)$.
(Notice that the density function on $\mathcal{P}(\boldsymbol{N})$ satisfies the conditions, except that it is not defined for *all* subsets of $\boldsymbol{N}$.)

It follows that $m(A) \geq 0$ for every A, and that $m(S - A) = m(S) - m(A)$. Property (c) is called *finite additivity*, and clearly

$$m(A_1 \cup \cdots \cup A_n) = m(A_1) + \cdots + m(A_n)$$

for any disjoint finite collection $\{A_1, \ldots, A_n\}$.

At this point, we can only give two trivial examples of measures:

1.11 Example Let S be a finite set, and let $m(A) = |A|$ for $A \subseteq S$. This is the *counting measure* on S.

1.12 Example Let S be a nonempty set and let $a \in S$. Let $m(A) = 1$ if $a \in A$, and $m(A) = 0$ if $a \notin A$. (A *trivial* measure.)

In the next section, we construct nontrivial measures on $\boldsymbol{N}$.

Exercises

1.1 If S is a finite nonempty set, then every filter on S is a principal filter.

1.2 Let S be an uncountable set, and let I be the collection of all $X \subset S$ such that $|X| \leq \aleph_0$. I is a nonprincipal ideal on S.

1.3 Let S be an infinite set and let $Z \subseteq S$ be such that both Z and $S - Z$ are infinite. The collection $I = \{X \subseteq S \mid X - Z \text{ is finite}\}$ is a nonprincipal ideal.

1.4 If a set $A \subseteq S$ has more than one element, then the principal filter generated by A is not maximal.

1.5 If $\mathcal{F}$ is a nonempty set of filters on S, then $\bigcap\{F \mid F \in \mathcal{F}\}$ is a filter on S.

1.6 The filter constructed in the proof of Lemma 1.7 is the smallest filter on S that includes the collection G.

1.7 Let A be the set of all natural numbers that are divisible by a given number $p > 0$. Show that $d(A) = 1/p$.

1.8 Prove that the set $\{2^n \mid n \in \boldsymbol{N}\}$ has density 0.

1.9 Construct a set A of natural numbers such that

$$\limsup_{n\to\infty} A(n)/n = 1 \text{ and } \liminf_{n\to\infty} A(n)/n = 0.$$

2. Ultrafilters

2.1 Definition A filter U on S is an *ultrafilter* if for every $X \subseteq S$, either $X \in U$ or $S - X \in U$.

The dual notion is a *prime ideal:*

2.2 Definition An ideal I on S is a *prime ideal* if for every $X \subseteq S$, either $X \in I$ or $S - X \in I$.

2.3 Lemma *A filter F on S is an ultrafilter if and only if it is a maximal filter on S.*

Proof. If F is an ultrafilter, then it is a maximal filter, because if $F' \supset F$ is a filter on S, then there is $X \subseteq S$ in $F' - F$. But because F is an ultrafilter, $S - X$ is in F, and hence in F', and we have $\emptyset = X \cap (S - X) \in F'$, a contradiction.

Conversely, let F be a filter but not an ultrafilter. There is some $X \subseteq S$ such that neither X nor $S - X$ is in F. Let $G = F \cup \{X\}$. We claim that G has the finite intersection property.

If $X_1, \ldots, X_n$ are in F, then $Y = X_1 \cap \cdots \cap X_n \in F$, and $Y \cap X \neq \emptyset$ because otherwise $(S - X) \supseteq Y$ and so $S - X \in F$, contrary to the assumption. Hence $X_1 \cap \cdots \cap X_n \cap X \neq \emptyset$, which means that $G = F \cup \{X\}$ has the finite intersection property.

Thus there is a filter F' on S such that $F' \supseteq F \cup \{X\}$. But that means that F is not a maximal filter. □

We have seen earlier that there are principal filters that are maximal. In other words, there exist principal ultrafilters. Do there exist nonprincipal ultrafilters?

Let S be an infinite set, and let F be the filter of cofinite subsets of S. If U is an ultrafilter and U extends F, then U cannot be principal. Thus, to find a nonprincipal ultrafilter it is enough to find an ultrafilter that extends the filter of cofinite sets.

Conversely, if U is a nonprincipal ultrafilter, then it extends the filter of cofinite sets because every $X \in U$ is infinite (Exercise 2.2).

The next theorem states that *any* filter can be extended to an ultrafilter. However, the proof uses the Axiom of Choice, and in fact it is known that the theorem cannot be proved in Zermelo-Fraenkel set theory alone.

2.4 Theorem *Every filter on a set S can be extended to an ultrafilter on S.*

The proof uses Zorn's Lemma. In order to apply it we need the following fact:

2.5 Lemma *If C is a set of filters on S and if for every $F_1, F_2 \in C$, either $F_1 \subseteq F_2$ or $F_2 \subseteq F_1$, then the union of C is also a filter on S.*

Proof. This is a matter of simple verification of (a), (b), and (c) in Definition 1.1. □

Proof of Theorem 2.4. Let F_0 be a filter on S; we find a filter $F \supseteq F_0$ that is maximal.

Let P be the set of all filters F on S such that $F \supseteq F_0$, and let us consider the partially ordered set $(P, \subset)$. By Lemma 2.5, every chain C in P has an upper bound, namely $\bigcup C$. Thus Zorn's Lemma is applicable, and $(P, \subset)$ has a maximal element U. Clearly, U is a maximal filter on S and $U \supseteq F_0$. By Lemma 2.3, U is an ultrafilter. □

There is a natural relation between ultrafilters and measures. Let us call a measure m on S *two-valued* if it only takes values 0 and 1: for every $A \subseteq S$, either $m(A) = 0$ or $m(A) = 1$.

2.6 Theorem *(a) If m is a two-valued measure on S then $U = \{A \subseteq S \mid m(A) = 1\}$ is an ultrafilter.*

(b) If U is an ultrafilter on S, then the function m on $\mathcal{P}(S)$ defined by $m(A) = 1$ if $A \in U$ and $m(A) = 0$ if $A \notin U$ is a two-valued measure on S.

Proof. Compare the definitions of ultrafilter and measure. Note that if A and B are disjoint, then at most one of them is in an ultrafilter (or has measure 1). □

We now present one of many applications of ultrafilters. This particular application is a generalization of the concept of limits of sequences of real numbers.

2.7 Definition Let U be an ultrafilter on $\boldsymbol{N}$, and let $\langle a_n \rangle_{n=0}^{\infty}$ be a bounded sequence of real numbers. We say that a real number a is the *U-limit* of the sequence,

$$a = \lim_U a_n,$$

if for every $\varepsilon > 0$, $\{n \mid |a_n - a| < \varepsilon\} \in U$.

First we observe that if a U-limit exists then it is unique. For, assume that $a < b$ and both a and b are U-limits of $\langle a_n \rangle_{n=0}^{\infty}$. Let $\varepsilon = (b-a)/2$. Then the sets $\{n \mid |a_n - a| < \varepsilon\}$ and $\{n \mid |a_n - b| < \varepsilon\}$ are disjoint and so cannot both be in U.

If U is a principal ultrafilter, $U = \{A \mid n_0 \in A\}$ for some n_0, then $\lim_U a_n = a_{n_0}$, for any sequence $\langle a_n \rangle_{n=0}^{\infty}$. This is because for every $\varepsilon > 0$, $\{n \mid |a_n - a_{n_0}| < \varepsilon\} \supseteq \{n_0\} \in U$.

If $\langle a_n \rangle_{n=0}^{\infty}$ is convergent and $\lim_{n\to\infty} a_n = a$, then for every nonprincipal ultrafilter U, $\lim_U a_n = a$. This is because for every $\varepsilon > 0$, there exists some k such that $\{n \mid |a_n - a| < \varepsilon\} \supseteq \{n \mid n \geq k\}$, and $\{n \mid n \geq k\} \in U$.

The important property of ultrafilter limits is that the U-limit exists for every bounded sequence:

2.8 Theorem *Let U be an ultrafilter on $\boldsymbol{N}$ and let $\langle a_n \rangle_{n=0}^{\infty}$ be a bounded sequence of real numbers. Then $\lim_U a_n$ exists.*

Proof. As $\langle a_n \rangle_{n=0}^{\infty}$ is bounded, there exist numbers a and b such that $a < a_n < b$ for all n. For every $x \in [a, b]$, let

$$A_x = \{n \mid a_n < x\}.$$

Clearly, $A_a = \emptyset$, $A_b = \boldsymbol{N}$, and $A_x \subseteq A_y$ whenever $x \leq y$. Therefore $A_a \notin U$, $A_b \in U$, and if $A_x \in U$ and $x \leq y$, then $A_y \in U$. Now let

$$c = \sup\{x \mid A_x \notin U\};$$

we claim that $c = \lim_U a_n$. As for every $\varepsilon > 0$, $A_{c-\varepsilon} \notin U$ while $A_{c+\varepsilon} \in U$, and because $A_{c+\varepsilon} = A_{c-\frac{\varepsilon}{2}} \cup \{n \mid c - \varepsilon < a_n < c + \varepsilon\}$, it follows that $\{n \mid |a_n - c| < \varepsilon\} \in U$. □

As an application of ultrafilter limits, we construct a nontrivial measure on $\boldsymbol{N}$.

2.9 Theorem *There exists a measure m on $\boldsymbol{N}$ such that $m(A) = d(A)$ for every set A that has density.*

Proof. Let U be a nonprincipal ultrafilter on $\boldsymbol{N}$. For every set $A \subseteq \boldsymbol{N}$, let

$$m(A) = \lim_U \frac{A(n)}{n}$$

where $A(n) = |A \cap n|$. Clearly, if A has density then $m(A) = d(A)$. Also, it is easy to verify that $m(\emptyset) = 0$ and $m(\boldsymbol{N}) = 1$, and that $A \subseteq B$ implies $m(A) \leq m(B)$. If A and B are disjoint then $(A \cup B)(n) = A(n) + B(n)$, and the additivity of m follows from this property of ultrafilter limits:

$$\lim_U (a_n + b_n) = \lim_U a_n + \lim_U b_n.$$

We leave the proof, which is analogous to that for ordinary limits, as an exercise (Exercise 2.5). □

Exercises

2.1 If U is an ultrafilter on S, then $\mathcal{P}(S) - U$ is a prime ideal.

2.2 If U is a nonprincipal ultrafilter, then every $X \in U$ is infinite.

2.3 Let U be an ultrafilter on S. Show that the collection V of sets $X \subseteq S \times S$ defined by $X \in V$ if and only if $\{a \in S \mid \{b \in S \mid (a, b) \in X\} \in U\} \in U$ is an ultrafilter on $S \times S$.

2.4 Let U be an ultrafilter on S and let $f : S \to T$. Show that the collection V of sets $X \subseteq T$ defined by $X \in V$ if and only if $f^{-1}[X] \in U$ is an ultrafilter on T.

2.5 Prove that $\lim_U (a_n + b_n) = \lim_U a_n + \lim_U b_n$.

3. Closed Unbounded and Stationary Sets

In this section we introduce an important filter on a regular uncountable cardinal — the filter generated by the closed unbounded sets. Although the results of this section can be formulated and proved for any regular uncountable cardinal, we restrict our investigations to the least uncountable cardinal $\aleph_1$. (See Exercises 3.5–3.9.)

3.1 Definition A set $C \subseteq \omega_1$ is *closed unbounded* if
(a) C is unbounded in ω_1, i.e., $\sup C = \omega_1$.
(b) C is *closed*, i.e., every increasing sequence

$$\alpha_0 < \alpha_1 < \cdots < \alpha_n < \cdots \qquad (n \in \omega)$$

of ordinals in C has its supremum $\sup\{\alpha_n \mid n \in \omega\} \in C$.

An important, albeit simple, fact about closed unbounded sets is that they have the finite intersection property; this is a consequence of the following:

3.2 Lemma *If C_1 and C_2 are closed unbounded subsets of ω_1, then $C_1 \cap C_2$ is closed unbounded.*

Proof. It is easy to see that $C_1 \cap C_2$ is closed; if $\alpha_1 < \alpha_2 < \cdots$ is a sequence in both C_1 and C_2, then $\alpha = \sup\{\alpha_n \mid n \in \omega\} \in C_1 \cap C_2$.

To see that $C_1 \cap C_2$ is unbounded, let $\gamma < \omega_1$ be arbitrary and let us find α in $C_1 \cap C_2$ such that $\alpha > \gamma$. Let us construct an increasing sequence

$$\alpha_0 < \beta_0 < \alpha_1 < \beta_1 < \cdots < \alpha_n < \beta_n < \cdots$$

of countable ordinals as follows: Let α_0 be the least ordinal in C_1 above γ. Then let β_0 be the least $\beta > \alpha_0$ in C_2. Then $\alpha_1 \in C_1$, $\beta_1 \in C_2$, and so on.

Let α be the supremum of $\{\alpha_n\}_{n\in\omega}$; it is also the supremum of $\{\beta_n\}_{n\in\omega}$. The ordinal α is in both C_1 and C_2 and therefore $\alpha \in C_1 \cap C_2$. □

The set ω_1 of all countable ordinals is closed and unbounded. (However, here we use a consequence of the Axiom of Choice, namely the regularity of ω_1: the supremum of every countable sequence of countable ordinals is a countable ordinal.) Another example of a closed unbounded set is the set of all limit countable ordinals (Exercise 3.2). For other, less obvious examples see Exercises 3.3 and 3.4.

As the collection of all closed unbounded subsets of ω_1 has the finite intersection property, it generates a filter on ω_1:

$$F = \{X \subseteq \omega_1 \mid X \supseteq C \text{ for some closed unbounded } C\}. \tag{3.3}$$

The filter (3.3) is called the *closed unbounded filter* on ω_1.

3.4 Lemma *If $\{C_n\}_{n\in\omega}$ is a countable collection of closed unbounded sets, then $\bigcap_{n=0}^{\infty} C_n$ is closed and unbounded. Consequently, if $\{X_n\}_{n\in\omega}$ is a countable collection of sets in the closed unbounded filter F, then $\bigcap_{n=0}^{\infty} X_n \in F$.*

Proof. It is easy to verify that the intersection $C = \bigcap_{n=0}^{\infty} C_n$ is closed. To prove that C is unbounded, let $\gamma < \omega_1$; we find $\alpha \in C$ greater than γ.

We note that $C = \bigcap_{n=0}^{\infty} D_n$ where for each n, $D_n = C_0 \cap \cdots \cap C_n$; the sets D_n are closed unbounded and form a decreasing sequence: $D_0 \supseteq D_1 \supseteq D_2 \supseteq \cdots$.

Let $\langle \alpha_n \mid n \in \omega \rangle$ be the following sequence of countable ordinals: $\gamma < \alpha_0 < \alpha_1 < \cdots$, and for each n, α_{n+1} is the least ordinal in D_n above α_n. Let α be the supremum of $\{\alpha_n\}_{n\in\omega}$. To show that $\alpha \in C$, we prove that $\alpha \in D_n$ for each n. But for any n, α is the supremum of $\{\alpha_k \mid k \geq n+1\}$, and all the α_k for $k \geq n+1$ are in D_n because $D_n \supseteq D_{n+1} \supseteq \cdots \supseteq D_k \supseteq \cdots$ $(k > n)$. □

Closely related to closed unbounded sets are stationary sets. *Stationary sets* are those sets $S \subseteq \omega_1$ which do not belong to the ideal dual to the closed unbounded filter. A reformulation of this leads us to the following definition.

3.5 Definition A set $S \subseteq \omega_1$ is *stationary* if for every closed unbounded set C, $S \cap C$ is nonempty.

Clearly, every closed unbounded set is stationary, and if S is stationary and $S \subseteq T \subseteq \omega_1$, then T is also stationary. Later in this section we present an example of a stationary set that does not have a closed unbounded subset. But first we prove the following theorem.

A function f with domain $S \subseteq \omega_1$ is *regressive* if $f(\alpha) < \alpha$ for all $\alpha \neq 0$.

3.6 Theorem *A set $S \subseteq \omega_1$ is stationary if and only if every regressive function $f : S \to \omega_1$ is constant on an unbounded set. In fact, f has a constant value on a stationary set.*

This theorem is a good example of how an uncountable aleph such as $\aleph_1$ differs substantially from $\aleph_0$. On uncountable cardinals there is no analogue of the function $f : \omega \to \omega$ defined by $f(n) = n - 1$ for $n > 0$, $f(0) = 0$, which has each value only finitely many times. A consequence of Theorem 3.6 is that on ω_1, a function that satisfies $f(\alpha) < \alpha$ repeats some value uncountably often, unless the domain of f is "small," that is, not stationary.

One direction of the theorem states that if $S \subseteq \omega_1$ is not stationary, then there is a function on S such that $f(\alpha) < \alpha$ for all $\alpha \neq 0$, and f takes each value at most countably many times. Such a function is constructed in the following example.

3.7 Example Let $A \subseteq \omega_1$ be a nonstationary set; hence there is a closed unbounded C such that $A \cap C = \emptyset$. Let $f : A \to \omega_1$ be defined as follows:

$$f(\alpha) = \sup(C \cap \alpha) \quad (\alpha \in A).$$

Because $f(\alpha) \in C$ (see Exercise 3.1) and $C \cap A = \emptyset$, we have $f(\alpha) < \alpha$. And for each $\gamma < \omega_1$, when $\alpha \in A$ is greater than the least element of C above γ, then $f(\alpha) > \gamma$ and so f does not have the same value for uncountably many α's.

To prove the other direction of the theorem we need a lemma.

3.8 Lemma *Let $\{C_\xi \mid \xi < \omega_1\}$ be a collection of closed unbounded sets. The set $C \subseteq \omega_1$ defined by*

(3.9) $$\alpha \in C \text{ if and only if } \alpha \in C_\xi \text{ for all } \xi < \alpha \quad (\alpha \in \omega_1)$$

called the diagonal intersection *of the C_ξ, is closed unbounded.*

Proof. First we prove that C is closed. So let $\alpha_0 < \alpha_1 < \cdots < \alpha_n < \cdots$ be an increasing sequence of elements of C and let α be its supremum. To prove that α is in C we show that $\alpha \in C_\xi$ for all $\xi < \alpha$.

Let $\xi < \alpha$. Then there is some k such that $\xi < \alpha_k$, and hence $\xi < \alpha_n$ for all $n \geq k$. Since each α_n is in C, it satisfies (3.9), and so for each $n \geq k$ we have $\alpha_n \in C_\xi$. But C_ξ is closed and so α, the supremum of $\{\alpha_n\}_{n \geq k}$, is in C_ξ.

Next we prove that C is unbounded. Let $\gamma < \omega_1$ and let us find $\alpha \in C$ greater than γ. We construct an increasing sequence $\alpha_0 < \alpha_1 < \alpha_2 < \cdots$ as follows:

Let $\alpha_0 = \gamma$. The set $\bigcap_{\xi < \alpha_0} C_\xi$ is closed unbounded (by Lemma 3.4) and so we let α_1 be its least element above α_0. Then we let α_2 be the least element of $\bigcap_{\xi < \alpha_1} C_\xi$ above α_1, and in general,

(3.10) $$\alpha_n < \alpha_{n+1} \in \bigcap_{\xi < \alpha_n} C_\xi.$$

The let α be the supremum of $\{\alpha_n\}_{n \in \omega}$, and let us prove that $\alpha \in C$.

We are to show that $\alpha \in C_\xi$ for all $\xi < \alpha$. So let $\xi < \alpha$. There is k such that $\xi < \alpha_k$, and for all $n \geq k$ we have $\alpha_{n+1} \in C_\xi$, by (3.10). But α is the supremum of $\{\alpha_n\}_{n > k}$ and hence is in C_ξ. □

Proof of Theorem 3.6. Let S be a stationary set and let $f : S \to \omega_1$ be a regressive function. For each $\gamma < \omega_1$, let $A_\gamma = f^{-1}[\{\gamma\}]$; we show that for some γ the set A_γ is stationary.

Assuming otherwise, no A_γ is stationary and so for each $\gamma < \omega_1$ there is a closed unbounded set C_γ such that $A_\gamma \cap C_\gamma = \emptyset$. Let C be the diagonal intersection of the C_γ, that is,

(3.11) $$\alpha \in C \text{ if and only if } \alpha \in C_\gamma \text{ for all } \gamma < \alpha.$$

If $\alpha \in C_\gamma$, then $\alpha \notin A_\gamma$ and so $f(\alpha) \neq \gamma$. Thus (3.11) means that if $\alpha \in C$, then $f(\alpha) \neq \gamma$ for all $\gamma < \alpha$; in other words, $f(\alpha)$ is not less than α. But C is closed unbounded and hence it intersects the stationary set S. But for all $\alpha \in S$, $f(\alpha)$ is less than α. A contradiction. □

As an application of Theorem 3.6 we construct a stationary set $S \subseteq \omega_1$, whose complement is also stationary. Note however that the construction uses the Axiom of Choice. (It is known that the Axiom of Choice is indispensable in this case.)

3.12 Example A stationary set whose complement is stationary.

Let C be the set of all countable limit ordinals. For each $\alpha \in C$ there exists an increasing sequence $x_\alpha = \langle x_{\alpha n} \mid n \in \omega \rangle$ with limit α. For each n, we let $f_n : C \to \omega_1$ be the function $f_n(\alpha) = x_{\alpha n}$.

For each n, because $f_n(\alpha) < \alpha$ on C, there exists γ_n such that the set $S_n = \{\alpha \in C \mid f_n(\alpha) = \gamma_n\}$ is stationary.

We claim that at least one of the sets S_n has a stationary complement. If not, then each S_n contains a closed unbounded set, so their intersection does too. Hence $\bigcap_{n=0}^{\infty} S_n$ contains an ordinal α that is greater than the supremum of the set $\{\gamma_n\}_{n \in \omega}$. But in that case the sequence $x_\alpha = \langle x_{\alpha n} \mid n \in \omega \rangle = \langle f_n(\alpha) \mid n \in \omega \rangle = \langle \gamma_n \mid n \in \omega \rangle$ does not converge to α, a contradiction. □

As another application of Theorem 3.6 we prove a combinatorial theorem known as the Δ-*lemma*. Although it is possible to prove the Δ-lemma directly, the present proof illustrates the power of Theorem 3.6.

3.13 Theorem *Let $\{A_i \mid i \in I\}$ be an uncountable collection of finite sets. Then there exists an uncountable $J \subseteq I$ and a set A such that for all distinct $i, j \in J$, $A_i \cap A_j = A$.*

Proof. We may assume that $I = \omega_1$, and since the union of $\aleph_1$ finite sets has size $\aleph_1$, we may also assume that all the A_i are subsets of ω_1. Clearly, uncountably many A_i have the same size and so we assume that we have a collection $\{A_\alpha \mid \alpha < \omega_1\}$ of subsets of ω_1, each of size n, for some fixed number n.

Let C be the set of all $\alpha < \omega_1$ such that $\max A_\xi < \alpha$ whenever $\xi < \alpha$. The set C is closed unbounded (compare with Exercise 3.4). For each $k \leq n$, let $S_k = \{\alpha \in C \mid |A_\alpha \cap \alpha| = k\}$; there is at least one k for which S_k is stationary. For each $m = 1, \ldots, k$, let $f_m(\alpha) =$ the mth element of A_α; we have $f_m(\alpha) < \alpha$ on S_k. By k applications of Theorem 3.6, we obtain a stationary set $T \subseteq S_k$ and a set A (of size k) such that $A_\alpha \cap \alpha = A$ for all $\alpha \in T$.

Now when $\alpha < \beta$ are in T, then $A_\alpha \subset \beta$ (because $\beta \in C$) and $A_\alpha \cap \alpha = A_\beta \cap \beta = A$, and it follows that $A_\alpha \cap A_\beta = A$ [because $(A_\alpha - \alpha) \cap A_\beta = \emptyset$]. Thus $\{A_\alpha \mid \alpha \in T\}$ is an uncountable subcollection of $\{A_\alpha \mid \alpha < \omega_1\}$ which satisfies the theorem. □

Exercises

3.1 An unbounded set $C \subseteq \omega_1$ is closed if and only if for every $X \subset C$, if $\sup X < \omega_1$, then $\sup X \in C$.

3.2 The set of all countable limit ordinals is closed unbounded.

If X is a set of ordinals, then α is a *limit point* of X if for every $\gamma < \alpha$ there is $\beta \in X$ such that $\gamma < \beta < \alpha$. A countable α is a limit point of X if and only if there exists a sequence $\alpha_0 < \alpha_1 < \cdots$ in X such that $\sup\{\alpha_n \mid n \in \omega\} = \alpha$. Every closed unbounded $C \subseteq \omega_1$ contains all its countable limit points.

3.3 If X is an unbounded subset of ω_1, then the set of all countable limit points of X is a closed unbounded set.

If $f : \omega_1 \to \omega_1$ is an increasing function, then $\alpha < \omega_1$ is a *closure point* of f if $f(\xi) < \alpha$ whenever $\xi < \alpha$.

3.4 The set of all closure points of every increasing function $f : \omega_1 \to \omega_1$ is closed unbounded.

Let κ be a regular uncountable cardinal. A set $C \subseteq \kappa$ is *closed unbounded* if $\sup C = \kappa$ and $\sup(C \cap \alpha) \in C$ for all $\alpha < \kappa$.

3.5 If C_1 and C_2 are closed unbounded, then $C_1 \cap C_2$ is closed unbounded.

The *closed unbounded filter* on κ is the filter generated by the closed unbounded sets.

3.6 If $\lambda < \kappa$ and each C_α, $\alpha < \lambda$, is closed unbounded, then $\bigcap_{\alpha<\lambda} C_\alpha$ is closed unbounded.

A set $S \subseteq \kappa$ is *stationary* if $S \cap C \neq \emptyset$ for every closed unbounded set $C \subseteq \kappa$.

3.7 The set $\{\alpha < \kappa \mid \alpha \text{ is a limit ordinal and } \operatorname{cf}(\alpha) = \omega\}$ is stationary. If $\kappa > \aleph_1$, then the set $\{\alpha < \kappa \mid \operatorname{cf}(\alpha) = \omega_1\}$ is stationary.

3.8 If each C_α, $\alpha < \kappa$, is closed unbounded, then the *diagonal intersection* $\{\alpha < \kappa \mid \alpha \in C_\xi \text{ for all } \xi < \alpha\}$ is closed unbounded.

A function with $\operatorname{dom} f \subseteq \kappa$ is *regressive* if $f(\alpha) < \alpha$ for all $\alpha \in \operatorname{dom} f$, $\alpha \neq 0$.

3.9 A set $S \subseteq \kappa$ is stationary if and only if every regressive function on S is constant on an unbounded set.

4. Silver's Theorem

Using the techniques introduced in Sections 2 and 3 we now prove the following result on the Generalized Continuum Hypothesis for singular cardinals.

4.1 Silver's Theorem *Let $\aleph_\lambda$ be a singular cardinal such that $\operatorname{cf}\lambda > \omega$. If for every $\alpha < \lambda$, $2^{\aleph_\alpha} = \aleph_{\alpha+1}$, then $2^{\aleph_\lambda} = \aleph_{\lambda+1}$.*

We prove Silver's Theorem for the special case $\aleph_\lambda = \aleph_{\omega_1}$, using the theory of stationary subsets of $\aleph_1$. The full result can be proved in a similar way, using the general theory of stationary subsets of $\kappa = \operatorname{cf}\lambda$ (see Exercise 4.1). Thus assume that $2^{\aleph_\alpha} = \aleph_{\alpha+1}$ for all $\alpha < \omega_1$. One consequence of this assumption, to be used repeatedly in this section, is that $\aleph_\alpha^{\aleph_1} < \aleph_{\omega_1}$ for all $\alpha < \omega_1$ (see Theorem 3.12 in Chapter 9).

Let f and g be two functions on ω_1. The functions f and g are *almost disjoint* if there is some $\alpha < \omega_1$ such that $f(\beta) \neq g(\beta)$ for all $\beta \geq \alpha$.

4.2 Lemma *Let $\{A_\alpha \mid \alpha < \omega_1\}$ be a family of sets such that $|A_\alpha| \leq \aleph_\alpha$ for every $\alpha < \omega_1$, and let F be a family of almost disjoint functions,*

$$F \subset \prod_{\alpha<\omega_1} A_\alpha.$$

Then $|F| \leq \aleph_{\omega_1}$.

Proof. Without loss of generality we may assume that $A_\alpha \subseteq \omega_\alpha$, for each $\alpha < \omega_1$. Let S_0 be the set of all limit ordinals $0 < \alpha < \omega_1$. For $f \in F$ and $\alpha \in S_0$, let $f^*(\alpha)$ denote the least β such that $f(\alpha) < \omega_\beta$. As $f^*(\alpha) < \alpha$ for every $\alpha \in S_0$, there exists, by Theorem 3.6, a stationary set $S \subset S_0$ such that f^* is constant on S. Therefore $f \restriction S$ is a function from S into ω_β, for some $\beta < \omega_1$. Let us denote this function $f \restriction S$ by $\varphi(f)$.

If f and g are two different functions in F, then $\varphi(f)$ and $\varphi(g)$ are also different: even if their domains are the same, say S, then $f \restriction S \neq g \restriction S$ because f and g are almost disjoint. Thus φ is a one-to-one mapping with domain F. The values of φ range over functions defined on a subset S of ω_1 into some $\omega_\beta < \omega_{\omega_1}$. Thus we have

$$|F| \leq 2^{\aleph_1} \cdot \sum_{\beta<\omega_1} \aleph_\beta^{\aleph_1} \leq \aleph_{\omega_1}.$$

□

A slight modification of Lemma 4.2 gives the following:

4.3 Lemma *Let $F \subset \prod_{\alpha<\omega_1} A_\alpha$ be a family of almost disjoint functions, such that the set $T = \{\alpha < \omega_1 \mid |A_\alpha| \leq \aleph_\alpha\}$ is stationary. Then $|F| \leq \aleph_{\omega_1}$.*

Proof. In the proof of Lemma 4.2, let S_0 be the stationary set $S_0 = \{\alpha \in T \mid \alpha \text{ is a limit ordinal}\}$. The rest of the proof is the same. □

Using Lemma 4.3, we easily get the following:

4.4 Lemma *Let f be a function on ω_1 such that $f(\alpha) < \aleph_{\alpha+1}$ for all $\alpha < \omega_1$. Let F be a family of almost disjoint functions on ω_1, and let*

$$F_f = \{g \in F \mid \textit{for some stationary set } T \subseteq \omega_1,\ g(\alpha) < f(\alpha) \textit{ for all } \alpha \in T\}.$$

Then $|F_f| \leq \aleph_{\omega_1}$.

Proof. For a fixed stationary set T, the set $\{g \in F \mid g(\alpha) < f(\alpha)$ for all $\alpha \in T\}$ has cardinality at most $\aleph_{\omega_1}$, by Lemma 4.3. Thus $|F| \le 2^{\aleph_1} \cdot \aleph_{\omega_1} = \aleph_{\omega_1}$. □

We now prove the crucial lemma for Silver's Theorem:

4.5 Lemma *Let $\{A_\alpha \mid \alpha < \omega_1\}$ be a family of sets such that $|A_\alpha| \le \aleph_{\alpha+1}$ for every $\alpha < \omega_1$, and let F be a family of almost disjoint functions,*

$$F \subset \prod_{\alpha<\omega_1} A_\alpha.$$

Then $|F| \le \aleph_{\omega_1+1}$.

Proof. Let U be an ultrafilter on ω_1 that extends the closed unbounded filter. Thus every set $S \in U$ is stationary.

Without loss of generality we may assume that $A_\alpha \subseteq \omega_{\alpha+1}$, for each $\alpha < \omega_1$. Let us define a relation $<$ on the set F as follows:

$$f < g \quad \text{if and only if} \quad \{\alpha < \omega_1 \mid f(\alpha) < g(\alpha)\} \in U.$$

We claim that $<$ is a linear ordering of F. If $f < g$ and $g < h$ then $f < h$, because

$$\{\alpha \mid f(\alpha) < h(\alpha)\} \supseteq \{\alpha \mid f(\alpha) < g(\alpha)\} \cap \{\alpha \mid g(\alpha) < h(\alpha)\} \in U.$$

If $f, g \in F$ and $f \ne g$, then $\{\alpha \mid f(\alpha) = g(\alpha)\}$ is at most countable and therefore not in U, and so one of the sets $\{\alpha \mid f(\alpha) < g(\alpha)\}$, $\{\alpha \mid g(\alpha) < f(\alpha)\}$ belongs to U. Thus either $f < g$ or $g < f$. It follows that $<$ is a linear ordering on F.

Now, if $f, g \in F$ and $g < f$, then $g \in F_f$ where

$$F_f = \{g \in F \mid \text{for some stationary } T,\ g(\alpha) < f(\alpha) \text{ for all } \alpha \in T\},$$

and by Lemma 4.4, $|F_f| \le \aleph_{\omega_1}$. Thus for every $f \in F$, $|\{g \in F \mid g < f\}| \le \aleph_{\omega_1}$.

As $<$ is a linear ordering of the set F, it follows from Theorem 1.14 in Chapter 8 that $|F| \le \aleph_{\omega_1+1}$. □

Silver's Theorem (for $\lambda = \omega_1$) is now an easy consequence of Lemma 4.5:

Proof of Silver's Theorem. For each $\alpha < \omega_1$, let $A_\alpha = \mathcal{P}(\omega_\alpha)$; as $2^{\aleph_\alpha} = \aleph_{\alpha+1}$, we have $|A_\alpha| = \aleph_{\alpha+1}$. For every set $X \subseteq \omega_{\omega_1}$, let $f_X \in \prod_{\alpha<\omega_1} A_\alpha$ be the function defined by

$$f_X(\alpha) = X \cap \omega_\alpha.$$

If $X \ne Y$ then $f_X \ne f_Y$, and moreover, f_X and f_Y are almost disjoint. This is because there is some $\alpha < \omega_1$ such that $X \cap \omega_\beta \ne Y \cap \omega_\beta$ for all $\beta \ge \alpha$. Thus $F = \{f_X \mid X \in \mathcal{P}(\omega_{\omega_1})\}$ is a family of almost disjoint functions, and by Lemma 4.5, $|F| \le \aleph_{\omega_1+1}$. Therefore, $2^{\aleph_{\omega_1}} = \aleph_{\omega_1+1}$. □

Exercises

4.1 Prove Silver's Theorem for arbitrary λ of uncountable cofinality. [Throughout Section 4, replace ω_1 by $\kappa = \operatorname{cf}\lambda$, and replace the sequence $\langle \aleph_\alpha \mid \alpha < \omega_1 \rangle$ by a continuous increasing sequence of cardinals $\langle \lambda_\alpha \mid \alpha < \kappa \rangle$. Use Exercise 3.9.]

Chapter 12

Combinatorial Set Theory

1. Ramsey's Theorems

The classic example of the type of question we want to consider in this section is the well-known puzzle: Show that in any group of 6 people there are 3 who either all know each other or are strangers to each other. We are implicitly assuming that the relation "x knows y" is symmetric.

The argument goes as follows. Consider one of these people, say Joe. Of the remaining 5 people, either there are at least 3 who know Joe, or there are at least 3 who do not know Joe. Let us assume that Peter, Paul, and Mary know Joe. If two of them, say Peter and Paul, know each other, then we have 3 people who know each other (Joe, Peter, and Paul). Otherwise, Peter, Paul, and Mary are 3 people who are strangers to each other. If Peter, Paul, and Mary are 3 people who do not know Joe, the argument is similar.

We now restate this problem in a more abstract form that allows for ready generalizations.

Let S be a set. For $r \in \boldsymbol{N}$, $r \neq 0$, $[S]^r = \{X \subseteq S \mid |X| = r\}$ is the collection of all r-element subsets of S. (See Exercise 3.5 in Chapter 4.) Let $\{A_i\}_{i=0}^{s-1}$ be a partition of $[S]^r$ into s classes ($s \in \boldsymbol{N} - \{0\}$); i.e., $[S]^r = \bigcup_{i=0}^{s-1} A_i$ and $A_i \cap A_j = \emptyset$ for $i \neq j$. We say that a set $H \subseteq S$ is *homogeneous* for the partition if $[H]^r \subseteq A_i$ for some i; i.e., if all r-element subsets of H belong to the same class A_i of the partition.

It may help to think of the elements of the different classes as being colored by distinct colors. We thus have s colors (numbered $0, 1, \ldots, s-1$) and each r-element subset of S is colored by one of them. Homogeneous set are *monochromatic*; i.e., all r-element subsets of them are colored by the same color.

In this terminology, the introductory example shows that, for a set S with $|S| \geq 6$, every partition of $[S]^2$ into two classes has a homogeneous set H with $|H| \geq 3$. (The classes are $A_0 = \{\{x, y\} \in [S]^2 \mid x \text{ and } y \text{ know each other}\}$ and $A_1 = \{\{x, y\} \in [S]^2 \mid x \text{ and } y \text{ are strangers to each other}\}$.)

Put yet another way: pick a set S of at least 6 points and connect each pair of points by a line segment colored either red or blue (but not both). The resulting graph then contains a monochromatic triangle (i.e., all red or all blue).

Let κ, λ be cardinal numbers. We write $\kappa \to (\lambda)^r_s$ as a shorthand for the statement: for every set S with $|S| = \kappa$ and every partition of $[S]^r$ into s classes, there exists a homogeneous set H with $|H| \geq \lambda$. The negation of this statement is denoted $\kappa \not\to (\lambda)^r_s$.

We proved that $6 \to (3)^2_2$; it is an easy exercise to show that $5 \not\to (3)^2_2$. (See Exercise 1.1.)

More generally, $\kappa \to (\lambda_1, \ldots, \lambda_s)^r_s$ means: for every set S with $|S| = \kappa$ and every partition $\{A_i\}_{i=0}^{s-1}$ of $[S]^r$ into s classes there is some i and a set $H \subseteq S$ with $[H]^r \subseteq A_i$ and $|H_i| \geq \lambda_i$. Exercises 1.2 and 1.3 exhibit some very simple properties of these symbols.

The fundamental result about partitions on finite sets is the Finite Ramsey's Theorem. It asserts that if S is sufficiently large, any coloring of its r-element subsets will have monochromatic sets of the prescribed size k.

1.1 The Finite Ramsey's Theorem *For any positive natural numbers k, r, s there exists a natural number n such that $n \to (k)^r_s$.*

The readers interested primarily in infinite sets can skip the proof without harm.

Proof. We first consider $s = 2$ and prove that for all $r, p, q \in \boldsymbol{N} - \{0\}$ there exists $n \in \boldsymbol{N}$ for which $n \to (p, q)^r_2$ by induction on r.

Let $r = 1$. Then it suffices to take $n = p + q - 1$: it is clear that if $|S| = n$ and $[S]^1 = S = A_0 \cup A_1$, we cannot have both $|A_0| \leq p - 1$ and $|A_1| \leq q - 1$. If $|A_0| \geq p$ let $H = A_0$; otherwise, let $H = A_1$.

We now assume that the statement is true for r (and any p, q) and prove it for $r + 1$. We denote the least n for which $n \to (p, q)^r_2$ by $R(p, q; r)$. The proof now proceeds by induction on $(p + q)$.

Consider a set S with $|S| = n > 0$ (as yet unspecified) and a partition $[S]^{r+1} = A_0 \cup A_1$ where $A_0 \cap A_1 = \emptyset$. The statement is true if $p \leq r$ or $q \leq r$: if $p < r$ ($q < r$, respectively) we can take as H any subset of S with $|H| = p$ ($|H| = q$, respectively); if $p = q = r$ then either $A_0 \neq \emptyset$ and any $H \in A_0$ will do, or $A_1 \neq \emptyset$ and any $H \in A_1$ will do. So we assume that $p > r$, $q > r$ and the statement is true for p', q' if $p' + q' < p + q$.

Fix $a \in S$, let $S^a = S - \{a\}$ and define a partition $\{B_0, B_1\}$ of $[S^a]^r$ by $X \in B_0$ (B_1, respectively) if and only if $\{a\} \cup X \in A_0$ (A_1, respectively) (for $X \in [S^a]^r$).

If we choose n so large that $n - 1 = R(p', q'; r)$ (p', q' as yet unspecified) then there will be a set $H^a \subseteq S^a$ so that either

(i) $[H^a]^r \subseteq B_0$ and $|H^a| \geq p'$, or

(ii) $[H^a]^r \subseteq B_1$ and $|H^a| \geq q'$.

Either way, all $(r+1)$-element sets of the form $\{a\} \cup X$, $X \in [H^a]^r$ are thus guaranteed to be colored by the same color, so it remains to concern ourselves with $(r+1)$-element subsets of H^a.

Assume (i) occurs. By the inductive assumption, there exists n for which $n \to (p-1,q)_2^{r+1}$ holds; let $p' = R(p-1,q;r+1)$ be the least such n. It then follows that either there is $H' \subseteq H^a$ with $|H'| \geq p-1$ and $[H']^{r+1} \subseteq A_0$; in this case we let $H = H' \cup \{a\}$ and notice that $|H| \geq p$ and $[H]^{r+1} \subseteq A_0$. Or, there is $H'' \subseteq H^a$ with $|H''| \geq q$ and $[H'']^{r+1} \subseteq A_1$; in this case we let $H = H''$ and notice that $|H| \geq q$ and $[H]^{r+1} \subseteq A_1$.

The case (ii) is handled similarly, letting $q' = R(p,q-1;r+1)$, the least n for which $n \to (p,q-1)_2^{r+1}$ holds.

In conclusion, if we let $n = R(p',q';r)$ for $p' = R(p-1,q;r+1)$ and $q' = R(p,q-1;r+1)$ then $n \to (p,q)_2^{r+1}$.

This concludes the proof for $s = 2$. In particular, taking $p = q = k$ we have $n \to (k)_2^r$ for all k, r. The proof of the Finite Ramsey's Theorem can now be completed by induction on s. The case $s = 1$ is trivial and $s = 2$ is proved above.

Assume that m has the property $m \to (k)_s^r$. Let us consider a partition $\{A_i\}_{i=0}^s$ of $[S]^r$ into $(s+1)$ classes, where S is a set of an as yet unspecified cardinality n. Then $\{B_0, B_1\}$ defined by

$$B_0 = \bigcup_{i=0}^{s-1} A_i, \quad B_1 = A_s$$

is a partition of $[S]^r$ into 2 classes. If $n = R(l,l;r)$ (for as yet unspecified l) then there is $H' \subseteq S$, $|H'| \geq l$, such that either $[H']^r \subseteq B_0$ or $[H']^r \subseteq B_1$. We take $l = \max\{m,k\}$. In the first case, $\{A_i\}_{i=0}^{s-1}$ partitions $[H']^r$ into s classes, so our choice of l guarantees existence of $H \subseteq H'$ such that $|H| \geq k$ and $[H]^r \subseteq A_i$ for some $i \in \{0,\ldots,s-1\}$. In the second case, our choice of l allows us to let $H = H'$ and conclude $|H| \geq k$, $[H]^r \subseteq A_s$. Either way, we are done. □

We remark that the exact determination of the numbers $R(p,q;r)$ is difficult and only a few are known despite much effort and extensive use of computers. Here is a complete list, as of now (1998): $R(3,l;2)$ for $l = 3,4,5,6,7,8,9$ has values $6,9,14,18,23,28,36$, respectively; $R(4,4;2) = 18$, $R(4,5;2) = 25$, $R(3,3,3;2) = 17$, and $R(4,4;3) = 13$. They tend to grow rapidly with increasing $(p+q)$ and r. This is a subject of great interest in the area of mathematics known as (finite) combinatorics, but we now turn to analogous questions for infinite cardinal numbers. There the most important result is the (Infinite) Ramsey's Theorem.

1.2 Ramsey's Theorem $\aleph_0 \to (\aleph_0)_s^r$ *holds for all* $r, s \in \boldsymbol{N} - \{0\}$.

In words, if r-element subsets of an infinite set are colored by a finite number s of colors, then there is an infinite subset, all of whose r-element subsets are colored by the same color.

Proof. It suffices to consider $S = \boldsymbol{N}$ (see Exercise 1.2). We proceed by induction on r.

$r = 1$: If $[\boldsymbol{N}]^1 = \boldsymbol{N} = \bigcup_{i=0}^{s-1} A_i$ then at least one of the sets A_i must be infinite (Theorem 2.7 in Chapter 4) and we let H be such an A_i.

$r = 2$: This is a warm-up for the general induction step. Let $[\boldsymbol{N}]^2 = \bigcup_{i=0}^{s-1} A_i$. We construct sequences $\langle a_n \rangle_{n=0}^{\infty}$, $\langle i_n \rangle_{n=0}^{\infty}$, and $\langle H_n \rangle_{n=0}^{\infty}$ by recursion. We set $a_0 = 0$ and let $B_i^0 = \{b \in \boldsymbol{N} \mid b \neq a_0 \text{ and } \{a_0, b\} \in A_i\}$. Then $\{B_i^0\}_{i=0}^{s-1}$ is a partition of $\boldsymbol{N} - \{a_0\}$; we take i_0 to be the first i for which B_i^0 is infinite (refer to the case $r = 1$) and let $H_0 = B_{i_0}^0$. All the subsequent terms of the sequence $\langle a_n \rangle$ will be selected from H_0, thus guaranteeing that $\{a_0, a_n\} \in A_{i_0}$ for all $n > 0$. We select a_1 to be the first element of H_0 and let $B_i^1 = \{b \in H_0 \mid b \neq a_1 \text{ and } \{a_0, b\} \in A_i\}$. Again, $\{B_i^1\}_{i=0}^{s-1}$ is a partition of the infinite set $H_0 - \{a_1\}$ and we let i_1 be the first i for which B_i^1 is infinite, and $H_1 = B_{i_1}^1$. Proceeding in this manner one obtains the desired sequences. It is obvious from the construction that $\langle a_n \rangle_{n=0}^{\infty}$ is an increasing sequence and $\{a_n, a_m\} \in A_{i_n}$ for all $m > n$. The sequence $\langle i_n \rangle_{n=0}^{\infty}$ has values from the finite set $\{0, \ldots, s-1\}$, hence there exists j and an infinite set M such that $i_n = j$ for all $n \in M$. It remains to let $H = \{a_n \mid n \in M\}$. Then H is infinite and $[H]^2 \subseteq A_j$ because $\{a_n, a_m\} \in A_{i_n} = A_j$ for all $n, m \in M$, $n < m$.

The general case is similar. We assume the theorem is true for r and prove it for $r+1$. So let $[\boldsymbol{N}]^{r+1} = \bigcup_{i=0}^{s-1} A_i$. Consider an arbitrary $a \in \boldsymbol{N}$ and an arbitrary infinite $S \subseteq \boldsymbol{N}$ such that $a \notin S$. We define a partition $\{B_i\}_{i=0}^{s-1}$ of $[S]^r$ as follows: for $X \in [S]^r$, $X \in B_i$ if and only if $\{a\} \cup X \in A_i$ (i.e., we color the r-element subset X of S by the same color the $(r+1)$-element set $\{a\} \cup X$ was colored by originally). By the inductive assumption, there is an infinite $H \subseteq S$ such that $[H]^r \subseteq B_i$ for some i. We select one such $i = i(a, S)$ and $H = H(a, S)$. (We can always use the least such i, but we use the Axiom of Choice [$\mathcal{P}(\boldsymbol{N})$ can be well-ordered] to select a particular H. The use of the Axiom of Choice can be avoided, at the cost of additional complications.) We note that all sets of the form $\{a\} \cup X$ where $X \in [H]^r$ belong to A_i.

The rest of the proof imitates the case $r = 2$ closely. We construct sequences $\langle a_n \rangle_{n=0}^{\infty}$, $\langle i_n \rangle_{n=0}^{\infty}$, and $\langle H_n \rangle_{n=0}^{\infty}$ by recursion. We let $a_0 = 0$, $i_0 = i(0, \boldsymbol{N} - \{0\})$, $H_0 = H(0, \boldsymbol{N} - \{0\})$. Having constructed a_n, i_n, and H_n, we let a_{n+1} be the least element of H_n, and $i_{n+1} = i(a_{n+1}, H_n - \{a_{n+1}\})$, $H_{n+1} = H(a_{n+1}, H_n - \{a_{n+1}\})$. The point is again to guarantee that all sets of the form $\{a_n, a_{k_1}, \ldots, a_{k_r}\}$ where $n < k_1, \ldots, n < k_r$ (and $k_1, \ldots, k_r$ are mutually distinct) belong to A_{i_n}.

Again, there is $j \in \{0, 1, \ldots, s-1\}$ such that $M = \{m \in \boldsymbol{N} \mid i_m = j\}$ is infinite. We let $H = \{a_m \mid m \in M\}$ and notice that H is infinite and $[H]^{r+1} \subseteq A_j$. □

We conclude this section with two simple applications of Ramsey's Theorem.

1.3 Corollary *Every infinite ordered set $(P, \leq)$ contains an infinite subset S such that either any two distinct elements of S are comparable (i.e., S is a chain) or any two distinct elements of S are incomparable.*

Proof. Apply Ramsey's Theorem to the partition $\{A_0, A_1\}$ of $[P]^2$ where

$$A_0 = \{\{x, y\} \in [P]^2 \mid x \text{ and } y \text{ are comparable}\},$$
$$A_1 = \{\{x, y\} \in [P]^2 \mid x \text{ and } y \text{ are incomparable}\}.$$

□

1.4 Corollary *Every infinite linearly ordered set contains a subset similar to either $(\boldsymbol{N}, <)$ or $(\boldsymbol{N}, >)$.*

Proof. Let $(P, \preccurlyeq)$ be an infinite linearly ordered set and let $\leq$ be some well-ordering of P. We partition $[P]^2$ as follows:

$$A_0 = \{\{x, y\} \in [P]^2 \mid x < y \text{ and } x \prec y\};$$
$$A_1 = \{\{x, y\} \in [P]^2 \mid x < y \text{ and } x \succ y\}.$$

Let H be an infinite homogeneous set provided by Ramsey's Theorem. The relation $<$ well-orders H; let $\tilde{H}$ be the initial segment of H of order type ω. If $[H]^2 \subseteq A_0$ then $(\tilde{H}, \prec)$ is similar to $(\boldsymbol{N}, <)$; if $[H]^2 \subseteq A_1$ then $(\tilde{H}, \succ)$ is similar to $(\boldsymbol{N}, <)$. □

Exercises

1.1 Show that $5 \not\to (3)^2_2$.

1.2 Prove that the words "every set S" in the definition of $\kappa \to (\lambda)^r_s$ can be replaced by "some set S."

1.3 Assume that $\kappa \to (\lambda)^r_s$ holds. Prove

(a) If $\kappa' \geq \kappa$ then $\kappa' \to (\lambda)^r_s$.

(b) If $\lambda' \leq \lambda$ then $\kappa \to (\lambda')^r_s$.

(c) If $s' \leq s$ then $\kappa \to (\lambda)^r_{s'}$.

(d) If $r' \leq r$ then $\kappa \to (\lambda)^{r'}_s$.

Prove analogous statements for $\kappa \to (\lambda_1, \ldots, \lambda_s)^r_s$. Also show that $\kappa \to (\lambda_1, \ldots, \lambda_s)^r_s$ holds if and only if $\kappa \to (\lambda_{\pi(1)}, \ldots, \lambda_{\pi(s)})^r_s$ holds, where π is any one-to-one mapping of $\{1, \ldots, s\}$ onto itself.

1.4 Show that for an infinite cardinal κ, $\kappa \to (\kappa)^1_\lambda$ holds if and only if $\lambda < \mathrm{cf}(\kappa)$.

1.5 Let $[S]^{<\omega} = \bigcup_{n=0}^{\infty} [S]^n$. Show that there is a partition $\{A_0, A_1\}$ of $[\boldsymbol{N}]^{<\omega}$ such that $[H]^{<\omega} \cap A_0 \neq \emptyset$ and $[H]^{<\omega} \cap A_1 \neq \emptyset$ for every infinite $H \subseteq \boldsymbol{N}$. [*Hint*: Put $X \in A_0$ if and only if $|X| \in X$.]

2. Partition Calculus for Uncountable Cardinals

Ramsey's Theorem asserts that any partition of r-element subsets of an infinite set S into a finite number of classes has an *infinite* homogeneous set. The next natural question is, how large does S have to be in order that every partition has an *uncountable* homogeneous set. By analogy with Ramsey's Theorem one might expect $\aleph_1 \to (\aleph_1)^r_s$. However, as the next theorem shows, this is false.

2.1 Theorem $2^{\aleph_0} \not\to (\aleph_1)^2_2$.

Proof. The argument is quite similar to the one used to deduce Corollary 1.4. Let $\lambda = 2^{\aleph_0}$ be the cardinality of the continuum. The set $\boldsymbol{R}$ of all real numbers has $|\boldsymbol{R}| = \lambda$ and is linearly ordered by $\leq$. Let $\preccurlyeq$ be some well-ordering of $\boldsymbol{R}$ of order type λ. We partition $[\boldsymbol{R}]^2$ as follows:

$$A_0 = \{\{x, y\} \in [\boldsymbol{R}]^2 \mid x \prec y \text{ and } x < y\},$$
$$A_1 = \{\{x, y\} \in [\boldsymbol{R}]^2 \mid x \prec y \text{ and } x > y\}.$$

Let H be an uncountable homogeneous set for this partition. That means that either

(i) for all $x, y \in H$, $x \prec y$ implies $x < y$, or

(ii) for all $x, y \in H$, $x \prec y$ implies $x > y$.

We show that this is impossible.

Let us assume (i) holds. Let $\varphi : \mu \to H$ be an isomorphism of $(\mu, \in)$ and $(H, \prec)$, where μ is an ordinal (necessarily, $\mu \leq \lambda$ and μ uncountable). We note that $\xi < \eta < \mu$ implies $\varphi(\xi) \prec \varphi(\eta)$ and hence $\varphi(\xi) < \varphi(\eta)$. Thus $\{(\varphi(\xi), \varphi(\xi + 1)) \mid \xi < \mu\}$ is an uncountable collection of mutually disjoint open intervals in $\boldsymbol{R}$, an impossibility by Theorem 3.2 in Chapter 10.

Case (ii) is similar. □

For a slightly different proof see Exercise 2.1. There is a weaker positive result.

2.2 Erdős-Dushnik-Miller Partition Theorem $\aleph_1 \to (\aleph_1, \aleph_0)^2_2$.

In words, for every partition of pairs of elements of an uncountable set into two classes, if one class has no infinite homogeneous set, then the other one has to have an uncountable homogeneous set.

Proof. Let $\{A, B\}$ be a partition of $[\omega_1]^2$. For $\alpha \in \omega_1$ let $B(\alpha) = \{\beta \in \omega_1 \mid \beta \neq \alpha \text{ and } \{\alpha, \beta\} \in B\}$. There are two possibilities.

1. For every uncountable $X \subseteq \omega_1$ there exists some $\alpha \in X$ for which $|B(\alpha) \cap X| = \aleph_1$. In this case we construct a countable homogeneous set for B by recursion. We let $X_0 = \omega_1$ and let α_0 be the first element of X_0 for which $|B(\alpha_0) \cap X_0| = \aleph_1$. We next let $X_1 = B(\alpha_0) \cap X_0$ and let α_1 be the first element of X_1 for which $|B(\alpha_1) \cap X_1| = \aleph_1$. In general, at stage $n + 1$ we let $X_{n+1} = B(\alpha_n) \cap X_n \subset X_n$ and let α_{n+1} be the first element of X_{n+1} for which $|B(\alpha_{n+1}) \cap X_{n+1}| = \aleph_1$. It is clear that $H = \{\alpha_n \mid n \in \boldsymbol{N}\}$ is a countable set and $[H]^2 \subseteq B$.

2. In the opposite case, there is an uncountable set $X \subseteq \omega_1$ such that $|B(\alpha) \cap X| \leq \aleph_0$ for all $\alpha \in X$. This time we construct an uncountable homogeneous set for A by transfinite recursion of length ω_1. Having defined a one-to-one sequence $\langle \alpha_\nu \mid \nu < \lambda \rangle$ where $\alpha_\nu \in X$ for each $\nu < \lambda < \omega_1$, we observe that

the set $\bigcup_{\nu<\lambda} B(\alpha_\nu) \cap X$ of those $\beta \in X$ for which some $\{\alpha_\nu, \beta\} \in B$ (for some $\nu < \lambda$) is at most countable (it is a countable union of at most countable sets). Therefore, $X - \bigcup_{\nu<\lambda}(B(\alpha_\nu) \cup \{\alpha_\nu\})$ is uncountable, and we let α_λ be its first element. Clearly $\alpha_\lambda \neq \alpha_\nu$ and $\{\alpha_\nu, \alpha_\lambda\} \in A$, for all $\nu < \lambda$. The set $H = \{\alpha_\nu \mid \nu < \omega_1\}$ satisfies $|H| = \aleph_1$ and $[H]^2 \subseteq A$. □

Actually, the theorem holds for any infinite cardinal κ (in place of $\aleph_1$), i.e., $\kappa \to (\kappa, \aleph_0)^2_2$. The proof for regular κ is a straightforward modification of the one just given for $\kappa = \aleph_1$, and is left as an exercise.

To guarantee that every partition of $[S]^2$ has an uncountable homogeneous set, the underlying set S must be more than just uncountable.

2.3 Erdős-Radó Partition Theorem $(2^{\aleph_0})^+ \to (\aleph_1)^2_{\aleph_0}$.

Assuming the validity of the Continuum Hypothesis, this reduces to $\aleph_2 \to (\aleph_1)^2_{\aleph_0}$.

Proof. We denote $(2^{\aleph_0})^+$ by λ and let $\{A_n\}_{n\in \boldsymbol{N}}$ be a partition of $[\lambda]^2$. For any pair $\{\alpha, \beta\} \in [\lambda]^2$ we let $n(\alpha, \beta)$ be the unique n for which $\{\alpha, \beta\} \in A_n$ ($n(\alpha, \beta)$ is the "color" of $\{\alpha, \beta\}$).

For every $\alpha < \lambda$ we construct a transfinite sequence f_α recursively as follows:

$f_\alpha(0) = 0$;

If $\langle f_\alpha(\eta) \mid \eta < \xi\rangle$ is defined ($\xi < \alpha$) and if there exists some $\sigma < \alpha$ such that $\sigma \neq f_\alpha(\eta)$ and $n(f_\alpha(\eta), \sigma) = n(f_\alpha(\eta), \alpha)$ holds for all $\eta < \xi$, then we let $f_\alpha(\xi)$ to be the least such σ; otherwise we stop.

We note that $\operatorname{dom} f_\alpha$ is an initial segment of α (or α itself) and f_α is an increasing sequence of elements of α.

We claim that for some $\alpha < \lambda$, $|\operatorname{dom} f_\alpha| \geq \aleph_1$. Assume, to the contrary, that $|\operatorname{dom} f_\alpha| \leq \aleph_0$ for all $\alpha < \lambda$. Consider the sequences $g_\alpha : \operatorname{dom}(f_\alpha) \to \boldsymbol{N}$ defined by $g_\alpha(\eta) = n(f_\alpha(\eta), \alpha)$. We note that $g_\alpha = g_\beta$ implies $f_\alpha = f_\beta$: clearly $\operatorname{dom} f_\alpha = \operatorname{dom} g_\alpha = \operatorname{dom} g_\beta = \operatorname{dom} f_\beta$, and if $f_\alpha(\eta) = f_\beta(\eta)$ for all $\eta < \xi$, then $n(f_\alpha(\eta), \sigma) = n(f_\alpha(\eta), \alpha)$ holds if and only if $n(f_\beta(\eta), \sigma) = n(f_\beta(\eta), \beta)$ (for $\eta < \xi$), showing that $f_\alpha(\xi) = f_\beta(\xi)$ also, by definition of f. Our assumption implies that $\operatorname{dom} g_\alpha$ is a countable ordinal; hence there are only $\aleph_1 \cdot (\aleph_0^{\aleph_0}) = 2^{\aleph_0}$ possibly distinct g_α's. Therefore there exist $\beta < \alpha < \lambda$ such that $g_\beta = g_\alpha$, hence $f_\beta = f_\alpha$. As then $n(f_\alpha(\eta), \beta) = n(f_\beta(\eta), \beta) = g_\beta(\eta) = g_\alpha(\eta) = n(f_\alpha(\eta), \alpha)$ for all $\eta \in \gamma = \operatorname{dom} f_\alpha$, it follows that $f_\alpha(\gamma)$ is defined (as β). This is a contradiction that proves the claim.

We now fix α with $|\operatorname{dom} f_\alpha| \geq \aleph_1$ and let $X = \operatorname{ran} f_\alpha$. Then $|X| \geq \aleph_1$ and the set X has the property that, for any $\sigma, \tau, \tau' \in X$, $\sigma < \tau$, $\sigma < \tau'$, $n(\sigma, \tau) = n(\sigma, \tau')$; i.e., $\{\sigma, \tau\} \in A_n$ if and only if $\{\sigma, \tau'\} \in A_n$. We let $B_n = \{\sigma \in X \mid \{\sigma, \tau\} \in A_n$ for some (or all) $\tau \in X, \sigma < \tau\}$. The collection $\{B_n\}$ is a partition of X and $|X| \geq \aleph_1$ implies that for some $n \in \boldsymbol{N}$, $|B_n| \geq \aleph_1$. Clearly $[B_n]^2 \subseteq A_n$, so $H = B_n$ is the desired homogeneous set. □

It is not too hard to generalize this result. First, we define the transfinite sequence of *beths*, the cardinal numbers obtained by iterated exponentiation.

2.4 Definition

$$\beth_0 = \aleph_0;$$
$$\beth_{\alpha+1} = 2^{\beth_\alpha};$$
$$\beth_\lambda = \sup\{\beth_\alpha \mid \alpha < \lambda\} \text{ for limit } \lambda \neq 0.$$

A more general version of the Erdős-Radó Partition Theorem can now be stated.

2.5 Theorem $(\beth_n)^+ \to (\aleph_1)^{n+1}_{\aleph_0}$ *holds for all* $n \in \boldsymbol{N}$.

The proof is by induction on n, following closely the special case $n = 1$ in Theorem 2.3. There are analogous results for any infinite cardinal κ — see Exercise 2.5. Many results of similar nature, both positive and negative, have been obtained by researchers in the part of set theory known as infinitary combinatorics. One problem in particular has played a very important role in the development of modern set theory. It is the question of whether there are any uncountable cardinal numbers κ for which an analogue of Ramsey's Theorem holds.

2.6 Definition An uncountable cardinal κ is called *weakly compact* if $\kappa \to (\kappa)^r_s$ holds for all $r, s \in \boldsymbol{N} - \{0\}$.

2.7 Theorem *Weakly compact cardinals are strongly inaccessible.*

Proof. We have to prove that a weakly compact cardinal κ is regular and a strong limit cardinal.

(i) Regularity: Assume $\kappa = \bigcup_{\nu<\lambda} P_\nu$ where $\lambda < \kappa$ and $|P_\nu| < \kappa$ for all $\nu < \lambda$. Define a partition of $[\kappa]^2$ by:

$$\{\alpha, \beta\} \in A_0 \text{ if and only if there is } \nu < \lambda \text{ such that } \alpha \in P_\nu \text{ and } \beta \in P_\nu;$$
$$\{\alpha, \beta\} \in A_1 \text{ otherwise.}$$

Clearly $\{A_0, A_1\}$ cannot have a homogeneous set of cardinality κ.

(ii) Strong limit property: Assume $\lambda < \kappa \leq 2^\lambda$. By Exercise 2.1, $2^\lambda \not\to (\lambda^+)^2_2$, hence $\kappa \not\to (\lambda^+)^2_2$ and, as $\lambda^+ \leq \kappa$, $\kappa \not\to (\kappa)^2_2$.

□

Exercises

2.1 Prove that $2^\kappa \not\to (\kappa^+)^2_2$ for any infinite κ.
[*Hint*: Follow the proof of Theorem 2.1, but replace $(\boldsymbol{R}, \leq)$ by the set $\{0,1\}^\kappa$ ordered lexicographically (see Chapter 4, Section 4).]

2.2 Prove that $\kappa \to (\kappa, \aleph_0)^2_2$ holds for all infinite regular κ.

2.3 Prove: If X is an infinite set and $\leq$ and $\preccurlyeq$ are two well-orderings of X, then there is $Y \subseteq X$ with $|Y| = |X|$ such that $y_1 \leq y_2$ if and only if $y_1 \preccurlyeq y_2$ holds for all $y_1, y_2 \in Y$.
[*Hint*: Let $A = \{\{x, y\} \in [X]^2 \mid x < y \text{ and } x \prec y\}$ and $B = \{\{x, y\} \in [X]^2 \mid x < y \text{ and } x \succ y\}$. Apply the Erdős-Dushnik-Miller Theorem and show that B cannot have an infinite homogeneous set.]

2.4 Prove that $(\beth_n)^+ \to (\aleph_1)^{n+1}_{\aleph_0}$.
[*Hint*: Induction.]

2.5 For any cardinal κ define $\exp_0(\kappa) = \kappa$, $\exp_{n+1}(\kappa) = 2^{\exp_n(\kappa)}$. Prove: For any infinite κ, $(\exp_n(\kappa))^+ \to (\kappa^+)^{n+1}_{\kappa}$. In particular, $(2^\kappa)^+ \to (\kappa^+)^2_2$.

3. Trees

Like partitions, trees originated in finite combinatorics, but turned out to be of great interest to set theorists as well.

3.1 Definition A *tree* is an ordered set $(T, \leq)$ which has a least element and is such that, for every $x \in T$, the set $\{y \in T \mid y < x\}$ is well-ordered by $\leq$. See Figure 1.

Elements of T are called *nodes.* If $x, y \in T$ and $y < x$, we say that y is a *predecessor* of x and x is a *successor* of y. The unique least element of T is the *root.* By Theorem 3.1 in Chapter 6, the well-ordered set $\{y \in T \mid y < x\}$ of all predecessors of x is isomorphic to a unique ordinal number $h(x)$, the *height* of x. The set $T_\alpha = \{x \in T \mid h(x) = \alpha\}$ is the *αth level* of T. If $h(x)$ is a successor ordinal, x is called a *successor node*, otherwise it is a *limit node.* The least α for which $T_\alpha = \emptyset$ is called the *height* of the tree T, $h(T)$.

A *branch* in T is a maximal chain (i.e., a linearly ordered subset) in T. The order type of a branch b is called its *length* and is denoted $\ell(b)$. It is always an ordinal number less than or equal to the height of T. A branch whose length equals the height of the tree is called *cofinal.*

A subset T' of T is a *subtree* of T if for all $x \in T'$, $y \in T$, $y < x$ implies $y \in T'$. Then T' is also a tree (when ordered by $\leq$) and $T'_\alpha = T_\alpha \cap T'$ for all $\alpha < h(T')$. The set $T^{(\alpha)} = \bigcup_{\beta<\alpha} T_\beta$ is a subtree of T, for any $\alpha \leq h(T)$, and $h(T^{(\alpha)}) = \alpha$. If $x \in T_\alpha$ then $\{y \in T \mid y < x\}$ is a branch of $T^{(\alpha)}$ of length α; however, $T^{(\alpha)}$ may have other branches of length α as well (if α is a limit ordinal).

Finally, a set $A \subseteq T$ is an *antichain* in T if any two distinct elements of A are incomparable, i.e., $x, y \in A$, $x \neq y$ implies that neither $x < y$ nor $y < x$. The reader is strongly urged to work out Exercises 3.1 and 3.2, where some simple properties of these concepts are developed.

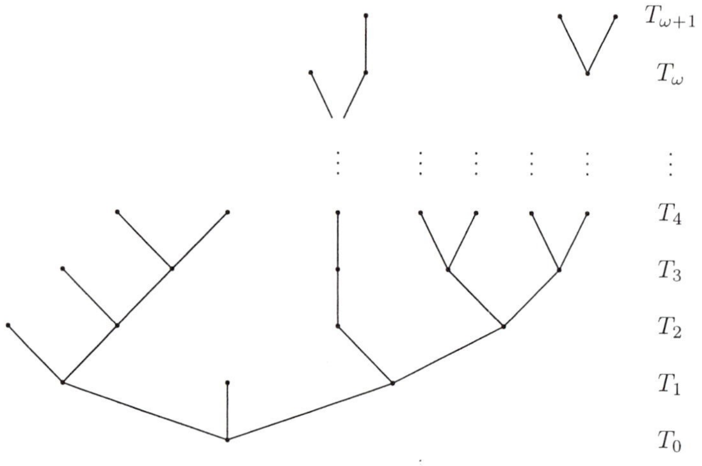

Figure 1

It is time for some examples of trees.

3.2 Example (a) Every well-ordered set $(W, \leq)$ is a tree. Hence trees can be viewed as generalizations of well-orderings. $h(W)$ is the order type of W; the only branch is W itself, and it is cofinal.

(b) Let λ be an ordinal number and A a nonempty set. Define $A^{<\lambda} = \bigcup_{\alpha<\lambda} A^\alpha$ to be the set of all transfinite sequences of elements of A of length less than λ. We let $T = A^{<\lambda}$ and order it by $\subseteq$; so $f \leq g$ for $f, g \in T$ means $f \subseteq g$, i.e., $f = g \restriction \operatorname{dom} f$. It is easy to verify that T is a tree. For $f \in T$, $h(f) = \alpha$ if and only if $f \in A_\alpha$; i.e., $T_\alpha = A^\alpha$. For $\alpha = \beta + 1$ and $f \in A^\alpha$, $f \restriction \beta$ is the immediate predecessor of f, and all $f \cup \{\langle \beta, a\rangle\}$, $a \in A$, are immediate successors of f. Branches in T are in one-to-one correspondence with functions from λ into A: if $F \in A^\lambda$ then $\{F \restriction \alpha \mid \alpha < \lambda\}$ is a branch in T. Conversely, if B is a branch in T then B is a compatible system of functions and $F = \bigcup B \in A^\lambda$. We note that all branches are cofinal.

(c) More generally, if $T \subseteq A^{<\lambda}$ is a subtree of $(A^{<\lambda}, \subseteq)$, branches in T are in one-to-one correspondence with those functions $F \in A^{<\lambda} \cup A^\lambda$ for which $F \restriction \alpha \in T$ for all $\alpha \in \operatorname{dom} F$ and either $F \notin T$, or $F \in T$ and F has no successors in T. We often identify branches and their corresponding functions in this situation.

(d) Let $A = \boldsymbol{N}$, $\lambda = \omega$. Let $T \subseteq \boldsymbol{N}^{<\omega}$ be the set of all finite decreasing sequences; i.e., $f \in T$ if and only if $f(i) > f(j)$ holds for all $i < j < \operatorname{dom} f \in \boldsymbol{N}$. Then T is a subtree of $(\boldsymbol{N}^{<\omega}, \subseteq)$. T is a tree of height ω which has no cofinal branches (see Exercise 2.8, Chapter 3).

(e) Let $(R, \leq)$ be a linearly ordered set. A *representation* of a tree $(T, \preccurlyeq)$ *by intervals* in $(R, \leq)$ is a one-to-one function Φ which assigns to each $x \in T$

an interval $\Phi(x)$ in $(R, \leq)$ so that, for $x, y \in T$,

(i) $x \preccurlyeq y$ if and only if $\Phi(x) \supseteq \Phi(y)$;

(ii) x and y are incomparable if and only if $\Phi(x) \cap \Phi(y) = \emptyset$.

This implies, in particular, that $(\Phi[T], \supseteq)$ is a tree isomorphic to $(T, \preccurlyeq)$. For example, the system $\langle D_s \mid s \in S\rangle$ constructed in Example 3.18, Chapter 10, is a representation of the tree $S = \mathrm{Seq}(\{0,1\}) = \{0,1\}^{<\omega}$ (ordered by $\subseteq$) by closed intervals on the real line.

The study of finite trees is one of the key concerns of combinatorics. We do not pursue it here, and turn our attention instead to infinite trees. We concern ourselves mainly with the question, under what circumstances does a tree have a cofinal branch. For trees whose height is a successor ordinal the answer is obvious: if $h(T) = \alpha + 1$ then $T_\alpha \neq \emptyset$ and $\{y \in T \mid y \leq x\}$ is a cofinal branch in T, for any $x \in T_\alpha$. Henceforth, we concentrate on trees of limit height. Example 3.2(d) shows that there are trees of height ω that have only finite branches. The next theorem, the most basic observation relevant to our question, shows that this cannot happen if the tree is sufficiently "slim."

3.3 König's Lemma *If T is a tree of height ω, all levels of which are finite, then T has a branch of length ω.*

Equivalently, every tree of height ω such that each node has finitely many immediate successors has an infinite branch.

Proof. We use recursion to construct an infinite sequence $\langle c_n \rangle_{n=0}^{\infty}$ of nodes of T so that, for each n, $\{a \in T \mid c_n \leq a\}$ is infinite.

We let c_0 be the root of T and note that $\{a \in T \mid c_0 \leq a\} = T$ is infinite. Given c_n such that $\{a \in T \mid c_n \leq a\}$ is infinite, we observe that

$$\{a \in T \mid c_n \leq a\} = \{c_n\} \cup \bigcup_{b \in S} \{a \in T \mid b \leq a\}$$

where S is the finite set of all immediate successors of c_n (Exercise 3.1(v)). Hence, for at least one $b \in S$, $\{a \in T \mid b \leq a\}$ is infinite, and we let c_{n+1} be one such b. It is easy to verify that $\{a \in T \mid a \leq c_n \text{ for some } n \in \boldsymbol{N}\}$ is a branch in T of length ω. □

There is an important point to be made about this recursive construction, and many similar ones to come. The Recursion Theorem, as formulated in Section 3 of Chapter 3, requires a function g, whose purpose is to "compute" c_{n+1} from c_n. In the previous proof, we did not explicitly specify such a g, and indeed some form of the Axiom of Choice is needed to do so. For example, let k be a choice function for $\mathcal{P}(T)$. Let S_c denote the set of all immediate successors of c in T. If we let $g(c, n) = k(\{b \in S_c \mid \{a \in T \mid b \leq a\} \text{ is infinite}\})$, then $c_{n+1} = g(c_n, n)$ defines $\langle c_n \mid n \in \boldsymbol{N}\rangle$ in conformity with the Recursion Theorem.

It is customary to omit this kind of detailed justification, and we do so from now on. The reader is encouraged to supply the details of the next few applications of the Recursion Theorem or the Transfinite Recursion Theorem as an exercise.

Returning to König's Lemma: there are several interesting ways to generalize it. For example, it is fairly easy to prove that any tree of height κ whose levels are finite has a branch of length κ (Exercise 3.3). We consider the following question: let T be a tree of height ω_1, each level of which is at most countable; does it have to have a branch of length ω_1? It turns out that the answer is "no."

3.4 Definition A tree of height ω_1 is called an *Aronszajn tree* if all its levels are at most countable and if it has no branch of length ω_1.

3.5 Theorem *Aronszajn trees of height ω_1 exist.*

Proof. We construct the levels T_α, $\alpha < \omega_1$, of an Aronszajn tree by transfinite recursion in such a way that

(i) $T_\alpha \subseteq \omega^\alpha$; $|T_\alpha| \leq \aleph_0$;

(ii) If $f \in T_\alpha$ then f is one-to-one and $(\omega - \operatorname{ran} f)$ is infinite;

(iii) If $f \in T_\alpha$ and $\beta < \alpha$ then $f \restriction \beta \in T_\beta$;

(iv) For any $\beta < \alpha$, any $g \in T_\beta$, and any finite $X \subseteq \omega - \operatorname{ran} g$, there is $f \in T_\alpha$ such that $f \supseteq g$ and $\operatorname{ran} f \cap X = \emptyset$.

Let us assume this has been done and let us show that $T = \bigcup_{\alpha<\omega_1} T_\alpha$ is then an Aronszajn tree. Clearly T is a tree (by (iii)), each level is at most countable (by (i)), and its height is ω_1 (by (iv), each $T_\alpha \neq \emptyset$). If B were a branch of length ω_1 in T, then $F = \bigcup B$ would be a one-to-one function from ω_1 into ω (by (ii)), a contradiction.

It remains to implement the construction of T_α's. We let $T_0 = \{\emptyset\}$. Assuming T_α satisfying (i)–(iv) has been constructed, we let $T_{\alpha+1} = \{g \cup \{\langle \alpha, a\rangle\} \mid g \in T_\alpha,\ a \in \omega - \operatorname{ran} g\}$. It is easy to check (i)–(iv) hold for $\alpha + 1$.

It remains to construct T_α for α limit. For any $g \in T_\beta$, $\beta < \alpha$, and any finite $X \subseteq (\omega - \operatorname{ran} g)$ we construct a particular $f = f(g, X)$ by recursion as follows. Fix an increasing sequence $\langle \alpha_n \rangle_{n=0}^{\infty}$ such that $\alpha_0 = \beta$ and $\sup\{\alpha_n \mid n \in \boldsymbol{N}\} = \alpha$. Let $f_0 = g \in T_{\alpha_0}$ and $X_0 = X \subseteq \omega - \operatorname{ran} f_0$. Having defined $f_n \in T_{\alpha_n}$ and finite $X_n \subseteq \omega - \operatorname{ran} f_n$, we first take some finite $X_{n+1} \supset X_n$, $X_{n+1} \subseteq \omega - \operatorname{ran} f_n$ (that is possible because the last set is infinite, by (ii)) and then select some $f_{n+1} \in T_{\alpha_{n+1}}$ such that $f_{n+1} \supseteq f_n$ and $X_{n+1} \cap \operatorname{ran} f_{n+1} = \emptyset$ (possible by (iv)). We let $f = \bigcup_{n=0}^{\infty} f_n$. Clearly $f : \alpha \to \omega$, f is one-to-one (because all f_n are), $\operatorname{ran} f \cap (\bigcup_{n=0}^{\infty} X_n) = \emptyset$, so $\omega - \operatorname{ran} f$ is infinite, and $\operatorname{ran} f \cap X = \emptyset$. Thus f satisfies (ii). For $\beta < \alpha$, $f \restriction \beta = f_n \restriction \beta$ when $\beta < \alpha_n$, so (iii) is satisfied as well.

We put this $f = f(g, X)$ into T_α for each $g \in \bigcup_{\beta<\alpha} T_\beta$ and each finite $X \subseteq \omega - \operatorname{ran} g$. Thus (iv) is satisfied. As $|\bigcup_{\beta<\alpha} T_\beta| \leq \sum_{\beta<\alpha} |T_\beta| \leq \aleph_0$ (by inductive assumption (i)) and the number of finite subsets of ω is countable, the set T_α is at most countable, and (i) is satisfied as well. □

More generally, a tree of height κ (κ an uncountable cardinal) is called an *Aronszajn tree* if all its levels have cardinality less than κ and there are no branches of length κ. The question of existence of such trees is very complicated and not yet fully resolved. It is easy to show that they always exist when κ is singular (Exercise 3.4). Of particular interest are uncountable cardinals for which an analogue of König's Lemma holds, i.e., no Aronszajn trees of height κ exist; such cardinals are said to have the *tree property*. It turns out that strongly inaccessible cardinals with the tree property are precisely the weakly compact cardinals defined in Section 2.

Exercises

3.1 Let $(T, \leq)$ be a tree. Prove

(i) The root is the unique $r \in T$ for which $h(r) = 0$. In particular, $T_0 \neq \emptyset$.

(ii) If $\alpha \neq \beta$ then $T_\alpha \cap T_\beta = \emptyset$.

(iii) $T = T^{(h(T))} = \bigcup_{\alpha<h(T)} T_\alpha$.

(iv) $x \in T$ is a successor node if and only if there exists $y \in T$ such that $y < x$ and there is no $z \in T$ for which $y < z < x$ holds. If x is a successor node then such y is unique; it is called the *immediate predecessor* of x, and x is called an *immediate successor* of y. (Each node has at most one immediate predecessor, but it may have many immediate successors.)

(v) If $x < y$ then there is a unique $z \in T$ such that $x < z \leq y$ and z is an immediate successor of x.

(vi) $h(T) = \sup\{\alpha + 1 \mid T_\alpha \neq \emptyset\} = \sup\{h(x) + 1 \mid x \in T\}$.

3.2 Prove:

(i) Every chain in T is well-ordered.

(ii) If b is a branch in T, $x \in b$, and $y < x$, then $y \in b$.

(iii) If b is a branch in T, then $|b \cap T_\alpha| = 1$ for $\alpha < \ell(b)$ and $|b \cap T_\alpha| = 0$ for $\alpha > \ell(b)$. Conclude that $\ell(b) \leq h(T)$.

(iv) Show that $h(T) = \sup\{\ell(b) \mid b \text{ is a branch in } T\}$.

(v) Show that each T_α ($\alpha < h(T)$) is an antichain in T.

3.3 Let $(T, \leq)$ be a tree of height κ (κ an infinite cardinal), all levels of which are finite. Prove that T has a branch of length κ.
[*Hint*: Let $U_\alpha = \{x \in T_\alpha \mid |\{y \in T \mid x \leq y\}| = \kappa\}$. Prove that $\{|U_\alpha| \mid \alpha < \kappa\} \subseteq \boldsymbol{N}$ is bounded; let m be its maximum ($m \geq 1$). Show that T has exactly m branches of length κ.]

3.4 Construct an Aronszajn tree of height $\aleph_\omega$. Generalize to an arbitrary singular cardinal κ.

4. Suslin's Problem

We proved in Chapter 4, Theorem 5.7, that the real numbers, with their usual ordering, are the unique (up to isomorphism) complete linearly ordered set without endpoints that has a countable dense subset. As an immediate consequence, every collection of mutually disjoint open intervals in $(\boldsymbol{R}, <)$ is at most countable (Theorem 3.2 in Chapter 10). Suslin asked in 1920 whether the ordering of real numbers is uniquely characterized by this weaker property.

4.1 Definition A *Suslin line* is a complete linearly ordered set without endpoints where every collection of mutually disjoint open intervals is at most countable, but where there is no countable dense subset.

The famous Suslin's Hypothesis asserts that there are no Suslin lines. It is now known that Suslin's Hypothesis can be neither proved nor refuted from the axioms of Zermelo-Fraenkel set theory with Choice. In this section we establish a relationship between Suslin lines and certain kinds of trees. In the next section we consider some additional axioms of combinatorial nature that lead to an answer to Suslin's problem.

Theorem 3.5 makes it clear that trees of height ω_1 with countable levels are not always "slim enough" to have a cofinal branch. However, we can impose a stronger condition: we can require that all antichains (not just the levels) be countable.

4.2 Definition A tree of height ω_1 is called a *Suslin tree* if all its antichains are at most countable and there are no branches of length ω_1.

Clearly a Suslin tree is an Aronszajn tree, but the converse need not be true. As the name indicates, there is a close connection between Suslin trees and Suslin lines. It is this connection that we proceed to establish.

Let $(S, <)$ be a Suslin line; we recall that every complete ordering is by definition dense. We construct a tree $T \subseteq \boldsymbol{N}^{<\omega_1}$ and its representation Φ by open intervals in S using transfinite recursion of length ω_1. More precisely, we construct the levels $T_\alpha \subseteq \boldsymbol{N}^\alpha$ and mappings Φ_α so that, for each $\alpha < \omega_1$,

$$(*)\quad \begin{cases} |T_\alpha| \leq \aleph_0,\ T^{(\alpha+1)} = \bigcup_{\beta\leq\alpha} T_\beta \text{ is a tree, and} \\ \Phi^{(\alpha+1)} = \bigcup_{\beta\leq\alpha} \Phi_\beta \text{ is a representation of } T^{(\alpha+1)} \text{ by intervals on } (S,<). \end{cases}$$

We let $T_0 = \{\emptyset\}$, $\Phi(\emptyset) = S$. Assume $T_\alpha \subseteq \boldsymbol{N}^\alpha$ has been constructed and satisfies (*). For each $f \in T_\alpha$ and $n \in \boldsymbol{N} - \{0\}$ we put $f_n = f \cup \{\langle \alpha, n\rangle\}$ into $T_{\alpha+1}$. Let $\Phi_\alpha(f) = (a, b)$; using density of $(S, <)$ we pick an increasing sequence $a = a_1 < a_2 < a_3 < \cdots < b$ and let $\Phi_{\alpha+1}(f_n) = (a_n, a_{n+1})$ for $n \geq 1$. If $a_0 = \sup\{a_n\}_{n=1}^\infty < b$ we also put $f_0 = f \cup \{\langle\alpha, 0\rangle\}$ into $T_{\alpha+1}$ and set $\Phi_{\alpha+1}(f_0) = (a_0, b)$. It is clear that the condition (*) holds for $\alpha + 1$.

Now let $\alpha < \omega_1$ be a limit ordinal. We assume that T_β and Φ_β satisfying (*) have been constructed for all $\beta < \alpha$. It is then clear that $T^{(\alpha)} = \bigcup_{\beta<\alpha} T_\beta$ is a tree, $|T^{(\alpha)}| \le \aleph_0$, and $\Phi^{(\alpha)} = \bigcup_{\beta<\alpha} \Phi_\beta$ is a representation of $T^{(\alpha)}$ by intervals in S.

Let f be a branch in $T^{(\alpha)}$; we can think of it as an element of $\boldsymbol{N}^\alpha$. Then $\Phi^{(\alpha)}(f \restriction \beta) = (a_\beta, b_\beta)$ is an interval in S and $\beta < \gamma < \alpha$ implies $a_\beta \le a_\gamma < b_\gamma \le b_\beta$. Put $a = \sup_{\beta<\alpha} a_\beta$, $b = \inf_{\beta<\alpha} b_\beta$ (here we use the completeness of S); clearly $a \le b$. If in fact $a < b$, we put f into T_α and define $\Phi_\alpha(f) = (a, b)$. It is easy to verify that $T^{(\alpha+1)} = T^{(\alpha)} \cup T_\alpha$ is a tree and $\Phi^{(\alpha+1)} = \Phi^{(\alpha)} \cup \Phi_\alpha$ is its representation (for example, if $f, g \in T_\alpha$ are incompatible, we consider the first β where $f(\beta) \ne g(\beta)$; we then have $f \restriction (\beta+1) \ne g \restriction (\beta+1)$, so $\Phi^{(\alpha)}(f \restriction (\beta+1)) \cap \Phi^{(\alpha)}(g \restriction (\beta+1)) = \emptyset$ by inductive assumption, and so $\Phi(f) \cap \Phi(g) = \emptyset$). The set T_α is at most countable, because otherwise $\{\Phi^{(\alpha)}(f) \mid f \in T_\alpha\}$ would be an uncountable collection of mutually disjoint intervals in S. Thus (*) is satisfied at stage α.

This completes the recursive construction. We let $T = \bigcup_{\alpha<\omega_1} T_\alpha$ and $\Phi = \bigcup_{\alpha<\omega_1} \Phi_\alpha$. Clearly T is a tree and Φ represents it by intervals in $(S, <)$. We claim that T is a Suslin tree. It is clear that T has no uncountable antichains: the image of any antichain in T by the one-to-one function Φ is a collection of mutually disjoint intervals in S and so it is at most countable.

We next show that T has no branches of length ω_1.

4.3 Claim *Let $(T, \le)$ be a tree where each node has at least two immediate successors. If T has no uncountable antichains then T has no branches of length $\ge \omega_1$.*

Proof. Let $\langle x_\alpha \mid \alpha < \omega_1\rangle$ be a chain in T, where $x_\alpha \in T_\alpha$. The set of immediate successors of x_α has at least 2 elements, so let $y_{\alpha+1}$ be an immediate successor of x_α different from $x_{\alpha+1}$. Then $\{y_{\alpha+1} \mid \alpha < \omega_1\}$ is an antichain in T. □

It remains to show that T has height ω_1, i.e., that each $T_\alpha \ne \emptyset$. We proceed by contradiction. Let $\overline{\alpha} < \omega_1$ be the first ordinal for which $T_{\overline{\alpha}} = \emptyset$; it is clear from the construction that $\overline{\alpha}$ has to be a limit ordinal. Let C be the set of all endpoints of all intervals $\Phi(g)$, $g \in T^{(\overline{\alpha})}$; then C is at most countable, and cannot be dense in S. In other words, there is an interval (c, d) disjoint with C. We use it to construct a branch in $T^{(\overline{\alpha})}$ as follows: we let $F(0) = \emptyset$ and note $(c, d) \subseteq S = \Phi(F(0))$. Given $F(\alpha) \in T_\alpha$ such that $(c, d) \subseteq \Phi(F(\alpha)) = (a, b)$ and the sequence $a = a_1 < a_2 < a_3 < \cdots < a_0 \le b$ used to construct $\Phi_{\alpha+1}$, we note that $C \cap (c, d) = \emptyset$ implies that there is a unique $n \in \boldsymbol{N}$ such that $(c, d) \subseteq (a_n, a_{n+1})$ (or $(c, d) \subseteq (a_0, b)$, in case $n = 0$); we let $F(\alpha+1) = F(\alpha) \cup \{\langle \alpha, n\rangle\}$ for that n. At α limit, $\bigcap_{\beta<\alpha} \Phi(F(\beta)) \supseteq (c, d)$ by the inductive assumption, so $g = \bigcup_{\beta<\alpha} F(\beta) \in T_\alpha$ and $\Phi(g) \supseteq (c, d)$, and we let $F(\alpha) = g$.

The resulting branch $f = \bigcup_{\alpha<\overline{\alpha}} F(\alpha)$ clearly again has $\bigcap_{\alpha<\overline{\alpha}} \Phi(f \restriction \alpha) \supseteq (c, d)$, so $f \in T_{\overline{\alpha}}$ according to the limit step of the recursive construction of $T_{\overline{\alpha}}$. Hence $T_{\overline{\alpha}} \ne \emptyset$, a contradiction.

The Suslin tree just constructed has two additional properties, both obvious from the proof:

(i) The set of all immediate successors of each node is countable.

(ii) If $x, y \in T_\alpha$ for limit $\alpha < h(T)$ and $\{z \in T \mid z < x\} = \{z \in T \mid z < y\}$, then $x = y$.

We call trees with properties (i) and (ii) *regular.*

The arguments just completed prove the following theorem.

4.4 Theorem *If a Suslin line exists then there exists a regular Suslin tree.*

The theorem has a converse.

4.5 Theorem *If a regular Suslin tree exists then there is a Suslin line.*

The existence of Suslin lines is thus equivalent to the existence of regular Suslin trees. It is in fact equivalent to the existence of Suslin trees, but we omit the unappealing proof of this fact.

Proof. Let $(T, \preccurlyeq)$ be a regular Suslin tree; without loss of generality we can assume that $T \subseteq A^{\omega_1}$ for some set A, and $\preccurlyeq$ is $\supseteq$ (Exercise 4.3). For each $f \in T$ the set S_f of immediate successors of f is countable, and we fix a dense linear ordering $\prec_f$ of it, without endpoints. We note that the length of any branch b in T must be a limit ordinal (because each node in T has successors) less than ω_1 (because T is Suslin). We let B be the set of all branches of T, and order it "lexicographically." More precisely, we view B as a subset of $\bigcup\{T^\alpha \mid \alpha < \omega_1,\ \alpha \text{ limit}\}$ and for $b, b' \in B$, $b \neq b'$, we put

$$b < b' \quad \text{if and only if} \quad b \restriction (\alpha + 1) \prec_{b\restriction\alpha} b' \restriction (\alpha + 1)$$

where α is the least ordinal such that $b(\alpha) \neq b'(\alpha)$. (We note that $b \subset b'$ or $b' \subset b$ is impossible: branches are maximal chains.) Arguments similar to those used in the proof of Theorem 4.7 in Chapter 4 show that $<$ is a linear ordering. If $b < b'$ and α is as above, there exist $g \in S_{b\restriction\alpha}$ such that $b \restriction (\alpha+1) \prec_{b\restriction\alpha} g \prec_{b\restriction\alpha} b' \restriction (\alpha + 1)$ (by density of $\prec_{b\restriction\alpha}$). Any branch $b'' \supseteq g$ then satisfies $b < b'' < b'$, so $<$ is dense on B. Similar arguments show B has no least or greatest element.

We claim that every system of mutually disjoint intervals in B is at most countable. Assume to the contrary that $\{(b_i, b_i') \mid i < \omega_1\}$ is a disjoint system. Let α_i be the first ordinal where $b_i(\alpha_i) \neq b_i'(\alpha_i)$ and let $g_i \in S_{b_i\restriction\alpha_i}$ be such that $b_i \restriction (\alpha_i + 1) \prec g_i \prec b_i' \restriction (\alpha_i + 1)$. It is easy to check that $\{g_i \mid i < \omega_1\}$ is an antichain in T, a contradiction.

Finally, we show that B does not have a countable dense subset. Let $C = \{b_n \mid n \in \boldsymbol{N}\}$ be one. Let $\alpha = \sup\{\ell(b_n) \mid n \in \boldsymbol{N}\}$ be the supremum of the lengths of the branches in C; as $|C| \leq \aleph_0$ and each $\ell(b_n) < \omega_1$, regularity of ω_1 implies $\alpha < \omega_1$ as well. But $T_\alpha \neq \emptyset$, so take some $f \in T_\alpha$, two of its immediate successors, say $f_1 \prec_f f_2$, and some branches $b_1 \supseteq f_1$, $b_2 \supseteq f_2$. We then have $b_1 < b_2$, but the interval (b_1, b_2) is clearly disjoint with C.

In summary, $(B, <)$ has all the properties of a Suslin line except completeness. We let $(\overline{B}, <)$ be the Dedekind completion of $(B, <)$ as constructed in

Section 5 of Chapter 4. It is an easy exercise (Exercise 4.4) to verify that $(\overline{B}, <)$ is a Suslin line. □

Exercises

4.1 Let $(S, <_1)$ be a Suslin line, $(\{a\}, <_2)$ a one-element set with its unique ordering, and $(\boldsymbol{R}, <_3)$ the real line. Show that the sum of the ordered sets $(S, <_1)$, $(\{a\}, <_2)$, and $(\boldsymbol{R}, <_3)$ is a Suslin line.

4.2 Let $(S, <)$ be a Suslin line; prove that $|S| \leq 2^{\aleph_0}$.
[*Hint*: Let T and Φ be as in the proof of Theorem 4.4; by letting $\overline{\alpha} = \omega_1$ in the argument showing that T has height ω_1 we get a set C of $|C| = \aleph_1$ which is dense in S. Show that each $x \in S$ is a limit of a sequence of elements of C; then apply Exercise 3.1 in Chapter 9.]

4.3 Show that a tree has property (ii) from the definition of regularity if and only if it is isomorphic to a tree of transfinite sequences (as in Example 3.2(c)).

4.4 Let $(B, <)$ be a dense linear ordering without endpoints that has all the properties of a Suslin line, except completeness. Show that its Dedekind completion $(\overline{B}, <)$ is a Suslin line.
[*Hint*: If $(\overline{B}, <)$ had a countable dense subset, then it would be isomorphic to the real line. But every infinite subset of $\boldsymbol{R}$ has a countable dense subset; so B would have a countable dense subset.]

4.5 A regular Suslin tree $(T, \leq)$ is *normal* if for all $\alpha < \beta < \omega_1$ and all $x \in T_\alpha$ there exists $y \in T_\beta$ such that $x < y$. A Suslin line $(L, \leq)$ is *proper* if no open interval in L has a countable dense subset. Show that a Suslin tree is normal if and only if the Suslin line constructed from it as in the proof of Theorem 4.5 is proper.

4.6 Prove: If a Suslin line exists, then a proper Suslin line exists.
[*Hint*: Define an equivalence relation on the Suslin line $(L, \leq)$ by $x \sim y$ if and only if the interval with endpoints x, y contains a countable dense subset. Show that each $[x]$ contains a countable dense subset. Let $\tilde{L} = \{[x] \mid x \in L\}$; let $[x] \preccurlyeq [y]$ if and only if $x \leq y$ and show that $(\tilde{L}, \preccurlyeq)$ is a proper Suslin line.]

5. Combinatorial Principles

Suslin's Hypothesis can be neither proved nor refuted from the axioms of ZFC. This was shown in the sixties by constructing models where the axioms of ZFC are satisfied and Suslin's Hypothesis fails, as well as such models where it is true. The study of models of set theory requires familiarity with formal logic and is beyond the scope of this book, except for a brief introduction in Chapter 15. However, early in the development of the subject researchers isolated a number of general principles of combinatorial nature which hold in one or another model of ZFC and have, as a consequence, a definite answer to Suslin's Problem, as well as to a number of other questions. If one is willing to accept such a principle as an additional axiom of set theory, one can then, by purely classical methods, without deep study of logic, prove from it results unavailable in ZFC alone. This has become the way to proceed in general topology and some parts of abstract algebra. We demonstrate this approach here by two examples, Jensen's Principle Diamond and Martin's Axiom. We want to stress that these axioms do not have the same epistemological status as the axioms of ZFC. At the present state of set theory there are no compelling reasons to believe that one or another of them is intuitively true. It is merely known that they do not lead to contradictions, assuming ZFC does not. Their significance lies in systematizing the morass of consistency results obtained by the technique of models.

The first combinatorial principle we consider is Jensen's Principle $\diamond$.

Principle $\diamond$ There exists a sequence $\langle W_\alpha \mid \alpha < \omega_1 \rangle$ such that, for each $\alpha < \omega_1$, $W_\alpha \subseteq \mathcal{P}(\alpha)$, W_α is at most countable, and for any $X \subseteq \omega_1$ the set $\{\alpha < \omega_1 \mid X \cap \alpha \in W_\alpha\}$ is stationary.

Principle $\diamond$ is known to hold in the constructible model (see Chapter 15, Section 2). Here we prove two of its consequences; others can be found in the exercises.

5.1 Theorem *If $\diamond$ holds, then $2^{\aleph_0} = \aleph_1$.*

Proof. If $X \subseteq \omega \subseteq \omega_1$ then $S_X = \{\alpha < \omega_1 \mid X \cap \alpha \in W_\alpha\}$ is stationary. For $\alpha \in S_X$, $\alpha \geq \omega$ this implies $X = X \cap \alpha \in W_\alpha$. So $\mathcal{P}(\omega) \subseteq \bigcup_{\alpha<\omega_1} W_\alpha$, but the cardinality of the last set is at most $\sum_{\alpha<\omega_1} \aleph_0 = \aleph_0 \cdot \aleph_1 = \aleph_1$. □

5.2 Theorem *If $\diamond$ holds, then Suslin lines exist.*

The key idea of the proof is as follows. As usual, we construct $T = \bigcup_{\alpha<\omega_1} T_\alpha$ by transfinite recursion. We want to make sure that T has no uncountable antichains. As $|T| = \aleph_1$, there are $2^{\aleph_1} > \aleph_1$ subsets of T, hence more than $\aleph_1$ sets $X \subseteq T$ which our construction has to prevent from becoming antichains. The crux of the matter is being able to take care of more than $\aleph_1$ sets in $\aleph_1$ steps. It is here that $\diamond$ is used. Roughly speaking, we proceed in such a way that those $A \in W_\alpha$ that are maximal antichains in the subtree $T^{(\alpha)}$ constructed before stage α can no longer grow — they remain maximal antichains for the rest of the construction of T. $\diamond$ implies that when X is a maximal antichain in T, then $X \cap T^{(\alpha)}$ is a maximal antichain in $T^{(\alpha)}$ and $X \cap T^{(\alpha)} \in W_\alpha$, for some $\alpha < \omega_1$. As $X \cap T^{(\alpha)}$ subsequently does not grow, we must have $X = X \cap T^{(\alpha)}$ so in particular X is at most countable.

We now proceed with the details. One minor problem arises because $\diamond$ applies to subsets of ω_1, while we plan to construct T as a subset of $\omega^{<\omega_1}$. The following lemma takes care of this.

5.3 Lemma *$\diamond$ implies that there exists a sequence $\langle Z_\alpha \mid \alpha < \omega_1 \rangle$ such that, for each $\alpha < \omega_1$, $Z_\alpha \subseteq \omega^{<\alpha}$, Z_α is at most countable, and for any $X \subseteq \omega^{<\omega_1}$ the set $\{\alpha < \omega_1 \mid X \cap \omega^{<\alpha} \in Z_\alpha\}$ is stationary.*

Proof. We note that $|\omega^{<\omega_1}| = |\bigcup_{\alpha<\omega_1} \omega^\alpha| = \sum_{\alpha<\omega_1} 2^{\aleph_0} = \aleph_1$ because $2^{\aleph_0} = \aleph_1$ by Theorem 5.1. Let F be a one-to-one mapping of $\omega^{<\omega_1}$ onto ω_1. We claim that $S_F = \{\alpha < \omega_1 \mid \sup F[\omega^{<\alpha}] = \alpha\}$ is closed unbounded.

It is clear that S_F is closed. Given $\beta < \omega_1$ we define recursively: $\alpha_0 = \beta$, $\alpha_{n+1} = \sup F[\omega^{<\alpha_n}]$, and set $\alpha = \sup\{\alpha_n \mid n < \omega\}$. It is clear that $\beta \leq \alpha < \omega_1$ and $\alpha \in S_F$.

We now set $Z_\alpha = \{F^{-1}[A] \cap \omega^{<\alpha} \mid A \in W_\alpha\}$ if $\alpha \in S_F$, $Z_\alpha = \emptyset$ otherwise. Clearly $Z_\alpha \subseteq \omega^{<\alpha}$ and is at most countable. If $X \subseteq \omega^{<\omega_1}$ then $F[X] \subseteq \omega_1$, so $S = \{\alpha < \omega_1 \mid F[X] \cap \alpha \in W_\alpha\}$ is stationary. Then $S \cap S_F$ is also stationary and $\alpha \in S \cap S_F$ implies $Z_\alpha \ni F^{-1}[F[X] \cap \alpha] \cap \omega^{<\alpha} = X \cap \omega^{<\alpha}$. □

Proof of Theorem 5.2. We construct a particularly nice Suslin tree.

5.4 Definition (see Exercise 4.5) A regular tree $(T, \leq)$ is *normal* if for all $\alpha < \beta < h(T)$ and all $x \in T_\alpha$, there exists $y \in T_\beta$ such that $x < y$.

We construct T_α by transfinite recursion, so that $|T_\alpha| \leq \aleph_0$ and $T^{(\alpha+1)} = \bigcup_{\beta\leq\alpha} T_\beta$ is normal, for all $\alpha < \omega_1$. We put $T_0 = \{\emptyset\} \subseteq \omega^0$. Given $T_\alpha \subseteq \omega^\alpha$ such that $|T_\alpha| \leq \aleph_0$ and $T^{(\alpha+1)}$ is normal, we let $T_{\alpha+1} = \{f \cup \{\langle\alpha, n\rangle\} \mid f \in T_\alpha,\ n \in \omega\}$ and note that $|T_{\alpha+1}| \leq \aleph_0$ and $T^{(\alpha+2)}$ is normal.

Now let α be a limit ordinal. By the inductive assumption, all T_β, $\beta < \alpha$, and hence also $T^{(\alpha)} = \bigcup_{\beta<\alpha} T_\beta$, are normal, and $|T^{(\alpha)}| \leq \sum_{\beta<\alpha} |T_\beta| \leq \aleph_0$.

Let $\{C_n \mid n \in \boldsymbol{N}\}$ be the at most countable collection of those elements of Z_α which happen to be maximal antichains in $T^{(\alpha)}$.

5.5 Claim *For each $f \in T^{(\alpha)}$ there is a branch b of length α such that $f \subseteq b$ and for each $n \in \boldsymbol{N}$, there is some $g \in C_n$ such that $b \supseteq g$.*

Proof. We fix an increasing sequence of ordinals $\langle\alpha_n \mid n \in \omega\rangle$ such that $\alpha_0 = \operatorname{dom} f$ and $\sup_{n\in\omega} \alpha_n = \alpha$. We construct b by recursion.

Let $b_0 = f$. Given b_n with $\operatorname{dom} b_n \geq \alpha_n$, there is some $g \in C_n$ comparable with b_n; otherwise, $C_n \cup \{b_n\}$ would be an antichain in $T^{(\alpha)}$, contradicting the maximality of C_n. If $\alpha_{n+1} \leq \operatorname{dom} g$ we let $b_{n+1} = g$; otherwise we take some $b_{n+1} \in T_{\alpha_{n+1}}$ such that $b_{n+1} \supseteq g \cup b_n$ (it exists by normality of $T^{(\alpha)}$). Letting $b = \bigcup_{n\in\omega} b_n$ we have $\operatorname{dom} b = \sup\{\alpha_n \mid n \in \omega\} = \alpha$, and the other properties of b required by the Claim are clearly satisfied. □

Returning to the proof of Theorem 5.2, for each $f \in T^{(\alpha)}$ we choose one branch b_f as in Claim 5.5, and we let $T_\alpha = \{b_f \mid f \in T^{(\alpha)}\}$. It is clear that $T^{(\alpha+1)}$ is a normal tree and $|T^{(\alpha+1)}| \leq \aleph_0$. The key observation is: each C_n remains a maximal antichain in $T^{(\alpha+1)}$. This is so because each $b \in T_\alpha$ is comparable with some $g \in C_n$ (in fact, $g \subseteq b$).

This completes the recursive construction. We let $T = \bigcup_{\alpha<\omega_1} T_\alpha$ and note that T is a normal tree of height ω_1. In view of Claim 4.3, it remains to show that T has no antichains of cardinality $\aleph_1$. Let X be such an antichain; we can assume X is maximal.

5.6 Claim $S_X = \{\alpha < \omega_1 \mid X \cap T^{(\alpha)}$ *is a maximal antichain in* $T^{(\alpha)}\}$ *is closed unbounded.*

Proof. Let $\beta < \omega_1$ be arbitrary. We construct a sequence $\langle \alpha_n \mid n \in \omega \rangle$ by recursion: $\alpha_0 = \beta$. Given α_n, the set $T^{(\alpha_n)}$ is at most countable, and for every $f \in T^{(\alpha_n)}$ there is some $g_f \in X$ comparable with f (otherwise, X would not be a maximal antichain). We let $\alpha_{n+1} = \sup(\{(\operatorname{dom} g_f) + 1 \mid f \in T^{(\alpha_n)}\} \cup \{\alpha_n\})$. If $\alpha = \sup\{\alpha_n \mid n \in \omega\}$ we have $\beta \leq \alpha < \omega_1$ and each $f \in T^{(\alpha)}$ is comparable with some element of $X \cap T^{(\alpha)}$, so $X \cap T^{(\alpha)}$ is a maximal antichain in $T^{(\alpha)}$. This shows S_X is unbounded. It is easy to see that S_X is closed. □

We now complete the proof of Theorem 5.2. Lemma 5.3 provides $\alpha \in S_X$, α limit, such that $X \cap \omega^{<\alpha} \in Z_\alpha$. So $X \cap T^{(\alpha)}$ is a maximal antichain in $T^{(\alpha)}$ and belongs to Z_α. By our construction and the key observation, $X \cap T^{(\alpha)}$ remains a maximal antichain in $T^{(\alpha+1)}$. But then it remains a maximal antichain in T! Indeed, if $f \in T - T^{(\alpha+1)}$ then $f \restriction \alpha \in T^{(\alpha+1)}$ and hence $f \supseteq f \restriction \alpha \supseteq g$ for some $g \in X \cap T^{(\alpha)}$, i.e., f is comparable with some $g \in X \cap T^{(\alpha)}$. It follows that $X = X \cap T^{(\alpha)}$, and in particular, $|X| \leq |T^{(\alpha)}| \leq \aleph_0$. □

The second combinatorial principle we study in this section has nonexistence of Suslin lines as one of its many consequences. Before stating it we need to introduce some terminology.

5.7 Definition Let $(P, \leq)$ be an ordered set. We say that a set $C \subseteq P$ is *cofinal* in P if for every $p \in P$ there is $q \in C$ such that $p \leq q$. A set $D \subseteq P$ is *directed* if for all $d_1, d_2 \in D$ there is $d \in D$ such that $d_1 \leq d$, $d_2 \leq d$. A set $A \subseteq P$ is a *lower set* if $a \in A$, $p \in P$, $p \leq a$ implies $p \in A$. Let $\mathcal{C}$ be a collection of cofinal subsets of P. A set $G \subseteq P$ is called $\mathcal{C}$*-generic* if G is a directed lower set and $G \cap C \neq \emptyset$ for each $C \in \mathcal{C}$.

5.8 Example Let $(T, \leq)$ be a tree. $D \subseteq T$ is directed if and only if D is a chain. Assume T is normal. Then $T(\alpha) = \bigcup_{\alpha \leq \beta} T_\beta = \{y \in T \mid$ there exists $x \in T_\alpha$ such that $x \leq y\}$ is cofinal in T, for each $\alpha < h(T)$. A set $G \subseteq T$ is $\mathcal{C}$-generic for $\mathcal{C} = \{T(\alpha) \mid \alpha < h(T)\}$ if and only if G is a branch of length $h(T)$ in T.

The following easy theorem is a fundamental fact about generic sets.

5.9 Theorem *Let* $\mathcal{C}$ *be a collection of cofinal subsets of* P *with* $|\mathcal{C}| \leq \aleph_0$. *Then for every* $p \in P$ *there exists a* $\mathcal{C}$*-generic set* G *such that* $p \in G$.

Proof. Let $\mathcal{C} = \{C_n \mid n \in \boldsymbol{N}\}$. We construct a sequence $\langle p_n \mid n \in \boldsymbol{N} \rangle$ recursively. We let $p_0 = p$; given p_n, we let $p_{n+1} = q$ for some $q \in C_n$ such that $p_n \leq q$. Finally we let $G = \{r \in P \mid r \leq p_n$ for some $n \in \boldsymbol{N}\}$. This makes G a lower set and G is clearly directed and $\mathcal{C}$-generic. □

As an application we prove a special case (for $\boldsymbol{R}$) of the Baire Category Theorem.

5.10 Corollary *The intersection of any at most countable collection of open dense sets in* $\boldsymbol{R}$ *is dense.*

Proof. Let $\mathcal{O}$ be an at most countable collection of open dense sets in $\boldsymbol{R}$. Let (a,b) be an open interval. We consider the set P of all closed intervals $[\alpha,\beta] \subseteq (a,b)$, $\alpha,\beta \in \boldsymbol{R}$, $\alpha < \beta$, and order P by reverse inclusion $\supseteq$. If $O \subseteq \boldsymbol{R}$ is open dense, we let $C_O = \{[\alpha,\beta] \in P \mid [\alpha,\beta] \subseteq O\}$; it is easy to verify that C_O is cofinal in P. We let $\mathcal{C} = \{C_O \mid O \in \mathcal{O}\}$; by Theorem 5.9, there is a $\mathcal{C}$-generic set G. In particular, G is directed by $\supseteq$. It follows that G is a collection of nonempty closed and bounded intervals with the finite intersection property. By Theorem 3.4 in Chapter 10, $\bigcap G \neq \emptyset$. Clearly, any $x \in \bigcap G$ belongs to $(a,b) \cap \bigcap \mathcal{O}$. □

The interesting question is whether $\mathcal{C}$-generic sets exist for uncountable $\mathcal{C}$. The next example shows they do not, in general.

5.11 Example Let $T = \omega_1^{<\omega}$, ordered by $\subseteq$ as usual. So T is the tree of all finite sequences of countable ordinals. For each $\alpha < \omega_1$ we let $C_\alpha = \{f \in T \mid \alpha \in \operatorname{ran} f\}$. C_α is cofinal in T: if $f \in T$, $\operatorname{dom} f = n$, then $g = f \cup \{\langle n,\alpha\rangle\} \in C_\alpha$ and $f \subseteq g$. Let G be $\mathcal{C}$-generic for $\mathcal{C} = \{C_\alpha \mid \alpha < \omega_1\}$. G is a collection of finite sequences and, being directed, any two finite sequences in G are compatible. Therefore $F = \bigcup G$ is a function from a subset of ω into ω_1. As $|\operatorname{ran} F| \leq \aleph_0$, there exists $\gamma \in \omega_1 - \operatorname{ran} F$, but this contradicts $C_\gamma \cap G \neq \emptyset$.

As in our study of trees, we have to restrict ourselves to sufficiently "slim" ordered sets to have a hope of positive results.

5.12 Definition Let $(P,\leq)$ be an ordered set. We say that $p,q \in P$ are *compatible* if there is $r \in P$ such that $p \leq r$, $q \leq r$; otherwise they are *incompatible*. An *antichain* in P is a subset $A \subseteq P$ such that $p,q \in A$, $p \neq q$ implies p and q are incompatible.

We note that when $(P,\leq)$ is a tree, $p,q \in P$ are compatible if and only if they are comparable, and so our definition of antichains agrees with the previous one for trees.

An ordered set $(P,\leq)$ satisfies the *countable antichain condition* if every antichain in P is at most countable.

Let κ be an infinite cardinal. We can now state *Martin's Axiom* for κ.

Martin's Axiom MA_κ If $(P,\leq)$ is an ordered set satisfying the countable antichain condition, then, for every collection $\mathcal{C}$ of cofinal subsets of P with $|\mathcal{C}| \leq \kappa$, there exists a $\mathcal{C}$-generic set.

We note that $\mathrm{MA}_{\aleph_0}$ is true by Theorem 5.9, so $\mathrm{MA}_{\aleph_1}$ is the first interesting case. We prove two of its consequences here, and consider some further examples in the exercises.

5.13 Theorem MA_κ *implies that the intersection of any collection* $\mathcal{O}$ *of open dense sets in* $\boldsymbol{R}$ *of* $|\mathcal{O}| \leq \kappa$ *is dense.*

Proof. Exactly the same as the proof of Corollary 5.10, with the reference to Theorem 5.9 replaced by a reference to MA_κ. P satisfies the countable antichain condition on account of Theorem 3.2 in Chapter 10. □

5.14 Corollary MA_κ *implies* $2^{\aleph_0} > \kappa$.

Proof. For each $x \in \boldsymbol{R}$, $O_x = \boldsymbol{R} - \{x\}$ is open dense. If $2^{\aleph_0} \leq \kappa$ then $\mathcal{O} = \{O_x \mid x \in \boldsymbol{R}\}$ is a collection of open dense sets in $\boldsymbol{R}$ with $|\mathcal{O}| \leq \kappa$, but $\bigcap \mathcal{O} = \emptyset$ is not dense in $\boldsymbol{R}$. □

In particular, $\mathrm{MA}_{\aleph_1}$ implies $2^{\aleph_0} > \aleph_1$, the negation of the Continuum Hypothesis.

5.15 Theorem $\mathrm{MA}_{\aleph_1}$ *implies that there are no Suslin lines.*

Proof. Let T be a regular Suslin tree of height ω_1. We let $\tilde{T} = \{t \in T \mid \{r \in T \mid t \leq r\}$ is uncountable$\}$. It is clear that $\tilde{T}$ is a subtree of T and $\tilde{T}$ satisfies the condition from Definition 5.4 (although $\tilde{T}$ need not be regular, even if T is!); in particular, $|\tilde{T}| = \aleph_1$. For each $\alpha < \omega_1$, $\tilde{T}(\alpha) = \{y \in \tilde{T} \mid x \leq y$ for some $x \in \tilde{T}_\alpha\}$ is cofinal in $\tilde{T}$. Let G be $\mathcal{C}$-generic for $\mathcal{C} = \{\tilde{T}(\alpha) \mid \alpha < \omega_1\}$. Then G is a cofinal branch through $\tilde{T}$, hence through T, contradicting the assumption that T is Suslin. □

Martin's Axiom (MA) is the statement that MA_κ holds for all infinite $\kappa < 2^{\aleph_0}$. If $2^{\aleph_0} = \aleph_1$, then $\kappa < 2^{\aleph_0}$ implies $\kappa = \aleph_0$ and MA holds. However, it is consistent to have MA and $2^{\aleph_0} > \aleph_1$. Thus MA can be regarded as a generalization of the Continuum Hypothesis. From Theorem 5.13 we see immediately that, if MA holds, then the intersection of any collection $\mathcal{O}$ of open dense subsets of $\boldsymbol{R}$ of $|\mathcal{O}| < 2^{\aleph_0}$ is dense. MA also implies that the union of less than $2^{\aleph_0}$ sets of Lebesgue measure 0 has measure 0, the Lebesgue measure is κ-additive for all $\kappa < 2^{\aleph_0}$ (not just countably additive), and many other results about $\boldsymbol{R}$, topological spaces, and sets in general.

Exercises

5.1 Show that $\diamondsuit$ is equivalent to the statement: There exists a sequence $\langle W'_\alpha \mid \alpha < \omega_1\rangle$ such that, for each $\alpha < \omega_1$, $W'_\alpha \subseteq \alpha^\alpha$, W'_α is at most countable, and for any $f : \omega_1 \to \omega_1$ the set $\{\alpha < \omega_1 \mid f \restriction \alpha \in W'_\alpha\}$ is stationary.

5.2 Assuming $\diamondsuit$ show that there is a Suslin tree $(T, \leq)$ such that the only automorphism of $(T, \leq)$ is the identity mapping (a *rigid* Suslin tree).
[*Hint*: Imitate the proof of Theorem 5.2; at limit stages α make sure that no automorphism h of $T^{(\alpha)}$ such that $h \in Z_\alpha$ can be extended to an automorphism of $T^{(\alpha+1)}$.]

5.3 Assuming $\diamondsuit$ show that there are $2^{\aleph_1}$ mutually non-isomorphic normal Suslin trees.

5.4 Assuming $\diamondsuit$ show that there exist $2^{\aleph_1}$ stationary subsets of ω_1 such that the intersection of any two of them is at most countable.
[*Hint*: Consider $S_X = \{\alpha < \omega_1 \mid X \cap \alpha \in W_\alpha\}$ for all $X \subseteq \omega_1$.]

5.5 Show that MA_κ is equivalent to the statement: If $(P, \leq)$ is an ordered set satisfying the countable antichain condition, then for every collection $\mathcal{C}$ of maximal antichains with $|\mathcal{C}| \leq \kappa$ there exists a directed set G such that, for every $C \in \mathcal{C}$, there exist $c \in C$ and $a \in G$ so that $c \leq a$.

5.6 Show that there exist $2^{\aleph_0}$ subsets of ω such that the intersection of any two of them is finite.
[*Hint*: For $f \in \{0,1\}^\omega$ let $A_f = \{f \restriction n \mid n \in \omega\} \subseteq \omega^{<\omega}$. Of course $|\omega^{<\omega}| = |\omega|$.]

5.7 Assuming $2^{\aleph_0} = \aleph_1$ show that there exist $2^{\aleph_1}$ subsets of ω_1 such that the intersection of any two of them is at most countable.

Chapter 13

Large Cardinals

1. The Measure Problem

Theorem 2.14 in Chapter 8 shows that there exists no σ-additive translation invariant measure μ on the σ-algebra of all subsets of $\boldsymbol{R}$ such that $\mu([a,b]) = b-a$ for every interval $[a,b]$ in $\boldsymbol{R}$. This raises the question whether there exists any σ-additive measure on $\mathcal{P}(\boldsymbol{R})$, or for that matter on $\mathcal{P}(S)$ for any infinite set S. Of course, the counting measure, which allows the value $\mu(S) = \infty$, is a trivial example, so we formulate the problem differently. We allow a measure to have only finite values, and we further require (without loss of generality) that $\mu(S) = 1$.

1.1 Definition Let S be a nonempty set. A (*nontrivial probabilistic σ-additive*) *measure on* S is a function $\mu : \mathcal{P}(S) \to [0,1]$ such that
(a) $\mu(\emptyset) = 0$, $\mu(S) = 1$.
(b) If $X \subseteq Y$, then $\mu(X) \leq \mu(Y)$.
(c) If X and Y are disjoint, then $\mu(X \cup Y) = \mu(X) + \mu(Y)$.
(d) $\mu(\{a\}) = 0$ for every $a \in S$.
(e) If $\{X_n\}_{n=0}^{\infty}$ is a collection of mutually disjoint subsets of S, then

$$\mu(\bigcup_{n=0}^{\infty} X_n) = \sum_{n=0}^{\infty} \mu(X_n).$$

A consequence of (d) and (e) is that every at most countable subset of S has measure 0. Hence if there is a measure on S, then S is uncountable. It is clear that whether there exists a measure on S depends only on the cardinality of S: If S carries a measure and $|S'| = |S|$ then S' also carries a measure. In Section 2 of Chapter 11 we constructed a nontrivial finitely additive measure on $\boldsymbol{N}$, a function that satisfies (a)–(d) in Definition 1.1. The *measure problem* is the question whether there exists such a function on some S that is σ-additive. The measure problem is related to the question whether the Lebesgue measure can be extended to all sets of reals. Precisely: Does there exist a σ-additive measure $\mu : \mathcal{P}(\boldsymbol{R}) \to [0, \infty) \cup \{\infty\}$ such that for every Lebesgue measurable set X, $\mu(X)$ is the Lebesgue measure of X. If there is such an extension μ of the Lebesgue measure then the restriction of μ to $\mathcal{P}([0,1])$ is a nontrivial measure on $S = [0,1]$ (satisfying (a)–(e) in Definition 1.1). Conversely, it can be proved that if a nontrivial measure exists on a set S of cardinality $2^{\aleph_0}$ then the Lebesgue measure can be extended to a σ-additive measure $\mu : \mathcal{P}(\boldsymbol{R}) \to [0, \infty) \cup \{\infty\}$.

The measure problem, a natural question arising in abstract real analysis, is deeply related to the continuum problem, and surprisingly also to the subject we touched upon in Chapter 9 — inaccessible cardinals. This problem has become the starting point for investigation of *large cardinals*, a theory that we explore further in the next section.

1.2 Theorem *If there is a measure on $2^{\aleph_0}$, then the Continuum Hypothesis fails.*

Proof. Let us assume that $2^{\aleph_0} = \aleph_1$ and that there is a measure μ on the set $S = \omega_1$, a function on $\mathcal{P}(S)$ satisfying (a)–(e) of Definition 1.1. Let I denote the ideal of sets of measure 0:

$$I = \{X \subseteq S \mid \mu(X) = 0\}.$$

This ideal has the following properties:

$$\text{For every } x \in S,\ \{x\} \in I. \tag{1.3}$$

$$\text{If } X_n \in I \text{ for all } n \in \boldsymbol{N}, \text{ then } \bigcup_{n=0}^{\infty} X_n \in I. \tag{1.4}$$

$$\text{There is no uncountable mutually disjoint collection } \mathcal{S} \subseteq \mathcal{P}(S) \text{ such that } X \notin I \text{ for all } X \in \mathcal{S}. \tag{1.5}$$

Property (1.4) is an immediate consequence of Definition 1.1(e). As for (1.5), suppose that $\mathcal{S}$ is such a disjoint collection. For each n, let

$$\mathcal{S}_n = \{X \in \mathcal{S} \mid \mu(X) \geq \frac{1}{n}\}.$$

Because $\mu(S) = 1$, each $\mathcal{S}_n$ can only be finite, and because $\mathcal{S} = \bigcup_{n=0}^{\infty} \mathcal{S}_n$, $\mathcal{S}$ is at most countable.

We now construct a "matrix" of subsets of S $\langle A_{\alpha n} \mid \alpha \in \omega_1,\ n \in \omega \rangle$.

For each $\xi < \omega_1$, there exists a function f_ξ on ω such that $\xi \subseteq \operatorname{ran} f_\xi$. Let us choose one f_ξ for each ξ, and let

$$A_{\alpha n} = \{\xi < \omega_1 \mid f_\xi(n) = \alpha\} \quad (\alpha < \omega_1,\ n < \omega).$$

The matrix $\langle A_{\alpha n} \rangle$ has the following properties:

$$\text{For every } n, \text{ if } \alpha \neq \beta, \text{ then } A_{\alpha n} \cap A_{\beta n} = \emptyset. \tag{1.6}$$

$$\text{For every } \alpha,\ S - \bigcup_{n=0}^{\infty} A_{\alpha n} \text{ is at most countable.} \tag{1.7}$$

We have (1.6) because $\xi \in A_{\alpha n} \cap A_{\beta n}$ would mean that $f_\xi(n) = \alpha$ and $f_\xi(n) = \beta$. The set in (1.7) is at most countable because it is included in the set $\alpha \cup \{\alpha\}$: If $\xi \notin A_{\alpha n}$ for all n, then $\alpha \notin \operatorname{ran} f_\xi$ and so $\xi \le \alpha$.

Let $\alpha < \omega_1$ be fixed. From (1.7) and (1.4) we conclude that not all the sets $A_{\alpha n}$, $n \in \omega$, are in the ideal I: ω_1 is the union of the sets $A_{\alpha n}$ and an at most countable set, which belongs to I by (1.3) and (1.4).

Thus for each $\alpha < \omega_1$ there exists some $n_\alpha \in \boldsymbol{N}$ such that $A_{\alpha n_\alpha} \notin I$. Because there are uncountably many $\alpha < \omega_1$ and only countably many $n \in \boldsymbol{N}$, there must exist some n such that the set $\{\alpha \mid n_\alpha = n\}$ is uncountable. Let

$$\mathcal{S} = \{A_{\alpha n} \mid n_\alpha = n\}.$$

This is an uncountable collection of subsets of S. By (1.6), the sets in $\mathcal{S}$ are mutually disjoint, and $A_{\alpha n} \notin I$ for each $A_{\alpha n} \in \mathcal{S}$. This contradicts (1.5).

We have a contradiction and therefore the assumption that $2^{\aleph_0} = \aleph_1$ must be false. This proves the theorem. □

The proof of Theorem 1.2 can be slightly modified to obtain the following result.

1.8 Theorem *If there is a measure on a set S, then some cardinal $\kappa \le |S|$ is weakly inaccessible.*

1.9 Corollary *If there is a measure on $2^{\aleph_0}$, then $2^{\aleph_0} \ge \kappa$ for some weakly inaccessible cardinal κ.*

To prove Theorem 1.8 we follow closely the proof of Theorem 1.2. We assume that for some set S there is a function $\mu : \mathcal{P}(S) \to [0,1]$ that satisfies Definition 1.1. We have shown that then there exists an ideal I on S that satisfies (1.3), (1.4), and (1.5). Let

(1.10) $\kappa =$ the least cardinal such that for some S of size κ there exists an ideal I on S with properties (1.3), (1.4), and (1.5) and

(1.11) I is an ideal on $S = \kappa$ with properties (1.3), (1.4), and (1.5).

(Of course, if such an ideal exists on some S of size κ, then there is one on κ.)

1.12 Lemma *For every $\lambda < \kappa$, if $\{X_\eta\}_{\eta<\lambda}$ are such that $X_\eta \in I$ for all $\eta < \lambda$, then $\bigcup_{\eta<\lambda} X_\eta \in I$.*

Proof. Otherwise, there exists some $\lambda < \kappa$ and $\{X_\eta\}_{\eta<\lambda}$ such that $X_\eta \in I$ for each η but $\bigcup_{\eta<\lambda} X_\eta \notin I$. We may assume that the X_η are mutually disjoint, because we can replace each X_η by $X'_\eta = X_\eta - \bigcup\{X_\nu \mid \nu < \eta\}$, and $\bigcup_{\eta<\lambda} X'_\eta = \bigcup_{\eta<\lambda} X_\eta$. Let

$$J = \{Y \subseteq \lambda \mid \bigcup_{\eta \in Y} X_\eta \in I\}.$$

We verify that J is a ideal on λ (e.g., $\lambda \notin J$ because $\bigcup_{\eta\in\lambda} X_\eta \notin I$). For each $\eta \in \lambda$, $\{\eta\} \in J$ because $X_\eta \in I$; thus J satisfies (1.3). Similarly, one verifies that J satisfies (1.4) and (1.5) as well. [For (1.5) we use the assumption that the X_η are mutually disjoint.]

Thus J is an ideal on $\lambda < \kappa$ with properties (1.3), (1.4), and (1.5), contrary to the assumption (1.10). □

1.13 Corollary *If $X \subset \kappa$ and $|X| < \kappa$, then $X \in I$.*

1.14 Corollary *κ is an uncountable regular cardinal.*

Proof. κ is uncountable because every at most countable subset of κ belongs to I. And κ is regular, because otherwise κ would be the union of less than κ sets of size $< \kappa$, and therefore would belong to the ideal, a contradiction. □

We can now finish the proof of Theorem 1.8.

1.15 Theorem *κ is weakly inaccessible.*

Proof. In view of Corollary 1.14 it suffices to prove that κ is a limit cardinal. Let us therefore assume that κ is a successor cardinal, $\kappa = \aleph_{\nu+1}$.

For each $\xi < \kappa$ we choose a function f_ξ on ω_ν such that $\xi \subseteq \operatorname{ran} f_\xi$ and let

$$A_{\alpha\eta} = \{\xi < \kappa \mid f_\xi(\eta) = \alpha\} \quad (\alpha < \omega_{\nu+1},\ \eta < \omega_\nu).$$

As in the proof of Theorem 1.2, one shows that the matrix $\langle A_{\alpha\eta}\rangle$ has the following properties:

(1.16) For every η, if $\alpha \neq \beta$, then $A_{\alpha\eta} \cap A_{\beta\eta} = \emptyset$.

(1.17) For every α, $|\kappa - \bigcup_{\eta<\omega_\nu} A_{\alpha\eta}| \leq \aleph_\nu$.

Still following the proof of Theorem 1.2, but using Lemma 1.12 in place of (1.4), we show that for each $\alpha < \omega_{\nu+1}$ there exists some $\eta < \omega_\nu$ such that $A_{\alpha\eta} \notin I$. And the same argument as before produces a collection $\mathcal{S}$, of size $\aleph_{\nu+1}$ (therefore uncountable), of mutually disjoint sets $\notin I$. This contradicts the property (1.5), and therefore κ cannot be a successor cardinal. □

A measure μ is *two-valued* if it takes only two values, 0 and 1. We return to two-valued measures in the next section, as a starting point for the theory of large cardinals.

1.18 Definition Let μ be a measure on S. A set $A \subseteq S$ is an *atom* if $\mu(A) > 0$ and if for every $X \subseteq A$, either $\mu(X) = 0$ or $\mu(A - X) = 0$. The measure μ is *atomless* if there exist no atoms.

We conclude this section with the proof of the following dichotomy due to Stanisław Ulam.

1.19 Theorem *If there exists a measure then either there exists a two-valued measure or there exists a measure on* $2^{\aleph_0}$.

1.20 Lemma *Let μ be an atomless measure on S.*
(a) For every $\varepsilon > 0$ and every $X \subseteq S$ with $\mu(X) > 0$ there is a $Y \subseteq X$ such that $0 < \mu(Y) \leq \varepsilon$.
(b) For every $X \subseteq S$ there is a $Y \subseteq X$ such that $\mu(Y) = \frac{1}{2}\mu(X)$.

Proof.
(a) Let $X_0 = X$, and for each n, we find an $X_{n+1} \subset X_n$ such that $0 < \mu(X_{n+1}) \leq \frac{1}{2}\mu(X_n)$. This is possible because X_n is not an atom and therefore there is an $X \subset X_n$ such that $\mu(X) > 0$ and $\mu(X_n - X) > 0$. As $\mu(X_n) = \mu(X) + \mu(X_n - X)$, it follows that either $X_{n+1} = X$ or $X_{n+1} = X_n - X$ has the desired property. Clearly, $0 < \mu(X_n) \leq \frac{1}{2^n}\mu(X)$ for every n, and part (a) follows.
(b) Let $X \subseteq S$ be such that $\mu(X) = m > 0$. By transfinite recursion on $\alpha < \omega_1$ we construct a disjoint family of subsets Y_α of X as follows: Let $Y_0 \subset X$ be such that $0 < \mu(Y_0) \leq m/2$; if $\mu(\bigcup_{\beta<\alpha} Y_\beta) < m/2$, we choose $Y_\alpha \subset X - \bigcup_{\beta<\alpha} Y_\beta$ such that $0 < \mu(Y_\alpha) \leq \frac{m}{2} - \mu(\bigcup_{\beta<\alpha} Y_\beta)$. It follows from (1.5) that there exists an α at which the construction stops, i.e., $\mu(\bigcup_{\beta<\alpha} Y_\beta) = m/2$.

□

Proof of Theorem 1.19. Assume that there exists a measure μ on a set S. If there exists an atom $A \subseteq S$, we define a two-valued measure ν on A as follows: $\nu(X) = \mu(X)/\mu(A)$ for all $X \subseteq A$.

If μ is atomless, we define a family $\{X_s \mid s \in \text{Seq}\}$ of subsets of S indexed by finite 0-1 sequences $s \in \text{Seq} = \bigcup_{n=0}^{\infty}\{0,1\}^n$. The sets X_s are defined by recursion on the length of s: For the empty sequence $\emptyset$, let $X_\emptyset = S$. Given X_s, we let $X_{s^\frown 0}$ and $X_{s^\frown 1}$ be subsets of X_s such that $X_{s^\frown 1} = X_s - X_{s^\frown 0}$ and $\mu(X_{s^\frown 0}) = \mu(X_{s^\frown 1}) = \frac{1}{2}\mu(X_s)$. Thus $\mu(X_s) = 1/2^n$ where n is the length of s.

Furthermore, for each $f \in \{0,1\}^\omega$, we let $X_f = \bigcap_{n=0}^{\infty} X_{f\restriction n}$. Note that if $f \neq g$ then $X_f \cap X_g = \emptyset$, and that $\mu(X_f) = 0$ for each f.

Now we define a measure ν on the set $\{0,1\}^\omega$ as follows:

$$\nu(Z) = \mu(\bigcup\{X_f \mid f \in Z\}) \quad (Z \subseteq \{0,1\}^\omega).$$

As μ is a measure, it is easily verified that ν has properties (a), (b), (c) and (e) of Definition 1.1. Property (d) follows from the fact that $\mu(X_f) = 0$ for each $f \in \{0,1\}^\omega$. □

Thus if there exists a measure then either there exists one on $2^{\aleph_0}$ in which case $2^{\aleph_0} \geq \kappa$ for some weakly inaccessible cardinal κ, or else there exists a two-valued measure, an alternative that we investigate in the next section.

2. Large Cardinals

In the preceding section we proved that if there exists a nontrivial σ-additive measure, then there exists a weakly inaccessible cardinal. This result is just one example of a vast body of results, known as the theory of large cardinals. In this section we give some more examples of large cardinal results and introduce the most studied prototype of large cardinals — *measurable cardinals.*

2.1 Lemma *If μ is a two-valued measure on S, then*

$$U = \{X \subseteq S \mid \mu(X) = 1\}$$

is a nonprincipal ultrafilter on S, and

$$\text{(2.2)} \qquad \textit{if } \{X_n\}_{n=0}^{\infty} \textit{ is such that } X_n \in U \textit{ for each } n, \textit{ then } \bigcap_{n=0}^{\infty} X_n \in U.$$

Proof. An easy verification. U is nonprincipal because μ is nontrivial, and satisfies (2.2) because μ is σ-additive. □

Property (2.2) is called *σ-completeness.* The converse of Lemma 2.1 is also true: if U is a σ-complete nonprincipal ultrafilter on S, then the function $\mu : \mathcal{P}(S) \to \{0, 1\}$ defined by

$$\mu(X) = \begin{cases} 1 & \text{if } X \in U, \\ 0 & \text{if } X \notin U \end{cases}$$

is a two-valued measure on S.

Thus the problem whether two-valued measures exist is equivalent to the problem of existence of nonprincipal σ-complete ultrafilters. We now investigate this question.

First we generalize the definition of σ-completeness.

2.3 Definition Let κ be an uncountable cardinal. A filter F on S is *κ-complete* if for every cardinal $\lambda < \kappa$, if $X_\alpha \in F$ for all $\alpha < \lambda$, then $\bigcap_{\alpha<\lambda} X_\alpha \in F$.

An ideal I on S is *κ-complete* if for every cardinal $\lambda < \kappa$, if $X_\alpha \in I$ for all $\alpha < \lambda$, then $\bigcup_{\alpha<\lambda} X_\alpha \in I$.

A filter is κ-complete if and only if its dual ideal is κ-complete. An $\aleph_1$-complete filter is also called *σ-complete* or *countably complete*; it means that $\bigcap_{n=0}^{\infty} X_n \in F$ whenever all $X_n \in F$. Similarly for ideals.

2.4 Lemma *If there exists a nonprincipal σ-complete ultrafilter then there exists an uncountable cardinal κ and a nonprincipal κ-complete ultrafilter on κ.*

Proof. Let κ be the least cardinal such that there exists a nonprincipal σ-complete ultrafilter on κ, and let U be such an ultrafilter. We wish to show that U is κ-complete (note that because U is σ-complete, κ must be uncountable).

Let I be the ideal dual to U: $I = \mathcal{P}(\kappa) - U$. I is a nonprincipal σ-complete prime ideal on κ; we show that I is κ-complete. If not, then there exists $\lambda < \kappa$ and $\{X_\eta\}_{\eta<\lambda}$ such that $X_\eta \in I$ for each η but $\bigcup_{\eta<\lambda} X_\eta \notin I$. We may assume that the X_η are mutually disjoint. Let

$$J = \{Y \subseteq \lambda \mid \bigcup_{\eta \in Y} X_\eta \in I\}.$$

J is a σ-complete prime ideal on λ; it is nonprincipal because $X_\eta \in I$ for each $\eta \in \lambda$. The dual of J is a nonprincipal σ-complete ultrafilter on λ.

But $\lambda < \kappa$ and that contradicts the assumption that κ is the least cardinal on which there is a nonprincipal σ-complete ultrafilter. Thus U is κ-complete. □

2.5 Definition A *measurable cardinal* is an uncountable cardinal κ on which there exists a nonprincipal κ-complete ultrafilter.

The discussion leading to Definition 2.5 shows that measurable cardinals are related to the measure problem investigated in Section 1. The existence of a measurable cardinal is equivalent to the existence of a nontrivial two-valued σ-additive measure.

We devote the rest of this section to measurable cardinals.

2.6 Theorem *Every measurable cardinal is strongly inaccessible.*

Proof. We recall that a cardinal is strongly inaccessible if it is regular, uncountable, and strong limit. Let κ be a measurable cardinal, and let U be a nonprincipal κ-complete ultrafilter on κ.

Let I be the prime ideal dual to the ultrafilter U. Every singleton belongs to I, and by κ-completeness, every $X \subset \kappa$ of cardinality $< \kappa$ belongs to I. If κ were singular, the set κ would also have to belong to I, by κ-completeness. Hence κ is regular.

Now let us assume that κ is not strong limit. Therefore, there is $\lambda < \kappa$ such that $2^\lambda \geq \kappa$, so there is a set $S \subseteq \{0,1\}^\lambda$ of cardinality κ. On S there is also a nonprincipal κ-complete ultrafilter, say V. For each $\alpha < \lambda$, exactly one of the two sets

$$\{f \in S \mid f(\alpha) = 0\} \quad \text{and} \quad \{f \in S \mid f(\alpha) = 1\} \tag{2.7}$$

belongs to V; let us call that set X_α. Thus for each $\alpha < \lambda$ we have $X_\alpha \in V$, and by κ-completeness, $X = \bigcap_{\alpha<\lambda} X_\alpha$ is also in V. But there is at most one function f in S that belongs to all the X_α: the value of f at α is determined by the choice of one of the sets in (2.7). Hence $|X| \leq 1$, a contradiction since V is nonprincipal. Hence κ is a strong limit cardinal, and therefore strongly inaccessible. □

One can prove more than just inaccessibility of measurable cardinals. We give several examples of properties of measurable cardinals that can be obtained by elementary methods.

Let us recall (Section 3 in Chapter 12) that an uncountable cardinal κ has the *tree property* if there exists no Aronszajn tree of height κ. The following theorem shows that every measurable cardinal has the tree property.

2.8 Theorem *Let κ be a measurable cardinal. If T is a tree of height κ such that each node has less than κ immediate successors, then T has a branch of length κ.*

Proof. For each $\alpha < \kappa$, let T_α be the set of all $s \in T$ of height α. Since κ is a strongly inaccessible cardinal, it follows, by induction on α, that $|T_\alpha| < \kappa$ for all $\alpha < \kappa$. Hence $|T| \leq \kappa$, and therefore $|T| = \kappa$. Let U be a nonprincipal κ-complete ultrafilter on T.

We find a branch of length κ as follows. By induction on α we show that for each α there exists a unique $s_\alpha \in T_\alpha$ such that

$$\{t \in T \mid s_\alpha \leq t\} \in U \tag{2.9}$$

and that $s_\alpha < s_\beta$ when $\alpha < \beta$. First, let s_0 be the root of T. Given s_α, and assuming (2.9), we note that the set $\{t \in T \mid s_\alpha \leq t\}$ is the disjoint union of $\{s_\alpha\}$ and the sets $\{t \in T \mid u \leq t\}$ where u ranges over all immediate successors of s_α. Since U is a κ-complete ultrafilter and s_α has fewer than κ immediate successors, there is a unique immediate successor $u = s_{\alpha+1}$ such that $\{t \in T \mid s_{\alpha+1} \leq t\} \in U$.

When η is a limit ordinal less than κ and $\{s_\alpha\}_{\alpha<\eta}$, $s_\alpha \in T_\alpha$, are such that $s_\alpha < s_\beta$ if $\alpha < \beta$, and all the s_α satisfy (2.9), then we have

$$S = \bigcap_{\alpha<\eta} \{t \in T \mid s_\alpha \leq t\} \in U, \tag{2.10}$$

because U is κ-complete. The set $S_\eta = \{s \in T_\eta \mid s \leq t \text{ for some } t \in S\}$ is nonempty and of size less than κ; it follows that there is a unique $s \in S_\eta$ for which $\{t \in T \mid s \leq t\} \in U$, and we let s_η be this s. It is clear that $b = \{s_\alpha \mid \alpha < \kappa\}$ is a branch in T, of length κ. □

In Section 2 of Chapter 12 we introduced weakly compact cardinals. We state without proof the following equivalence that gives another characterization of weakly compact cardinals.

2.11 Theorem *The following are equivalent, for every uncountable cardinal κ:*
(a) $\kappa \to (\kappa)^2_2$.
(b) $\kappa \to (\kappa)^r_s$ for all positive integers r and s.
(c) κ is strongly inaccessible and has the tree property.

Since measurable cardinals are strongly inaccessible and have the tree property, it follows that every measurable cardinal is weakly compact. This can also be proved directly, by showing $\kappa \to (\kappa)^r_s$. We conclude this introduction to large cardinals by presenting the proof of the special case for $r = s = 2$.

2.12 Theorem *If κ is a measurable cardinal, then every partition of $[\kappa]^2$ into two sets has a homogeneous set of cardinality κ.*

Proof. Let $\{P_1, P_2\}$ be a partition of $[\kappa]^2$. To find a homogeneous set, let U be a nonprincipal κ-complete ultrafilter on κ.

For each $\alpha \in \kappa$, let

$$S_\alpha^1 = \{\beta \in \kappa \mid \beta \neq \alpha \text{ and } \{\alpha, \beta\} \in P_1\},$$
$$S_\alpha^2 = \{\beta \in \kappa \mid \beta \neq \alpha \text{ and } \{\alpha, \beta\} \in P_2\}.$$

Exactly one of S_α^1, S_α^2 belongs to U. Let

$$Z_1 = \{\alpha \mid S_\alpha^1 \in U\}, \quad Z_2 = \{\alpha \mid S_\alpha^2 \in U\}.$$

Since $Z_1 \cup Z_2 = \kappa$, either Z_1 or Z_2 is in U. Let us assume that $Z_1 \in U$ and let us find $H \subseteq \kappa$ of cardinality κ such that $[H]^2 \subseteq P_1$.

We construct $H = \{\alpha_\xi \mid \xi < \kappa\}$ by recursion. At step γ, we have constructed an increasing sequence

$$\langle \alpha_\xi \mid \xi < \gamma \rangle$$

of elements of Z_1 such that for each $\xi < \eta < \gamma$,

$$\alpha_\eta \in S_{\alpha_\xi}^1. \tag{2.13}$$

Because $S_{\alpha_\xi}^1 \in U$ for all $\xi < \gamma$, the set

$$Z_1 \cap \bigcap_{\xi < \gamma} S_{\alpha_\xi}^1$$

is in U (by κ-completeness) and therefore contains some α greater than all α_ξ, $\xi < \gamma$. Let α_γ be the least such α. Then (2.13) holds for all $\xi < \eta < \gamma + 1$.

Let $H = \{\alpha_\xi \mid \xi < \kappa\}$. If $\xi < \eta$, then $\alpha_\eta \in S_{\alpha_\xi}^1$ and so $\{\alpha_\xi, \alpha_\eta\} \in P_1$. $[H]^2 \subseteq P_1$ and therefore H is homogeneous for the partition. □

Exercises

A strongly inaccessible cardinal κ is a *Mahlo cardinal* if the set of all regular cardinals $< \kappa$ is stationary.

2.1 If κ is Mahlo, then the set of all strongly inaccessible cardinals $< \kappa$ is stationary. [*Hint*: The set of all strong limit $\alpha < \kappa$ is closed unbounded.]

Let κ be a measurable cardinal. A nonprincipal κ-complete ultrafilter U on κ is *normal* if every regressive function f with $\operatorname{dom} f \in U$ is constant on some set $A \in U$.

In Exercises 2.2–2.5, let U be a nonprincipal κ-complete ultrafilter on κ. For f and g in κ^κ, let

$$f \equiv g \text{ if } \{\alpha < \kappa \mid f(\alpha) = g(\alpha)\} \in U.$$

2.2 $\equiv$ is an equivalence relation on κ^κ.

Let W be the set of all equivalence classes of $\equiv$ on κ^κ; let $[f]$ denote the equivalence class of f. Let

$$[f] < [g] \text{ if } \{\alpha < \kappa \mid f(\alpha) < g(\alpha)\} \in U.$$

2.3 $<$ is a linear ordering of W.

2.4 $<$ is a well-ordering of W. [*Hint*: Otherwise, there is a sequence $\langle f_n \mid n \in \boldsymbol{N}\rangle$ such that $[f_n] > [f_{n+1}]$. Let $X_n = \{\alpha \mid f_n(\alpha) > f_{n+1}(\alpha)\}$ and let $X = \bigcap_{n=0}^\infty X_n$. If $\alpha \in X$, then $f_0(\alpha) > f_1(\alpha) > \cdots$, a contradiction.]

2.5 Let $h : \kappa \to \kappa$ be the least function (in $(W, <)$) with the property that for all $\gamma < \kappa$, $\{\alpha < \kappa \mid h(\alpha) > \gamma\} \in U$. Let $V = \{X \subseteq \kappa \mid h^{-1}[X] \in U\}$. Show that V is a normal ultrafilter.

Thus for every measurable cardinal κ, there exists a normal ultrafilter on κ. In Exercises 2.6–2.9, U is a normal ultrafilter on κ.

2.6 Every set $A \in U$ is stationary. [*Hint*: Use Exercise 3.9 in Chapter 11 and the definition of normality.]

2.7 Let $\lambda < \kappa$ be a regular cardinal and let $E_\lambda = \{\alpha < \kappa \mid \operatorname{cf}(\alpha) = \lambda\}$. The set E_λ is not in U. [*Hint*: Assume that $E_\lambda \in U$. For each $\alpha \in E_\lambda$, let $\{x_{\alpha\xi} \mid \xi < \lambda\}$ be an increasing sequence with limit α. For each $\xi < \lambda$ there is y_ξ and $A_\xi \in U$ such that $x_{\alpha\xi} = y_\xi$ for all $\alpha \in A_\xi$. Let $A = \bigcap_{\xi<\lambda} A_\xi$. Then $A \in U$, but A contains only one element, namely $\sup\{y_\xi \mid \xi < \lambda\}$; a contradiction.]

2.8 The set of all regular cardinals $< \kappa$ is in U. [*Hint*: Otherwise, the set $S = \{\alpha < \kappa \mid \operatorname{cf}(\alpha) < \alpha\}$ is in U, and because cf is a regressive function on S, there exists $\lambda < \kappa$ such that $\{\alpha < \kappa \mid \operatorname{cf}(\alpha) = \lambda\} \in U$; a contradiction.]

2.9 Every measurable cardinal is a Mahlo cardinal. [*Hint*: Exercises 2.6 and 2.8.]

Chapter 14

The Axiom of Foundation

1. Well-Founded Relations

The notion of a well-ordering is one of the key concepts of set theory. It emerged in Chapter 3, when we introduced natural numbers; the fact that the natural numbers are well-ordered by size is essentially equivalent to the Induction Principle. We studied well-orderings in full generality in Chapter 6, and we saw many applications of them in subsequent chapters. It is thus of great interest to consider whether the concept allows further useful generalizations. It turns out that, for many purposes, the "ordering" stipulation is unimportant; it is the "well-" part, i.e., the requirement that every nonempty subset has a minimal element, that is crucial. This leads to the basic definition of this section.

1.1 Definition Let R be a binary relation in A, and let $X \subseteq A$. We say that $a \in X$ is an *R-minimal* element of X if there is no $x \in X$ such that xRa. R is *well-founded on A* if every nonempty subset of A has an R-minimal element. The set $\{x \in A \mid xRa\}$ is called the *R-extension* of a in A and is denoted $\mathrm{ext}_R(a)$. Thus a is an R-minimal element of X if and only if $\mathrm{ext}_R(a) \cap X = \emptyset$.

1.2 Example

(a) The empty relation $R = \emptyset$ is well-founded on any A, empty or not.

(b) Any well-ordering of A is well-founded on A. In particular, $\in_\alpha$ is well-founded on α, for any ordinal number α.

(c) Let $A = \mathcal{P}(\omega)$; then $\in_A$ is well-founded on A.

(d) $\in_A$ is well-founded on A if $A = V_n$ ($n \in \boldsymbol{N}$) or $A = V_\omega$ (see Exercise 3.3 in Chapter 6).

(e) If $(T, \leq)$ is a tree (Chapter 12, Section 3) then $<$ is well-founded on T.

The next two lemmas list some simple properties of well-founded relations.

1.3 Lemma *Let R be a well-founded relation on A.*
(a) R is antireflexive in A, i.e., aRa is false, for all $a \in A$.
(b) R is asymmetric in A, i.e., aRb implies that bRa does not hold.
(c) There is no finite sequence $\langle a_0, a_1, \ldots, a_n \rangle$ such that $a_1 R a_0$, $a_2 R a_1$, ..., $a_n R a_{n-1}$, $a_0 R a_n$.
(d) There is no infinite sequence $\langle a_i \mid i \in \boldsymbol{N} \rangle$ of elements of A such that $a_{i+1} R a_i$ holds for all $i \in \boldsymbol{N}$.

Proof.
(a) If aRa then $X = \{a\} \neq \emptyset$ does not have an R-minimal element.
(b) If aRb and bRa then $X = \{a, b\} \neq \emptyset$ does not have an R-minimal element.
(c) Otherwise, $X = \{a_0, \ldots, a_n\}$ has no R-minimal element.
(d) $X = \{a_i \mid i \in \boldsymbol{N}\}$ would have no R-minimal element if (d) failed.

□

1.4 Lemma* *Let R be a binary relation in A such that there is no infinite sequence $\langle a_i \mid i \in \boldsymbol{N} \rangle$ of elements of A for which $a_{i+1} R a_i$ holds for all $i \in \boldsymbol{N}$. Then R is well-founded on A.*

This is a converse to Lemma 1.3(d). In this chapter we do not assume the Axiom of Choice. The few results where it is needed are marked by an asterisk. Under the assumption of the Axiom of Choice, a relation R is well-founded if and only if there is no infinite "R-decreasing" sequence of elements of A.

Proof. If R is not well-founded on A, then A has a nonempty subset X with the property that, for every $a \in X$, there is some $b \in X$ such that bRa holds. Choose $a_0 \in X$. Then choose $a_1 \in X$ such that $a_1 R a_0$. Given $\langle a_0, \ldots, a_i \rangle$, choose $a_{i+1} \in X$ so that $a_{i+1} R a_i$. The resulting sequence $\langle a_i \mid i \in \boldsymbol{N} \rangle$ satisfies $a_{i+1} R a_i$ for all $i \in \boldsymbol{N}$. □

The importance of well-founded relations rests on the fact that, like well-orderings, they allow proofs by induction and constructions of functions by recursion.

1.5 The Induction Principle *Let $\mathbf{P}$ be some property. Assume that R is a well-founded relation on A and, for all $x \in A$,*

(*) *if $\mathbf{P}(y)$ holds for all $y \in \operatorname{ext}_R(x)$, then $\mathbf{P}(x)$.*

Then $\mathbf{P}(x)$ holds for all $x \in A$.

Proof. Otherwise, $X = \{x \in A \mid \mathbf{P}(x) \text{ does not hold}\} \neq \emptyset$. Let a be an R-minimal element of X. Then $\mathbf{P}(a)$ fails, but $\mathbf{P}(y)$ holds for all yRa, contradicting (*). □

1.6 The Recursion Theorem *Let $\mathbf{G}$ be an operation. Assume that R is a well-founded relation on A. Then there is a unique function f on A such that, for all $x \in A$,*

$$f(x) = \mathbf{G}(f \restriction \operatorname{ext}_R(x)).$$

The proof of Theorem 1.6 uses ideas similar to those employed to prove other Recursion Theorems, such as the original one in Section 3 of Chapter 3, and Theorem 4.5 in Chapter 6. It is convenient to state some definitions and prove a simple lemma first.

1.7 Definition A set $B \subseteq A$ is *R-transitive* in A if $\operatorname{ext}_R(x) \subseteq B$ holds for all $x \in B$. In other words, B is R-transitive if it has the property that $x \in B$ and yRx imply $y \in B$.

It is clear that the union and the intersection of any collection of R-transitive subsets of A is R-transitive.

1.8 Lemma *For every $C \subseteq A$ there is a smallest R-transitive $B \subseteq A$ such that $C \subseteq B$.*

Proof. Let $B_0 = C$, $B_{n+1} = \{y \in A \mid yRx$ holds for some $x \in B_n\}$, $B = \bigcup_{n=0}^{\infty} B_n$. It is clear that B is R-transitive and $C \subseteq B$. Moreover, for any R-transitive B', if $C \subseteq B'$ then each $B_n \subseteq B'$, by induction, and hence $B \subseteq B'$. □

Proof of the Recursion Theorem. Let $T = \{g \mid g$ is a function, $\operatorname{dom} g$ is R-transitive in A, and $g(x) = \mathbf{G}(g \restriction \operatorname{ext}_R(x))$ holds for all $x \in \operatorname{dom} g\}$.

We first show that T is a compatible system of functions.

Consider $g_1, g_2 \in T$ and assume that $X = \{x \in \operatorname{dom} g_1 \cap \operatorname{dom} g_2 \mid g_1(x) \neq g_2(x)\} \neq \emptyset$. Let a be an R-minimal element of X. Then yRa implies $y \in \operatorname{dom} g_1$, $y \in \operatorname{dom} g_2$, and $g_1(y) = g_2(y)$. Hence $g_1(a) = g_1 \restriction \operatorname{ext}_R(a) = g_2 \restriction \operatorname{ext}_R(a) = g_2(a)$, a contradiction with $a \in X$.

We now let $f = \bigcup T$. Clearly f is a function, $\operatorname{dom} f = \bigcup\{\operatorname{dom} g \mid g \in T\}$ is R-transitive, and, if $x \in \operatorname{dom} f$, then $x \in \operatorname{dom} g$ for some $g \in T$, $g \subseteq f$, so $f(x) = g(x) = \mathbf{G}(g \restriction \operatorname{ext}_R(x)) = \mathbf{G}(f \restriction \operatorname{ext}_R(x))$.

It remains to prove that $\operatorname{dom} f = A$. If not, there is an R-minimal element a of $A - \operatorname{dom} f$. Then $\operatorname{ext}_R(a) \subseteq \operatorname{dom} f$ and $D = \operatorname{dom} f \cup \{a\}$ is R-transitive. We define g by

$$\begin{aligned} g(x) &= f(x) \quad \text{for } x \in \operatorname{dom} f; \\ g(a) &= \mathbf{G}(f \restriction \operatorname{ext}_R(a)). \end{aligned}$$

Clearly $g \in T$, so $g \subseteq f$ and $a \in \operatorname{dom} f$, a contradiction.

Uniqueness of f follows by the same argument that was used to prove that T is a compatible system. □

In Chapter 6 we used transfinite induction and recursion to show that every well-ordering is isomorphic to a unique ordinal number (ordered by $\in$). In the rest of this section we generalize this important result to well-founded relations.

We recall Definition 2.1 in Chapter 6: a set T is *transitive* if every element of T is a subset of T.

1.9 Definition A transitive set T is *well-founded* if and only if the relation $\in_T$ is well-founded on T, i.e. for every $X \subseteq T$, $X \neq \emptyset$, there is $a \in X$ such that $a \cap X = \emptyset$.

Ordinal numbers, $\mathcal{P}(\omega)$, V_n ($n \in \boldsymbol{N}$), and V_ω are examples of transitive well-founded sets (Exercise 1.2). The next theorem contains the main idea.

1.10 Theorem *Let R be a well-founded relation on A. There is a unique function f on A such that*

$$f(x) = \{f(y) \mid y \in A \text{ and } yRx\} = f[\mathrm{ext}_R(x)]$$

holds for all $x \in A$. The set $T = \operatorname{ran} f$ is transitive and well-founded.

Proof. The existence and uniqueness of f follow immediately from the Recursion Theorem; it suffices to take

$$\boldsymbol{G}(z) = \begin{cases} \operatorname{ran} z & \text{if } z \text{ is a function;} \\ \emptyset & \text{otherwise.} \end{cases}$$

$T = \operatorname{ran} f$ is transitive. This is easy: if $t \in T$ then $t = f(x)$ for some $x \in A$, so $s \in t = \{f(y) \mid yRx\}$ implies $s = f(y)$ for some $y \in A$, so $s \in T$.

T is well-founded. If not, there is $S \subseteq T$, $S \neq \emptyset$, with the property that for every $t \in S$ there is $s \in S$ such that $s \in t$. Let $B = f^{-1}[S]$; $B \neq \emptyset$ because f maps onto T. If $x \in B$ then $f(x) \in S$ so there is $s \in S$ such that $s \in f(x) = \{f(y) \mid yRx\}$. Hence $s = f(y)$ for some yRx, and $y \in f^{-1}[S] = B$. We have shown that for every $x \in B$ there is some $y \in B$ such that yRx, a contradiction with well-foundedness of R. □

The mapping f is in general not one-to-one. For example, $f(x) = \emptyset$ whenever x is an R-minimal element of A (i.e., $\mathrm{ext}_R(x) = \emptyset$). Similarly, $f(x) = \{\emptyset\}$ whenever all elements of $\mathrm{ext}_R(x) \neq \emptyset$ are R-minimal, and so on.

1.11 Definition R is *extensional* on A if $x \neq y$ implies $\mathrm{ext}_R(x) \neq \mathrm{ext}_R(y)$, for all $x, y \in A$. Put differently, R is extensional on A if it has the property that, if for all $z \in A$, zRx if and only if zRy, then $x = y$.

1.12 Theorem *The function f from Theorem 1.10 is one-to-one if and only if R is extensional on A. If it is one-to-one, it is an isomorphism between (A, R) and $(T, \in_T)$.*

Proof. Assume that R is extensional on A but f is not one-to-one. Then $X = \{x \in A \mid \text{there exists } y \in A,\ y \neq x, \text{ such that } f(x) = f(y)\} \neq \emptyset$. Let a be an R-minimal element of X and let $b \neq a$ be such that $f(a) = f(b)$. By extensionality of R, there exists $c \in A$ such that either cRa but not cRb, or cRb but not cRa. We consider only the first case; the second case is similar. From cRa it follows that $f(c) \in f(a) = f(b)$. As $f(b) = \{f(z) \mid zRb\}$, there exists some dRb for which $f(c) = f(d)$. We have $c \neq d$ because cRb fails. But this means that $c \in X$, contradicting the choice of a as an R-minimal element of X.

Assume that R is not extensional on A. Then there exist $a, b \in A$, $a \neq b$, for which $\text{ext}_R(a) = \text{ext}_R(b)$. We then have $f(a) = f[\text{ext}_R(a)] = f[\text{ext}_R(b)] = f(b)$ showing that f is not one-to-one.

Finally we show that f one-to-one implies f is an isomorphism. Of course aRb implies $f(a) \in f(b)$, by definition of f. Conversely, if $f(a) \in f(b)$ then $f(a) = f(x)$ for some xRb. As f is one-to-one, we have $a = x$, so aRb. □

The corollary of Theorems 1.10 and 1.12 asserting that every extensional well-founded relation R on A is isomorphic to the membership relation on a uniquely determined transitive well-founded set T is known as Mostowski's Collapsing Lemma. It shows how to define a unique representative for each class of mutually isomorphic extensional well-founded relations, and thus generalizes Theorem 3.1 in Chapter 6 (see Exercise 1.5).

Exercises

1.1 Given (A, R) and $X \subseteq A$, say that $a \in X$ is an *R-least* element of X if a is an R-minimal element of X and aRb holds for all $b \in X$, $b \neq a$. Show: if every nonempty subset of A has an R-least element then (A, R) is a well-ordering.

1.2 Prove that $\in_A$ is well-founded on A when $A = \mathcal{P}(\omega)$, $A = V_n$ for $n \in \boldsymbol{N}$, and $A = V_\omega$.

1.3 Let (A, R) be well-founded. Show that there is a unique function ρ, defined on A and with ordinal numbers as values, such that for all $x \in A$, $\rho(x) = \sup\{\rho(y) + 1 \mid yRx\}$. $\rho(x)$ is called the *rank* of x in (A, R).

1.4 (a) Let $A = \alpha$, $R = \in_\alpha$, where α is an ordinal number. Prove that $\rho(x) = x$ for all $x \in A$.

(b) Let $A = V_\omega$, $R = \in_A$. Prove that $\rho(x) =$ the least n such that $x \in V_{n+1}$, i.e., $V_n = \{x \in V_\omega \mid \rho(x) < n\}$.

1.5 Let (A, R) be a well-ordering and let $f : A \to T$ be the function from Theorem 1.10. Prove that $T = \alpha$ is an ordinal number and $f = \rho$ is an isomorphism of (A, R) onto $(\alpha, \in_\alpha)$.

1.6 (a) Let A be a transitive well-founded set and let $R = \in_A$. If $f : A \to T$ is the function from Theorem 1.10 then $T = A$ and $f(x) = x$ for all $x \in A$.

(b) If A and B are transitive well-founded sets and $A \neq B$ then $(A, \in_A)$ and $(B, \in_B)$ are not isomorphic.

2. Well-Founded Sets

In our discussion of Russell's paradox in Section 1 of Chapter 1 we touched upon the question whether a set can be an element of itself. We left it unanswered and, in fact, an answer was never needed: all results we proved until now hold, whether or not such sets exist. Nevertheless, the question is philosophically interesting, and important for more advanced study of set theory. We now have the technical tools needed to examine it in some detail.

We start with a simple, but basic, observation.

2.1 Lemma *For any set X there exists a smallest transitive set containing X as a subset; it is called the* transitive closure *of X and is denoted* $\mathrm{TC}(X)$.

Proof. We let $X_0 = X$, $X_{n+1} = \bigcup X_n = \{y \mid y \in x \text{ for some } x \in X_n\}$ and $\mathrm{TC}(X) = \bigcup\{X_n \mid n \in \boldsymbol{N}\}$. It is clear that $X \subseteq \mathrm{TC}(X)$ and $\mathrm{TC}(X)$ is transitive.

It T is any transitive set such that $X \subseteq T$, we see by induction that $X_n \subseteq T$ for all $n \in \boldsymbol{N}$, and hence $\mathrm{TC}(X) \subseteq T$. □

2.2 Lemma *$y \in \mathrm{TC}(X)$ if and only if there is a finite sequence $\langle x_0, x_1, \ldots, x_n\rangle$ such that $x_0 = X$, $x_{i+1} \in x_i$ for $i = 0, 1, \ldots, n-1$, and $x_n = y$.*

Proof. Assume $y \in \mathrm{TC}(X) = \bigcup_{n=0}^{\infty} X_n$ and proceed by induction. If $y \in X_0$ it suffices to take the sequence $\langle X, y\rangle$. If $y \in X_{n+1}$ we have $y \in x \in X_n$ for some x. By inductive assumption, there is a sequence $\langle x_0, \ldots, x_n\rangle$ where $x_0 = X$, $x_{i+1} \in x_i$ for $i = 0, \ldots, n-1$, and $x_n = x$. We let $x_{n+1} = y$ and consider $\langle x_0, \ldots, x_n, x_{n+1}\rangle$.

Conversely, given $\langle x_0, \ldots, x_n\rangle$ where $x_0 = X$ and $x_{i+1} \in x_i$ for $i = 0, 1, \ldots,$ $n-1$, it is easy to see, by induction, that $x_i \in X_i \subseteq \mathrm{TC}(X)$, for all $i \leq n$. □

2.3 Definition A set X is called *well-founded* if $\mathrm{TC}(X)$ is a transitive well-founded set.

For transitive sets, this definition agrees with Definition 1.9 because $\mathrm{TC}(X) = X$ when X is transitive. The significance of well-foundedness is revealed by the following theorem.

2.4 Theorem

(a) *If X is well-founded then there is no sequence $\langle X_n \mid n \in \boldsymbol{N}\rangle$ such that $X_0 = X$ and $X_{n+1} \in X_n$ for all $n \in \boldsymbol{N}$.*

(b)* *If there is no sequence $\langle X_n \mid n \in \boldsymbol{N}\rangle$ such that $X_0 = X$ and $X_{n+1} \in X_n$ for all $n \in \boldsymbol{N}$ then X is well-founded.*

In particular, a well-founded set X cannot be an element of itself (let $X_n = X$ for all n) and one cannot have $X \in Y$ and $Y \in X$ for any Y (consider the sequence $\langle X, Y, X, Y, X, Y, \ldots, \rangle$) or any other "circular" situation.

Proof.

(a) Assume X is well-founded and $\langle X_n \mid n \in \boldsymbol{N}\rangle$ satisfies $X_0 = X$, $X_{n+1} \in X_n$ for all $n \in \boldsymbol{N}$. We have $X_0 = X \subseteq \mathrm{TC}(X)$ and, by induction, $X_n \in \mathrm{TC}(X)$ for all $n \geq 1$. The set $\{X_n \mid n \geq 1\} \subseteq \mathrm{TC}(X)$ does not have an $\in$-minimal element, contradicting well-foundedness of $\mathrm{TC}(X)$.

(b) Assume X is not well-founded, so $\mathrm{TC}(X)$ is not a transitive well-founded set as defined in 1.9. This means that there exists $Y \subseteq \mathrm{TC}(X)$, $Y \neq \emptyset$, with the property that for every $y \in Y$ there is $z \in Y$ such that $z \in y$. Choose some $y \in Y$. By Lemma 2.2 there is a finite sequence $\langle X_0, \ldots, X_n\rangle$ such that $X_0 = X$, $X_{i+1} \in X_i$ for $i = 0, \ldots, n-1$, and $X_n = y$. W extend it to an infinite sequence by recursion, using the Axiom of Choice. Choose X_{n+1} to be some $z \in Y$ such that $z \in y = X_n$ (so $X_{n+1} \in X_n \cap Y$). Given $X_{n+k} \in Y$ choose $X_{n+k+1} \in X_{n+k} \cap Y$ similarly. The resulting sequence $\langle X_n \mid n \in \boldsymbol{N}\rangle$ is as required by the Theorem.

□

It is time to reconsider our intuitive understanding of sets. We recall Cantor's original description: a set is a collection into a whole of definite, distinct objects of our intuition or our thought. It seems reasonable to interpret this as meaning that the objects have to exist (in our mind) before they can be collected into a set. Let us accept this position (it is not the only possible one, as we see in Section 3), and recall that sets are the only objects we are interested in. Suppose we want to form a set "for the first time." That means that there are as yet no suitable objects (sets) in our mind, and so the only collection we can form is the empty set $\emptyset$. But now we have something! $\emptyset$ is now a definite object in our mind, so we can collect the set $\{\emptyset\}$. At this stage there are two objects in our mind, $\emptyset$ and $\{\emptyset\}$, and we can collect various sets of these. In fact, there are four: $\emptyset$, $\{\emptyset\}$, $\{\{\emptyset\}\}$, and $\{\emptyset, \{\emptyset\}\}$. Next we form sets made of these four objects (there are eight of them), and so on. Below, we describe this procedure rigorously, and show that one obtains by it precisely all the well-founded sets.

2.5 Definition (The cumulative hierarchy of well-founded sets.)

$$\begin{aligned} V_0 &= \emptyset; \\ V_{\alpha+1} &= \mathcal{P}(V_\alpha) \quad \text{for all } \alpha; \\ V_\alpha &= \bigcup_{\beta<\alpha} V_\beta \quad \text{for all limit } \alpha \neq 0. \end{aligned}$$

We note that $V_1 = \{\emptyset\}$, $V_2 = \{\emptyset, \{\emptyset\}\}$, $V_3 = \{\emptyset, \{\emptyset\}, \{\{\emptyset\}\}, \{\emptyset, \{\emptyset\}\}\}$ and V_n for $n \leq \omega$ have been defined in Exercise 3.3 in Chapter 6, and some of their properties stated in Exercises 3.4 and 3.5 in Chapter 6 (see also Exercises 1.2 and 1.4(b) in this chapter).

2.6 Lemma
(a) If $x \in V_\alpha$ and $y \in x$ then $y \in V_\beta$ for some $\beta < \alpha$.
(b) If $\beta < \alpha$ then $V_\beta \subseteq V_\alpha$.
(c) For all α, V_α is transitive and well-founded.

Proof.
(a) We proceed by transfinite induction on α. The statement is clear when $\alpha = 0$ or $\alpha \neq 0$ limit. If $x \in V_{\alpha+1}$ then $x \subseteq V_\alpha$, so $y \in x$ implies $y \in V_\alpha$ and we let $\beta = \alpha$.
(b) Again we use transfinite induction on α. Only the successor stage is nontrivial. We show that $V_\alpha \subseteq V_{\alpha+1}$. By (a), if $x \in V_\alpha$ then $x \subseteq \bigcup_{\beta<\alpha} V_\beta$. By inductive assumption, $\bigcup_{\beta<\alpha} V_\beta \subseteq V_\alpha$. We conclude that $x \subseteq V_\alpha$ and hence $x \in V_{\alpha+1}$. It follows that $V_\beta \subseteq V_\alpha \subseteq V_{\alpha+1}$ holds for all $\beta \leq \alpha$.
(c) Combining (a) and (b) gives that $x \in V_\alpha$ implies $x \subseteq \bigcup_{\beta<\alpha} V_\beta \subseteq V_\alpha$, showing that V_α is transitive.
We next show that each V_α is well-founded. Let $Y \subseteq V_\alpha$, $Y \neq \emptyset$. Let β be the least ordinal for which $Y \cap V_\beta \neq \emptyset$; clearly $\beta \leq \alpha$. Take any $x \in Y \cap V_\beta$; by Lemma 2.6(a), $y \in x$ implies $y \in V_\gamma$ for some $\gamma < \beta$, hence $y \notin Y$. This shows that x is an $\in$-minimal element of Y. As V_α is transitive, it is well-founded.

□

2.7 Theorem *A set X is well-founded if and only if $X \in V_\alpha$ for some ordinal α.*

Proof. 1) Assume that X is well-founded. It is easy to check that $\mathrm{TC}(X) \cup \{X\} = \mathrm{TC}(\{X\})$ is a transitive well-founded set. Let $Y = \{x \in \mathrm{TC}(\{X\}) \mid x \in V_\alpha$ for some $\alpha\}$; it suffices to prove that $Y = \mathrm{TC}(\{X\})$. If not, there is an $\in$-minimal element a of $\mathrm{TC}(\{X\}) - Y$. For each $y \in a$ we then have $y \in Y$; let $f(y)$ be the least α such that $y \in V_\alpha$. The Axiom Schema of Replacement guarantees that f is a well-defined function on a. We let $\gamma = \sup f[a]$. Then $y \in V_{f(y)} \subseteq V_\gamma$ holds for each $y \in a$, so $a \subseteq V_\gamma$ and $a \in V_{\gamma+1}$, contradicting $a \notin Y$.

2) If $X \in V_\alpha$, we have $X \subseteq V_\alpha$ by transitivity of V_α, so also $\mathrm{TC}(X) \subseteq V_\alpha$. Any $Y \subseteq \mathrm{TC}(X)$, $Y \neq \emptyset$, has an $\in$-minimal element, because V_α is transitive and well-founded. We conclude that $\mathrm{TC}(X)$ is well-founded, and, by definition, so is X. □

Earlier we proposed an intuitive understanding of the word "collection" that assumes that the objects being collected exist "previously." There is thus a process of collecting more and more complicated sets stage by stage, described rigorously by the cumulative hierarchy V_α, and yielding all well-founded sets. On the other hand, a set which is not well-founded does not fit this understanding, because the existence of a sequence $\langle X_n \mid n \in \boldsymbol{N} \rangle$ where $X_0 \ni X_1 \ni X_2 \ni \cdots$ implies an infinite regress in time. These considerations suggest that, from our present position, only well-founded sets are "true" sets. The reasonableness of this attitude can be further confirmed by observing that all axioms of set theory we have considered (so far) remain true if the word "set" is replaced everywhere by the words "well-founded set." We illustrate this by a few examples, and leave the rest of the axioms as an exercise.

2.8 Example

(a) The Axiom of Extensionality. All elements of a well-founded set are well-founded ($X \in V_\alpha$ implies $X \subseteq V_\alpha$). Therefore, if well-founded sets X and Y have the same well-founded elements then $X = Y$.

(b) The Axiom of Powerset. The powerset $\mathcal{P}(S)$ of a well-founded set S is well-founded. (If $S \in V_\alpha$ then $S \subseteq V_\alpha$, so $\mathcal{P}(S) \subseteq \mathcal{P}(V_\alpha) = V_{\alpha+1}$ and $\mathcal{P}(S) \in V_{\alpha+2}$ is well-founded.) Thus for any well-founded set S there is a well-founded set P ($= \mathcal{P}(S)$) such that for any well-founded X, $X \in P$ if and only if $X \subseteq S$.

(c) The Axiom Schema of Comprehension. Let $\mathbf{P}(x)$ be any property of x (in which the word "set" has been everywhere replaced by "well-founded set"). If A is well-founded, $\{x \in A \mid \mathbf{P}(x)\} \subseteq A$ is also well-founded.

These and similar arguments can be converted into a rigorous proof of the fact that all the axioms of set theory we have adopted so far are consistent with the assumption that only well-founded sets exist. Doing so requires a rigorous formal analysis of what is meant by a property, a subject of mathematical logic that we cannot pursue here. Nevertheless, this fact, the sharpened intuitive understanding of the way sets can be collected in stages that we described above, and an observation that all particular sets needed in mathematics are well-founded (see Exercise 2.3), make most set theorists include the statement that all sets are well-founded among their axioms for set theory.

The Axiom of Foundation (Also called the Axiom of Regularity.) All sets are well-founded.

It should be stressed that, whether or not one accepts the Axiom of Foundation, makes no difference as far as the development of ordinary mathematics in set theory is concerned. Natural numbers, integers, real numbers and functions on them, and even cardinal and ordinal numbers have been defined, and their properties proved in this book, without any use of the Axiom of Foundation. As far as they are concerned, it does not make any difference whether or not there exist any non-well-founded sets. However, the Axiom of Foundation is very useful in investigations of models of set theory (see Chapter 15).

Exercises

2.1 For any set X, $\mathrm{TC}(\{X\})$ is the smallest transitive set containing X as an element.

2.2 (a) $V_\alpha \in V_{\alpha+1} - V_\alpha$.

(b) $\beta < \alpha$ implies $V_\beta \subset V_\alpha$.

(c) $\alpha \in V_{\alpha+1} - V_\alpha$ for all ordinals α.

2.3 Assume that X and Y are well-founded sets. Prove that $\{X, Y\}$, (X, Y), $X \cup Y$, $X \cap Y$, $X - Y$, $X \times Y$, X^Y, $\bigcup X$, $\operatorname{dom} X$, and $\operatorname{ran} X$ are well-founded. Prove that $\boldsymbol{N}$, $\boldsymbol{Z}$, $\boldsymbol{Q}$, and $\boldsymbol{R}$ are well-founded sets.

2.4 Show that the Axioms of Existence, Pair, Union, Infinity, and Choice, as well as the Axiom Schema of Replacement, remain true if the word "set" is replaced everywhere by the words "well-founded set." [*Hint*: Use Exercise 2.3.]

2.5 An Induction Principle. Let $\mathbf{P}$ be some property. Assume that, for all well-founded sets x,

$$(*) \qquad \text{if } \mathbf{P}(y) \text{ holds for all } y \in x, \text{ then } \mathbf{P}(x).$$

Conclude that $\mathbf{P}(x)$ holds for all well-founded x.

2.6 A Recursion Theorem. Let $\boldsymbol{G}$ be an operation. There is a unique operation $\boldsymbol{F}$ such that $\boldsymbol{F}(x) = \boldsymbol{G}(\boldsymbol{F} \restriction x)$ for all well-founded x, $\boldsymbol{F}(x) = \emptyset$ for all non-well-founded x.

2.7 Use Exercise 2.6 to show that there is a unique operation $\boldsymbol{\rho}$ (rank) such that $\boldsymbol{\rho}(x) = \sup\{\boldsymbol{\rho}(y) + 1 \mid y \in x\}$ for well-founded x, $\boldsymbol{\rho}(x) = \{\{\emptyset\}\}$ otherwise. Prove that $V_\alpha = \{x \mid \boldsymbol{\rho}(x) \in \alpha\}$ holds for all ordinals α.

2.8 The Axiom of Foundation can be used to give a rigorous definition of order types of linear orderings (Chapter 4, Section 4). If $\mathfrak{A} = (A, <)$ is a linear ordering, let α be the least ordinal number such that there is a linear ordering $\mathfrak{A}' = (A', <')$ isomorphic to $\mathfrak{A}$, and of rank $\boldsymbol{\rho}(\mathfrak{A}') = \alpha$. (Cf. Exercise 2.7.) We define $\tau(\mathfrak{A}) = \{\mathfrak{A}' \mid \mathfrak{A}' \text{ is isomorphic to } \mathfrak{A} \text{ and } \boldsymbol{\rho}(\mathfrak{A}') = \alpha\}$. This is a set because $\tau(\mathfrak{A}) \subseteq V_{\alpha+1}$. Prove that $\mathfrak{A}$ is isomorphic to $\mathfrak{B} = (B, \prec)$ if and only if $\tau(\mathfrak{A}) = \tau(\mathfrak{B})$. *Isomorphism types* of arbitrary structures can be defined in the same way.

3. Non-Well-Founded Sets

For reasons discussed in the previous section, most set theorists accept the Axiom of Foundation as part of their axiomatic system for set theory. Nevertheless, alternatives allowing non-well-founded sets are logically consistent, have a certain intuitive appeal, and lately have found some applications. We discuss two such "anti-foundation" axioms in this section.

An alternative intuitive view of sets is that the objects of which a set is composed have to be definite and distinct when the set is "finished," but not necessarily beforehand. They may be formed as part of the same process that leads to the collection of the set. *A priori*, this does not exclude the possibility that a set could become one of its own elements, perhaps even its only element.

How could such sets be obtained? It turns out that natural generalizations of results in Sections 1 and 2, in particular of Theorems 1.10 and 1.12, lead to non-well-founded sets. It is convenient to restate these results using some new terminology.

3.1 Definition A *graph* is a structure (A, R) where R is a binary relation in A. A *pointed graph* is (A, R, p) where (A, R) is a graph and $p \in A \neq \emptyset$. A *decoration* of (A, R) or (A, R, p) is a function f with $\operatorname{dom} f = A$ such that, for all $x \in A$,

$$f(x) = \{f(y) \mid yRx\}.$$

If f is a decoration of a pointed graph (A, R, p), the set $f(p)$ is its *value*.

In this terminology, Theorems 1.10 and 1.12 imply, respectively:

Every well-founded graph has a unique decoration.

Every well-founded extensional graph has an injective (one-to-one) decoration.

Moreover, a set is well-founded if and only if it is the value of a decoration of some well-founded graph.

To prove this last remark, note that, if (A, R, p) is a well-founded graph and f a decoration, $f(p) \in \operatorname{ran} f = T$ where T is transitive and well-founded by Theorem 1.10, so $f(p) \subseteq T$ is well-founded. Conversely, given a well-founded set X, let $A = \mathrm{TC}(\{X\})$, $R = \in_A$, $p = X$, and $f = \mathrm{Id}_A$. It is easy to verify that (A, R, p) is a well-founded extensional pointed graph, f is an injective decoration of it, and $f(p) = X$ (see Exercise 2.1).

Let us now consider decorations of non-well-founded graphs. In the picture of (A, R, p), elements of A are represented by dots, aRb is depicted as an arrow from b to a, and the "point" p is circled.

3.2 Example

(a) See Figure 1(a). Let $A = \{a\}$ be a singleton, $R = \{(a, a)\}$, and $p = a$. Let S be the value of some decoration f of (A, R, p). Then $S = \{S\}$.

(b) See Figure 1(b). Let $A = \{a, b\}$ where $a \neq b$, $R = \{(a, b), (b, a)\}$, and $p = a$. If T is the value of some decoration of (A, R, p) then $T = \{\{T\}\}$.

(c) For a more complicated example of a non-well-founded set, and a hint of possible applications, we consider a task that arises in the study of programming languages. One has a set D of "data" and a set P of "programs." A program $\pi \in P$ accepts data as "inputs" and returns data as "outputs"; mathematically, it is just a function from D to D. This is straightforward when $D \cap P = \emptyset$, but interesting complications arise when programs accept other programs (or even themselves) as input, a situation quite common in programming.
As a simple specific example, we let $P = \{\pi\}$, $D = \{\emptyset, \pi\}$, and we stipulate that the program π accepts any input from D and returns it as output, without any modification; i.e., $\pi(\emptyset) = \emptyset$, $\pi(\pi) = \pi$. We then have $\pi = \{(\emptyset, \emptyset), (\pi, \pi)\} = \{\{\{\emptyset\}\}, \{\{\pi\}\}\}$. Such a set can be obtained as the value of any decoration of the graph depicted in Figure 1(c).

These examples show that we obtain non-well-founded sets if we postulate that (at least some) non-well-founded graphs have decorations. But first there is an interesting question to consider: what is the meaning of equality for non-well-founded sets? The Axiom of Extensionality, a *sine qua non* of set theory,

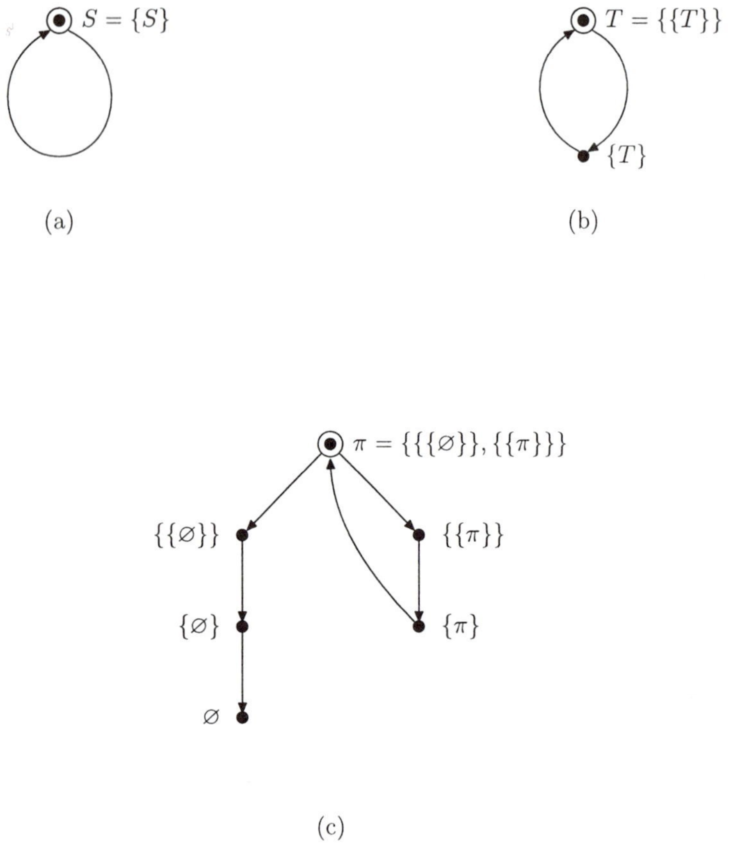

Figure 1

tells us that, for any sets X and Y, in order to show $X = Y$ it suffices to establish that X and Y have the same elements. For well-founded sets this is an intuitively effective procedure, because the elements of X and Y have been formed "previously." But if, for example, $X = \{X\}$ and $Y = \{Y\}$, trying to use Extensionality to decide whether $X = Y$ begs the question. Some stronger principle (compatible with the Axiom of Extensionality, of course) is needed to make the determination. The axiom we consider below implies existence of many non-well-founded sets, as well as a principle for deciding their equality. It is a natural generalization of Theorem 1.10, obtained by dropping the assumption that R is well-founded.

3.3 The Axiom of Anti-Foundation *Every graph has a unique decoration.*

It is known that this axiom is consistent with Zermelo-Fraenkel set theory (without the Axiom of Foundation).

3.4 Definition A set X is *reflexive* if $X = \{X\}$.

3.5 Theorem *The Axiom of Anti-Foundation implies that there exists a unique reflexive set.*

Proof. Let (A, R, p) be the pointed graph from Example 3.2(a). By the Axiom of Anti-Foundation it has a decoration f; $X = f(p)$ is a reflexive set. If Y is also a reflexive set, we define g on A by $g(a) = Y$. It is clear that g is also a decoration. By uniqueness, $f = g$ and hence $X = Y$. □

Next we develop a general criterion for determining whether two decorations have the same value. It involves an important notion of bisimulation, which originated in the study of ways how one (infinite) process can imitate another.

3.6 Definition Let (A_1, R_1) and (A_2, R_2) be graphs. For $B \subseteq A_1 \times A_2$ we define $B^+ \subseteq A_1 \times A_2$ by: $(a_1, a_2) \in B^+$ if for every $x_1 \in \text{ext}_{R_1}(a_1)$ there exists $x_2 \in \text{ext}_{R_2}(a_2)$, and for every $x_2 \in \text{ext}_{R_2}(a_2)$ there exists $x_1 \in \text{ext}_{R_1}(a_1)$, so that $(x_1, x_2) \in B$.

We say that B is a *bisimulation* between (A_1, R_1) and (A_2, R_2) if $B^+ \subseteq B$.

3.7 Lemma

(i) If $B \subseteq C \subseteq A_1 \times A_2$ then $B^+ \subseteq C^+$.

(ii) $\emptyset$ is a bisimulation.

(iii) The union of any collection of bisimulations is a bisimulation.

Proof. (i) and (ii) are obvious.

(iii) If $B = \bigcup_{i \in I} B_i$ and $B_i \subseteq B_i^+$ for all $i \in I$, then $B_i^+ \subseteq B^+$ for all $i \in I$ by (i), so $B = \bigcup_{i \in I} B_i \subseteq \bigcup_{i \in I} B_i^+ \subseteq B^+$. □

It follows immediately that, for any given (A_1, R_1), (A_2, R_2), there exists a *largest bisimulation*

$$\tilde{B} = \bigcup \{B \subseteq A_1 \times A_2 \mid B \text{ is a bisimulation}\}.$$

The next lemma connects this concept with decorations.

3.8 Lemma *Let f_1, f_2 be decorations of (A_1, R_1), (A_2, R_2), respectively. Set $B = \{(a_1, a_2) \in A_1 \times A_2 \mid f_1(a_1) = f_2(a_2)\}$. Then B is a bisimulation.*

Proof. We prove that $B = B^+$. We have

$$\begin{aligned}(a_1, a_2) \in B \text{ if and only if } & f_1(a_1) = f_2(a_2)\\ \text{if and only if } & \text{for every } x_1 \in \text{ext}_{R_1}(a_1),\ f_1(x_1) \in f_2(a_2) \text{ and vice versa}\\ \text{if and only if } & \text{for every } x_1 \in \text{ext}_{R_1}(a_1) \text{ there exists } x_2 \in \text{ext}_{R_2}(a_2) \text{ such that } f_1(x_1) = f_2(x_2) \text{ and vice versa}\\ \text{if and only if } & \text{for every } x_1 \in \text{ext}_{R_1}(a_1) \text{ there exists } x_2 \in \text{ext}_{R_2}(a_2) \text{ such that } (x_1, x_2) \in B \text{ and vice versa}\\ \text{if and only if } & (a_1, a_2) \in B^+.\end{aligned}$$

□

3.9 Definition Pointed graphs (A_1, R_1, p_1) and (A_2, R_2, p_2) are *bisimulation equivalent* if there is a bisimulation B between (A_1, R_1) and (A_2, R_2) such that $(p_1, p_2) \in B$. (Equivalently, if $(p_1, p_2) \in \tilde{B}$.)

The next theorem provides the promised criterion for equality.

3.10 Theorem *Let f_1 be a decoration of (A_1, R_1, p_1) and let f_2 be a decoration of (A_2, R_2, p_2). The Axiom of Anti-Foundation implies that f_1 and f_2 have the same value if and only if (A_1, R_1, p_1) and (A_2, R_2, p_2) are bisimulation equivalent.*

3.11 Corollary *Assuming the Axiom of Anti-Foundation, $X = Y$ if and only if the pointed graphs $(\mathrm{TC}(\{X\}), \in, X)$ and $(\mathrm{TC}(\{Y\}), \in, Y)$ are bisimulation equivalent.*

Before proving Theorem 3.10 we need a technical lemma.

3.12 Lemma

(a) *For $i = 1, 2$ let f_i be a decoration of (A_i, R_i, p_i). Let $\overline{A_i}$ be the smallest R_i-transitive subset of A_i such that $p_i \in \overline{A_i}$. Let $\overline{R_i} = R_i \cap (\overline{A_i} \times \overline{A_i})$ and $\overline{f_i} = f_i \restriction \overline{A_i}$. Then $\overline{f_i}$ is a decoration of $(\overline{A_i}, \overline{R_i}, p_i)$.*

(b) *Let B be a bisimulation between (A_1, R_1, p_1) and (A_2, R_2, p_2). Then $\overline{B} = B \cap (\overline{A_1} \times \overline{A_2})$ is a bisimulation between $(\overline{A_1}, \overline{R_1})$ and $(\overline{A_2}, \overline{R_2})$. If $(p_1, p_2) \in B$ then $\operatorname{dom} \overline{B} = \overline{A_1}$ and $\operatorname{ran} \overline{B} = \overline{A_2}$.*

Proof. We recall from the proof of Lemma 1.8 that $\overline{A_i} = \bigcup_{n=0}^{\infty} C_{i,n}$ where $C_{i,0} = \{p_i\}$, $C_{i,n+1} = \{y \in A_i \mid y R_i x \text{ holds for some } x \in C_{i,n}\} = \bigcup\{\text{ext}_{R_i}(x) \mid x \in C_{i,n}\}$. In particular, $\text{ext}_{R_i}(x) = \text{ext}_{\overline{R_i}}(x)$ for $x \in \overline{A_i}$. It follows immediately that $\overline{f_i}$ is a decoration and $\overline{B}$ is a bisimulation. The last statement implies that $\operatorname{dom} \overline{B}$ is an $\overline{R_1}$-transitive subset of $\overline{A_1}$, hence also an R_1-transitive subset of A_1. If $(p_1, p_2) \in B$ then $p_1 \in \operatorname{dom} \overline{B}$, and we conclude that $\overline{A_1} \subseteq \operatorname{dom} \overline{B}$. Similarly, $\overline{A_2} \subseteq \operatorname{ran} \overline{B}$. □

Proof of Theorem 3.10. Assume that $f_1(p_1) = f_2(p_2)$. Then $B = \{(a_1, a_2) \in A_1 \times A_2 \mid f_1(a_1) = f_2(a_2)\}$ is a bisimulation by Lemma 3.8, and $(p_1, p_2) \in B$. Hence (A_1, R_1, p_1) and (A_2, R_2, p_2) are bisimulation equivalent.

Conversely, let B be a bisimulation satisfying $(p_1, p_2) \in B$. From Lemma 3.12 we obtain a bisimulation $\overline{B}$ between $(\overline{A_1}, \overline{R_1})$ and $(\overline{A_2}, \overline{R_2})$, and decorations $\overline{f_1}$ and $\overline{f_2}$. We now define a graph (A, R) as follows:

$$A = \overline{B} = \{(a_1, a_2) \mid a_1 \in \overline{A_1},\ a_2 \in \overline{A_2},\ (a_1, a_2) \in \overline{B}\};$$
$$R = \{((b_1, b_2), (a_1, a_2)) \in A \times A \mid (b_1, a_1) \in \overline{R_1} \text{ and } (b_2, a_2) \in \overline{R_2}\}.$$

We define functions F_1 and F_2 on A by:

$$F_1((a_1, a_2)) = \overline{f_1}(a_1)$$
$$F_2((a_1, a_2)) = \overline{f_2}(a_2).$$

We note that $F_1((a_1, a_2)) = \overline{f_1}(a_1) = \{\overline{f_1}(b_1) \mid b_1 \overline{R_1} a_1\} = i\{F_1((b_1, b_2)) \mid b_1 \overline{R_1} a_1,\ b_2 \overline{R_2} a_2,\ (b_1, b_2) \in \overline{B}\} = \{F_1((b_1, b_2)) \mid (b_1, b_2) R (a_1, a_2)\}$, so F_1 is a decoration on (A, R). Similarly, F_2 is a decoration on (A, R). The Axiom of Anti-Foundation implies $F_1 = F_2$, so in particular $f_1(p_1) = \overline{f_1}(p_1) = F_1((p_1, p_2)) = F_2((p_1, p_2)) = \overline{f_2}(p_2) = f_2(p_2)$. □

Other "anti-foundation" axioms have been considered. For example, one can generalize Theorem 1.12 and obtain the following.

3.13 The Axiom of Universality *Every extensional graph has an injective decoration.*

The Axiom of Universality is also consistent, but it is incompatible with the Axiom of Anti-Foundation: as the next theorem shows, it provides many more non-well-founded sets.

3.14 Theorem *The Axiom of Universality implies that there exist collections of reflexive sets of arbitrary cardinality.*

Proof. Let A be any set, and let $R = \{(a, a) \mid a \in A\}$ be the identity relation on A. If $a \neq b$ then $\text{ext}_R(a) = \{a\} \neq \{b\} = \text{ext}_R(b)$, so (A, R) is extensional. If f is some injective decoration of (A, R) then $f(a) = \{f(a)\}$ holds for each $a \in A$, and $a \neq b$ implies $f(a) \neq f(b)$, by injectivity of f. Hence $\{f(a) \mid a \in A\}$ is a collection of reflexive sets of the same cardinality as A. □

Stronger results along these lines can be found in the exercises.

Exercises

3.1 Construct pointed graphs whose value S has the property
 (a) $S = \{\emptyset, S\}$;
 (b) $S = (\emptyset, S)$;
 (c) $S = \boldsymbol{N} \cup \{S\}$.

3.2 Show that $\tilde{B}^+ = \tilde{B}$ holds for the largest bisimulation $\tilde{B}$.

3.3 For given graphs (A_1, R_1), (A_2, R_2), (A_3, R_3) show

(i) Id_{A_1} is a bisimulation between (A_1, R_1) and (A_1, R_1).

(ii) If B is a bisimulation between (A_1, R_1) and (A_2, R_2) then B^{-1} is a bisimulation between (A_2, R_2) and (A_1, R_1).

(iii) If B is a bisimulation between (A_1, R_1) and (A_2, R_2), and C between (A_2, R_2) and (A_3, R_3), then $C \circ B$ is a bisimulation between (A_1, R_1) and (A_3, R_3).

Conclude that the notion of bisimulation equivalence is reflexive, symmetric, and transitive.

3.4 Show that any two of the following pointed graphs are bisimulation equivalent:

(i) Example 3.2(a);

(ii) Example 3.2(b);

(iii) $(\boldsymbol{N}, \{(n+1, n) \mid n \in \boldsymbol{N}\}, 0)$;

(iv) $(\boldsymbol{N}, >, 0)$.

3.5 Let (A, R) be a graph. Define by transfinite recursion:

$$\begin{aligned} W_0 &= \emptyset; \\ W_{\alpha+1} &= \{a \in A \mid \mathrm{ext}_R(a) \subseteq W_\alpha\}; \\ W_\alpha &= \bigcup_{\beta<\alpha} W_\beta \text{ for limit } \alpha \neq 0. \end{aligned}$$

Show that there exists λ such that $W_\lambda = W_{\lambda+1}$. Prove that $(W_\lambda, R \cap W_\lambda^2)$ is well-founded. We call W_λ the *well-founded part* of (A, R).

3.6 If W is the well-founded part of (A, R) and f_1, f_2 are decorations of (A, R) then $f_1 \restriction W = f_2 \restriction W$.

3.7 Let (A, R, p) be an extensional pointed graph, W its well-founded part, and $p \notin W$. Assuming the Axiom of Universality, show that there are sets of arbitrarily large cardinality, all elements of which are values of this graph.
[*Hint:* Take the union of an arbitrary collection of disjoint copies of (A, R); identify their well-founded parts, show that the resulting structure is extensional, and consider its injective decoration.]

Chapter 15

The Axiomatic Set Theory

1. The Zermelo-Fraenkel Set Theory With Choice

In the course of the previous fourteen chapters we introduced axioms which, taken together, constitute the Zermelo-Fraenkel set theory with Choice (ZFC). For the reader's convenience, we list all of them here.

The Axiom of Existence There exists a set which has no elements.

The Axiom of Extensionality If every element of X is an element of Y and every element of Y is an element of X, then $X = Y$.

The Axiom Schema of Comprehension Let $\mathbf{P}(x)$ be a property of x. For any A, there is B such that $x \in B$ if and only if $x \in A$ and $\mathbf{P}(x)$ holds.

The Axiom of Pair For any A and B, there is C such that $x \in C$ if and only if $x = A$ or $x = B$.

The Axiom of Union For any S, there is U such that $x \in U$ if and only if $x \in A$ for some $A \in S$.

The Axiom of Power Set For any S, there is P such that $X \in P$ if and only if $X \subseteq S$.

The Axiom of Infinity An inductive set exists.

The Axiom Schema of Replacement Let $\mathbf{P}(x, y)$ be a property such that for every x there is a unique y for which $\mathbf{P}(x, y)$ holds. For every A there is B such that for every $x \in A$ there is $y \in B$ for which $\mathbf{P}(x, y)$ holds.

The Axiom of Foundation All sets are well-founded.

The Axiom of Choice Every system of sets has a choice function.

(The reader might notice that some of the axioms are redundant. For example, the Axiom of Existence and the Axiom of Pair can be proved from the rest.)

We have shown in this book that the well-known concepts of real analysis (real numbers and arithmetic operations on them, limits of sequences, continuous functions, etc.) can be defined in set theory and their basic properties proved from Zermelo-Fraenkel axioms with Choice. A similar assertion can be made about any other branch of contemporary mathematics (except category theory). Fundamental objects of topology, algebra, or functional analysis (say, topological spaces, vector spaces, groups, rings, Banach spaces) are customarily defined to be sets of a specific kind. Topologic, algebraic, and analytic properties of these objects are then derived from the various properties of sets, which can be themselves in their turn obtained as consequences of the axioms of ZFC. Experience shows that all theorems whose proofs mathematicians accept on intuitive grounds can be in principle proved from the axioms of ZFC. In this sense, the axiomatic set theory serves as a satisfactory unifying foundation for mathematics.

Having ascertained that ZFC completely codifies current mathematical practice, one might wonder whether this is likely to be the case also for mathematical practice of the future. To put the question differently, can all true mathematical theorems (including those whose truth has not yet been demonstrated) be proved in Zermelo-Fraenkel set theory with Choice? Should the answer be "yes," we would know that all as yet open mathematical questions can be, at least in principle, decided (proved or disproved) from the axioms of ZFC alone. However, matters turned out differently.

Mathematicians have been baffled for decades by relatively easily formulated set-theoretic problems, which they were unable to either prove or disprove. A typical example of a problem of this kind is the Continuum Hypothesis (CH): Every set of real numbers is either at most countable or has the cardinality of the continuum. We showed in Chapter 9 that CH is equivalent to the statement $2^{\aleph_0} = \aleph_1$, but we proved neither $2^{\aleph_0} = \aleph_1$ nor $2^{\aleph_0} \neq \aleph_1$. Another similar problem is the Suslin's Hypothesis, and others arose in topology and measure theory (see Section 3). Persistent failures of all attempts to solve these problems led some mathematicians to suspect that they cannot be solved at all at the current level of mathematical art. The axiomatic approach to set theory makes it possible to formulate this suspicion as a rigorous, mathematically verifiable conjecture that, say, the Continuum Hypothesis cannot be decided from the axioms of ZFC (which codify the current state of mathematical art, as we have already discussed). This conjecture has been shown correct by work of Kurt Gödel and Paul Cohen.

First, Gödel demonstrated in 1939 that the Continuum Hypothesis cannot be disproved in ZFC (that is, one cannot prove $2^{\aleph_0} \neq \aleph_1$). Twenty-four years later, Cohen showed that it cannot be proved either. Their techniques were later used by other researchers to show that the Suslin's Hypothesis and many other problems are also undecidable in ZFC. We try to outline some of Gödel's and Cohen's ideas in Section 2, but readers more deeply interested in this matter should consult some of the more advanced texts of set theory.

The foregoing results put the classical open problems of set theory in a new perspective. Undecidability of the Continuum Hypothesis on the basis of our present understanding of sets as reflected by the axioms of ZFC means that some fundamental property of sets is still unknown. The task is now to find this property and formulate it as a new axiom which, when added to ZFC, would decide CH one way or another.

At one level, such axioms are very easy to come by. We could, for example, add to ZFC the axiom

$$\text{“}2^{\aleph_0} = \aleph_1.\text{”}$$

Unfortunately, one could instead add the axiom

$$\text{“}2^{\aleph_0} = \aleph_2\text{”}$$

and get a different, incompatible set theory. Or how about the axiom "$2^{\aleph_0} = \aleph_{\omega+17}$"? Cohen's work shows that this, also, produces a consistent set theory, incompatible with the previous two. Furthermore, it is possible to create two subvarieties of each of these three set theories by adding to them either the axiom asserting that Suslin's Hypothesis holds or the axiom asserting that it fails. In the opinion of some researchers, that is where the matter now stands. Similarly as in geometry, where there are, besides the classical euclidean geometry, also various noneuclidean ones (elliptic, hyperbolic, etc.), we have the cantorian set theory with $2^{\aleph_0} = \aleph_1$ as an axiom, and, besides it, various noncantorian set theories, with no logical reasons for preferring one to another.

Such an attitude is hardly completely satisfactory. To solve the continuum problem by arbitrarily adding $2^{\aleph_0} = \aleph_{\omega+17}$ as an axiom is clearly cheating. It is certainly not the way we have proceeded in the previous chapters. Whenever we accepted an axiom:

(a) It was intuitively obvious that sets, as we understand them, have the property postulated by the axiom (some doubts arose in the case of the Axiom of Choice, but those were discussed at length.)

(b) The axiom had important consequences both in set theory and in other mathematical disciplines; some of these consequences were in fact equivalent to it.

So far, there appears to be very little evidence that the axiom $2^{\aleph_0} = \aleph_{\omega+17}$ (or any other axiom of the form $2^{\aleph_0} = \aleph_\alpha$) satisfies either (a) or (b).

In fact, no axioms that would be intuitively obvious in the same way as the axioms of ZFC have been proposed. Perhaps our intuition in these matters has reached its limits. However, in recent years it has been discovered that a number of open questions, in particular in the field called descriptive set theory, has an intimate connection with various large cardinals, such as those we mentioned in Chapters 9 and 13. The typical pattern is that such questions can be answered one way assuming an appropriate large cardinal exists, and have an opposite answer assuming it does not, with the answer obtained with the help of large cardinals being preferable in the sense of being much more "natural," "profound," or "beautiful." The great amount of research into these matters performed over the last 40 years has produced the very rich, often very subtle and difficult theory of large cardinals, whose esthetic appeal makes it difficult not to believe that it describes true aspects of the universe of set theory. We discuss some of these results very briefly in Section 3.

2. Consistency and Independence

In order to understand methods for showing consistency and independence of the Continuum Hypothesis with respect to Zermelo-Fraenkel axioms for set theory, let us first investigate a similar, but much simpler, problem. In Chapter 2, we defined (strictly) *ordered sets* as pairs $(A, <)$ where A is a set and $<$ is an asymmetric and transitive binary relation in A. Equivalently, one might say that an ordered set is a structure $(A, <)$, which satisfies the following axioms.

The Axiom of Asymmetry There are no a and b such that $a < b$ and $b < a$.

The Axiom of Transitivity For all a, b, and c, if $a < b$ and $b < c$, then $a < c$.

We can also say that the Axiom of Asymmetry and the Axiom of Transitivity comprise an *axiomatic theory of order* and that ordered sets are *models* of this axiomatic theory. Finally, let us formulate yet another axiom.

The Axiom of Linearity For all a and b, either $a < b$ or $a = b$ or $b < a$.

Let us now ask whether the Axiom of Linearity can be proved or disproved in our axiomatic theory of order.

First, let us assume that the Axiom of Linearity can be proved in the theory of order. Then every model of the theory of order would have to satisfy also the Axiom of Linearity, a logical consequence of that theory. In a more familiar terminology, every ordering would have to be a linear ordering. But this is false; an example of a model for the theory of order in which the Axiom of Linearity does not hold is in Figure 1(a).

Next, let us assume that the Axiom of Linearity can be disproved in the theory of order. Then every model of the theory of order would have to satisfy also the negation of the Axiom of Linearity. In other words, every ordering

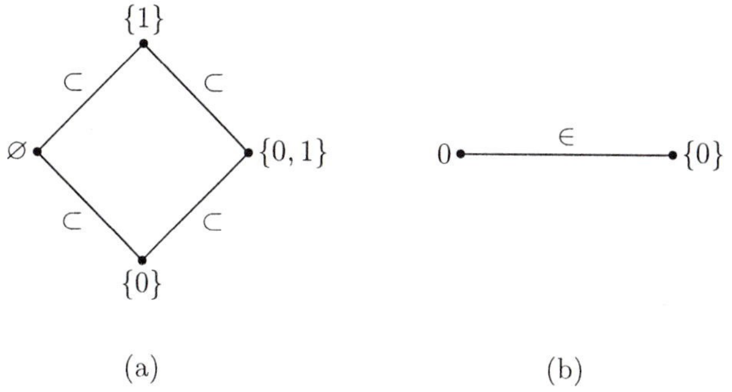

Figure 1: (a) $(\mathcal{P}(\{0,1\}), \subset)$; (b) $(\{0, \{0\}\}, \in)$.

would have to be nonlinear. But this is again false; see Figure 1(b) for an example of a model for the theory of order in which the Axiom of Linearity holds.

We conclude that the Axiom of Linearity is undecidable in the theory of order.

Let us now return to the question of undecidability of the Continuum Hypothesis in ZFC. By analogy with the preceding example, we see that in order to show that the Continuum Hypothesis cannot be proved in ZFC, one has to construct a model of ZFC in which the Continuum Hypothesis fails, and, similarly, in order to show that the Continuum Hypothesis cannot be disproved in ZFC, one has to construct a model of ZFC in which the Continuum Hypothesis holds. In the rest of this section, we outline constructions of just such models. But a technical point has to be clarified first. To describe a model for the theory of order, i.e., an ordered set, we have to specify the members of the model (by choosing the set A) and the meaning of the relation "less than" (by choosing a binary relation $<$ in A). Similarly, to describe a model for set theory, we have to specify the members of the model and the meaning of the relation "belongs to" in the model. But, in view of the paradoxes of set theory, it is too optimistic to expect that the members of a model for set theory can themselves be gathered into a set. (Actually, such models are not entirely impossible; their existence cannot be proved in ZFC, but can be proved if one extends ZFC by some large cardinal axiom — see Section 3).

To circumvent this difficulty, models for set theory have to be described by a pair of *properties*, $\mathbf{M}(x)$ and $\mathbf{E}(x, y)$. Here $\mathbf{M}(x)$ reads "x is a set of the model" and $\mathbf{E}(x, y)$ reads "x belongs to y in the sense of the model." A trivial example of a model for set theory is obtained by choosing $\mathbf{M}(x)$ to be the property "x is a set" and $\mathbf{E}(x, y)$ to be the property "$x \in y$." The model simply consists of all sets with the usual membership relation.

The first nontrivial model for set theory was constructed by Kurt Gödel in 1939 and became known as *the constructible model.* Gödel wanted to find a model in which the Continuum Hypothesis would hold, that is, in which $2^{\aleph_0} = \aleph_1$. Cantor's Theorem asserts that $2^{\aleph_0} \geq \aleph_1$; in other words $\aleph_1$ is the *smallest* cardinal to which the cardinality of the continuum can be equal. These considerations suggest a search for a model which would contain as few sets as possible. Gödel constructed such a model by transfinite recursion, in stages. At each stage, sets are put into the model *only if* their existence is guaranteed by one of the axioms of ZFC.

To begin with, the Axiom of Infinity and the Axiom Schema of Comprehension guarantee existence of the set of natural numbers, and so we let

$$L_0 = \omega.$$

Next, if $\mathbf{P}(n)$ is any property with parameters from ω, then the Axiom Schema of Comprehension postulates existence of the set $\{n \in \omega \mid \mathbf{P}(n)$ holds in $(\omega, \in)\}$, and we have to put it into the model. Therefore, we let

$$L_1 = \{X \subseteq L_0 \mid X = \{n \in L_0 \mid \mathbf{P}(n) \text{ holds in } (L_0, \in)\} \text{ for some property } P \text{ with parameters from } L_0\}.$$

In order to show that L_1 exists, it is necessary first to give formal definitions of logical concepts, such as "property" and "holds in," in set theory. Such definitions can be found in most textbooks of mathematical logic, but they are beyond the scope of this brief outline. Existence of L_1 itself then follows from the Axiom of Power Set and the Axiom Schema of Comprehension, so L_1 itself has to be in the model. Notice that $L_0 \subseteq L_1$: We can obtain $X = k$ for any $k \in L_0$ by taking "$n \in k$" as the property $\mathbf{P}(n)$.

Next we repeat the argument of the previous paragraph with L_1 in place of L_0. Since L_1 is in the model, all of its subsets which are definable in $(L_1, \in)$ by some property $\mathbf{P}$ must be put into the model, as well as the set L_2 of all such subsets.

In this fashion one defines L_3, L_4, ... At stage ω, we merely take the union of the previously constructed L_n:

$$L_\omega = \bigcup_{n \in \omega} L_n$$

(its existence follows from the Axiom Schema of Replacement and the Axiom of Union), and then continue as before.

The recursive definition of the operation L thus goes as follows:

$$\begin{aligned} L_0 &= \omega; \\ L_{\alpha+1} &= \{X \subseteq L_\alpha \mid X \text{ is definable in } (L_\alpha, \in) \text{ by some property } \mathbf{P} \text{ with} \\ &\qquad \text{parameters from } L_\alpha\}; \\ L_\alpha &= \bigcup_{\beta<\alpha} L_\beta \text{ if } \alpha \text{ is a limit ordinal, } \alpha > 0. \end{aligned}$$

It is easy to see that $L_\alpha \subseteq L_\beta$ whenever $\alpha < \beta$. A set is called *constructible* if it belongs to L_α for some ordinal α.

We are now ready to describe the constructible model. Sets of the model are going to be precisely the constructible sets [that is, $\mathbf{M}(x)$ is the property "x is constructible"]. The membership relation in the model is the usual one: If x and y are sets of the model, x belongs to y in the model if and only if $x \in y$ [that is, $\mathbf{E}(x, y)$ is the property "$x \in y$"].

Of course, it has to be verified that the constructible model is indeed a model for ZFC, that is, satisfies all of its axioms. As an illustration, we show that the Axiom of Pair holds in the constructible model: For any A and B in the model, there is C in the model such that, for all x in the model, x belongs to C in the model if and only if $x = A$ or $x = B$.

Taking into account the definition of the constructible model, this amounts to showing: For any constructible sets A and B, there is a constructible set C such that, for all constructible x, $x \in C$ if and only if $x = A$ or $x = B$.

Let constructible sets A and B be given; we first prove that $\{A, B\}$ is also a constructible set. Indeed, if $A \in L_\alpha$, $B \in L_\beta$, and $\gamma = \max\{\alpha, \beta\}$, then $A, B \in L_\gamma$ and the set $\{A, B\}$ is definable in $(L_\gamma, \in)$ as $\{x \in L_\gamma \mid x = A \text{ or } x = B\}$. We can conclude that $\{A, B\} \in L_{\gamma+1}$ and is, therefore, constructible. If one now sets $C = \{A, B\}$, then C is a constructible set which clearly has the required property.

The previous proof actually establishes a stronger result than mere validity of the Axiom of Pair in the constructible model; it demonstrates that the operation of unordered pair, when performed in the model on two sets from the model, results in the usual unordered pair of these two sets. We say that the operation of unordered pair is *absolute.* A similar detailed analysis of the set-theoretic concepts involved shows that all the remaining axioms of ZFC hold in the constructible model, and that many of the usual set-theoretic operations and notions are absolute. The notion of natural number is absolute (that is, a constructible set is a natural number in the sense of the constructible model if and only if it is really a natural number) and so is the notion of ordinal number. However, some concepts are not absolute; among the most important examples are the operation of power set and the notion of cardinal number. The reason for this is that, say, the power set of ω in the sense of the constructible model consists of all *constructible* subsets of ω, while the real power set of ω consists of all subsets of ω. So clearly the power set of ω in the model is a subset of the real power set of ω, but not necessarily vice versa. Indeed, it is this phenomenon which allows us to "cut down" the size of the continuum in the constructible model. Finally, the notion of constructibility is itself absolute: A set is constructible in the sense of the constructible model if and only if it is constructible. But all sets in the model are constructible! We can conclude that every set in the model is constructible in the sense of the model. This argument proves that the following statement holds in the constructible model.

The Axiom of Constructibility Every set is constructible.

To summarize, the constructible model satisfies all axioms of ZFC and, in addition, the Axiom of Constructibility. Consequently, the Axiom of Constructibility cannot be disproved in ZFC, and can be added to it without danger of producing contradictions. It turns out that there are many remarkable theorems one can prove in ZFC enriched by the Axiom of Constructibility; all of them then also hold in the constructible model, and thus cannot be disproved in ZFC. Gödel himself showed that the Axiom of Constructibility implies the Generalized Continuum Hypothesis:

$$2^{\aleph_\alpha} = \aleph_{\alpha+1} \quad \text{for all ordinals } \alpha.$$

Ronald Jensen used the Axiom of Constructibility to construct a linearly ordered set without endpoints in which every system of mutually disjoint intervals is at most countable, but which has no at most countable dense subset, and thus showed the failure of the Suslin's Hypothesis in this model. Many other deep results of a similar character have been obtained since. Here we merely indicate how the Axiom of Constructibility is used to show that $2^{\aleph_0} = \aleph_1$.

The proof is based on a fundamental result in mathematical logic, the Skolem-Löwenheim Theorem. This theorem asserts that for any structure $(A; R, \dots, F, \dots)$ where R is a binary relation, F is a unary function, etc., there is an *at most countable* $B \subseteq A$ such that any property with parameters from B holds in $(A; R, \dots, F, \dots)$ precisely when it holds in $(B; R \cap B^2, \dots, F \restriction B, \dots)$. (This is an abstract, generalized version of such theorems as "Every group has an at most countable subgroup," etc. It also generalizes Theorem 3.14 in Chapter 4.)

Let now $X \subseteq \omega$. The Axiom of Constructibility guarantees that $X \in L_{\alpha+1}$ for some, possibly uncountable, ordinal α. This means that there is a property $\mathbf{P}$ such that $n \in X$ if and only if $\mathbf{P}(n)$ holds in $(L_\alpha, \in)$. By the Skolem-Löwenheim Theorem, there is an *at most countable* set $B \subseteq L_\alpha$ such that $(B, \in)$ satisfies the same statements as $(L_\alpha, \in)$. In particular, $n \in X$ if and only if $\mathbf{P}(n)$ holds in $(B, \in)$. Moreover, the fact that a structure is of the form $(L_\beta, \in)$ for some ordinal β can itself be expressed by a suitable statement, which holds in $(L_\alpha, \in)$ and thus also in $(B, \in)$. From all this, one can conclude that $(B, \in)$ is (isomorphic to) a structure of the form $(L_\beta, \in)$ for some, necessarily at most countable, ordinal β. Since X is definable in $(B, \in)$, we get $X \in L_{\beta+1}$.

We can conclude that every set of natural numbers is constructed at some at most countable stage, i.e., $\mathcal{P}(\omega) \subseteq \bigcup_{\beta<\omega_1} L_{\beta+1}$. To complete the proof of $2^{\aleph_0} = \aleph_1$, we only need to show that the cardinality of the latter set is $\aleph_1$. This in turn follows if we show that L_γ is countable for all $\gamma < \omega_1$. Clearly, $L_0 = \omega$ is countable. The set L_1 consists of all subsets of L_0 definable in $(L_0, \in)$, but there are only countably many possible definitions (each definition is a finite sequence of letters from a finite alphabet of some formalized language, together with a finite sequence of parameters from the countable set L_0), and therefore only countably many definable subsets of L_0. We conclude that L_1 is countable and then proceed by induction, using the same idea at all successor stages, and the fact that a union of countably many countable sets is countable at limit stages.

The question of independence of the Continuum Hypothesis from the axioms of ZFC (that is, of showing that it cannot be proved in ZFC) remained open much longer. Finally, in 1963 Paul Cohen announced the discovery of a method which enabled him to construct a model of ZFC in which the Continuum Hypothesis failed. We devote the rest of this section to an outline of some of his ideas.

Let us again consider the universe of all sets as described by Zermelo-Fraenkel axioms with Choice. The only information about $2^{\aleph_0}$ we have been able to derive is provided by Cantor's Theorem: $2^{\aleph_0} > \aleph_0$ (more generally, $\mathrm{cf}(2^{\aleph_0}) > \aleph_0$, see Lemma 3.3 in Chapter 9). In particular, $2^{\aleph_0} = \aleph_1$ is a distinct possibility (and becomes a provable fact if the Axiom of Constructibility is also assumed). In general, it is, therefore, necessary to *add "new" sets* to our universe in order to get a model for $2^{\aleph_0} > \aleph_1$.

We concentrate our attention on the task of adding just one "new" set of natural numbers X. For the time being X is just a symbol devoid of content, a name for a set yet to be described. Let us see what one could say about it.

A key to the matter is a realization that one cannot expect to have *complete* information about X. If we found a property $\mathbf{P}$ which would tell us exactly which natural numbers belong to X, we could set $X = \{n \in \omega \mid \mathbf{P}(n)\}$ and conclude on the basis of the Axiom Schema of Comprehension that X exists in our universe, and so is not a "new" set. Cohen's basic idea was that partial descriptions of X are sufficient. He described the set X by a collection of "approximations" in much the same way as irrational numbers can be approximated by rationals.

Specifically, we call finite sequences of zeros and ones *conditions*; for example, $\emptyset$, $\langle 1\rangle$, $\langle 1,0,1\rangle$, $\langle 1,1,0\rangle$, $\langle 1,1,0,1\rangle$ are conditions. We view these conditions as providing partial information about X in the following sense: If the kth entry in a condition is 1, that condition determines that $k \in X$. If it is 0, the condition determines that $k \notin X$. For example, $\langle 1,1,0,1\rangle$ determines that $0 \in X$, $1 \in X$, $2 \notin X$, $3 \in X$ (but does not determine, say, $4 \in X$ either way.)

Next, it should be noted that adding one set X to the universe immediately gives rise to many other sets which were not in the universe originally, such as $\omega - X$, $\omega \times X$, X^2, $\mathcal{P}(X)$, etc. Each condition, by providing some information about X, enables us to make some conclusions also about these other sets, and about the whole expanded universe. Cohen writes $p \Vdash \mathbf{P}$ (*p forces* $\mathbf{P}$) to indicate that information provided by the condition p determines that the property $\mathbf{P}$ holds. For example, it is obvious that

$$\langle 1,1,0,1\rangle \Vdash (5,3) \in \omega \times X$$

(because, as we noted before, $\langle 1,1,0,1\rangle \Vdash 3 \in X$, and $5 \in \omega$ is true) or

$$\langle 1,1,0,1\rangle \Vdash \{2,3\} \notin \mathcal{P}(X)$$

(because $\langle 1,1,0,1\rangle \Vdash 2 \notin X$).

It should be noticed that conditions often clash: For example,

$$\langle 1,1,0\rangle \Vdash (0,1) \in X^2,$$

while

$$\langle 1,0,1\rangle \Vdash (0,1) \notin X^2.$$

To understand this phenomenon, the reader should think of $p \Vdash \mathbf{P}(X)$ as a *conditional* statement: *If* the set X is as described by the condition p, then X has the property $\mathbf{P}$. So $\langle 1,1,0\rangle \Vdash (0,1) \in X^2$ means: If $0 \in X$, $1 \in X$, and $2 \notin X$, then $(0,1) \in X^2$, while $\langle 1,0,1\rangle \Vdash (0,1) \notin X^2$ means: If $0 \in X$, $1 \notin X$, and $2 \in X$, then $(0,1) \notin X^2$. If we knew the set X, we would of course be able to determine whether it is as described by the condition $\langle 1,1,0\rangle$ or by the condition $\langle 1,0,1\rangle$ or perhaps by another of the remaining six conditions of length three. Since we cannot know the set X, we never know which of these conditions is "true"; thus we never are able to decide whether $(0,1) \in X^2$ or not. Nevertheless, it turns out that there are a great many properties of X which can be decided, because they must hold *no matter which* conditions are "true." As an illustration, let us show that every condition forces that X is infinite. If not, there is a condition p and a natural number k such that $p \Vdash$ "X has k elements.". For example, let $p = \langle 1,0,1\rangle$ and $k = 5$; we show that $\langle 1,0,1\rangle \Vdash$ "X has five elements" is impossible. Consider the condition $q = \langle 1,0,1,1,1,1,1\rangle$. First of all, the condition q contains all information supplied by p: $0 \in X$, $1 \notin X$, $2 \in X$. If the conclusion that X has five elements could be derived from p, it could be derived from q also. But this is absurd because clearly $q \Vdash$ "X has at least six elements"; namely, $q \Vdash 0 \in X$, $2 \in X$, $3 \in X$, $4 \in X$, $5 \in X$, $6 \in X$. The same type of argument leads to a contradiction for any p and k.

As a second and last illustration, we show that every condition forces that X is a "new" set of natural numbers. More precisely, if A is any set of natural numbers from the "original" universe (before adding X), then every condition forces that $X \neq A$. If not, there is a condition p such that $p \Vdash X = A$. Let us again assume that $p = \langle 1,0,1\rangle$; then $p \Vdash 0 \in X$, $1 \notin X$, $2 \in X$. Now there are two possibilities. If $3 \in A$, let $q_1 = \langle 1,0,1,0\rangle$. Since q_1 contains all information supplied by p, $q_1 \Vdash X = A$. But this is impossible, because $q_1 \Vdash 3 \notin X$, whereas $3 \in A$. If $3 \notin A$, let $q_2 = \langle 1,0,1,1\rangle$. We can again conclude that $q_2 \Vdash X = A$ and get a contradiction from $q_2 \Vdash 3 \in X$ and $3 \notin A$. Again, a similar argument works for any p.

Let us now review what has been accomplished by Cohen's construction. The universe of set theory has been extended by adding to it a "new," "imaginary" set X (and various other sets which can be obtained from X by set-theoretic operations). Partial descriptions of X by conditions are available. These descriptions are not sufficient to decide whether a given natural number belongs to X or not, but allow us, nevertheless, to demonstrate certain statements about X, such as that X is infinite and differs from every set in the original universe. Cohen has established that the descriptions by conditions are sufficient to show the validity of all axioms of Zermelo-Fraenkel set theory with Choice in the extended universe.

Although adding one set of natural numbers to the universe does not increase the cardinality of the continuum, one can next take the extended universe and, by repeating the whole construction, add to it another "new" set of natural numbers Y. If this procedure is iterated $\aleph_2$ times, the result is a model in which there are at least $\aleph_2$ sets of natural numbers, i.e., $2^{\aleph_0} \geq \aleph_2$. Alternatively, one can simply add $\aleph_2$ such sets at once by employing slightly modified conditions.

Cohen's method has been used to construct models in which $2^{\aleph_0} = \aleph_\alpha$ for any $\aleph_\alpha$ with $\mathrm{cf}(\aleph_\alpha) > \aleph_0$. It can also be used to build models in which Suslin's Hypothesis holds or fails to hold, models for Martin's Axiom MA_κ, as well as models for many other undecidable propositions of set theory.

The methods described in this section can also be applied to showing that the Axiom of Choice is neither provable nor refutable from the other axioms of Zermelo-Fraenkel set theory. It is not refutable because one can define the constructible model and prove that it is a model of ZFC using *only* the axioms of Zermelo-Fraenkel set theory *without Choice*. [A choice function for any system X of constructible sets can be *defined* roughly as follows: Select from each $A \in S$ ($A \neq \emptyset$) that element which is constructed at the earliest stage — that is, belongs to L_0 or to $L_{\alpha+1}$ for the least possible α. Should there be several, select the $\in$-least element of L_0 or the one whose definition in $(L_\alpha, \in)$ comes first in the alphabetic order of all possible definitions.] On the other hand, the Axiom of Choice is not provable in Zermelo-Fraenkel set theory either, because, as Cohen has shown, it is possible to extend the universe of set theory by adding to it a "new" set of real numbers without adding any well-ordering of this set, and thus obtain a model of set theory in which the set of real numbers cannot be well-ordered. Consistency of the Axiom of Foundation with the remaining axioms of ZFC is also established by the method of models. A model for set theory with the Axiom of Foundation is obtained by letting $\mathbf{M}(x)$ be the property "x is a well-founded set," and $\mathbf{E}(x, y)$ be "$x \in y$." The proof that this is indeed a model satisfying the Axiom of Foundation follows the lines of Example 2.8 in Section 2 of Chapter 14.

3. The Universe of Set Theory

In this last section we give some thought to possibilities of extending Zermelo-Fraenkel set theory with Choice by additional axioms. Our interest here is in axioms which could with some justification be regarded as true.

One possible candidate for such a new axiom has been introduced in Section 2; it is the Axiom of Constructibility. We have seen there that the Axiom of Constructibility has important consequences in set theory; for example, it implies the Generalized Continuum Hypothesis and existence of counterexamples to Suslin's Hypothesis. In recent years it has also been shown that the Axiom of Constructibility can be a powerful tool for other branches of abstract mathematics, and a series of important results, including some of great elegance and intuitive appeal, has been derived from it in model theory, general topology, and group theory. On the other hand, there do not seem to be any good intuitive reasons for belief that all sets are constructible; the ease with which the method of forcing from Section 2 establishes the possibility of existence of nonconstructible sets might suggest rather the opposite. Moreover, all of the aforementioned results can be proved equally well from other axioms, some of them weaker than the Axiom of Constructibility, and some of them actually contradicting it.

However, the most serious objections to accepting the Axiom of Constructibility at the same level as the axioms of ZFC stem from the fact that it yields some rather unnatural consequences in descriptive set theory, a branch of mathematics concerned with detailed study of the complexity of sets of real numbers. As descriptive set theory provides some of the most important insights into the foundations of set theory, we outline a few of its basic problems and results next.

We introduced Borel sets in Section 5 of Chapter 10 as sets of real numbers that are particularly simple. Their detailed study confirms this by showing that Borel sets exhibit particularly nice behavior: Every Borel set is either at most countable or it contains a perfect subset, and so, in accordance with the Continuum Hypothesis, it has either cardinality $\leq \aleph_0$ or $2^{\aleph_0}$. Similarly, every Borel set is Lebesgue measurable, and has many other nice properties. Descriptive set theory attempts to extend these results to more complex sets. How do we obtain such sets? Countable unions and intersections of Borel sets are Borel, so we cannot use them to obtain "new" sets. Continuous functions are well-understood, simple functions, but it turns out that an image of a Borel set by a continuous function (while hopefully still "simple" enough) need not be a Borel set. We define *analytic* sets as those sets that are images of a Borel set by a continuous function, and *coanalytic* sets as complements of analytic sets (in $\boldsymbol{R}$). The usual notation is $\boldsymbol{\Sigma}_1^1$ for the collection of all analytic sets and $\boldsymbol{\Pi}_1^1$ for the collection of all coanalytic sets. One can then proceed further and define recursively $\boldsymbol{\Sigma}_{n+1}^1$ as the collection of all continuous images of $\boldsymbol{\Pi}_n^1$ sets and $\boldsymbol{\Pi}_{n+1}^1$ as the collection of all complements of sets in $\boldsymbol{\Sigma}_{n+1}^1$. This is the *projective hierarchy*, and sets that belong to some $\boldsymbol{\Sigma}_n^1$ (or $\boldsymbol{\Pi}_n^1$) are the *projective sets*. We would expect projective sets to behave nicely, being still rather simple, and the classical descriptive set theory as developed by Nikolai Luzin and his students confirms this to an extent. For example, it is known that all analytic and coanalytic sets are Lebesgue measurable, and that all uncountable analytic sets contain a perfect subset. As to going further, in ZFC with the Axiom of Constructibility one gets discouraging results: there exist $\boldsymbol{\Sigma}_2^1$ (or $\boldsymbol{\Pi}_2^1$) sets that are not Lebesgue measurable, and there exist uncountable coanalytic ($\boldsymbol{\Pi}_1^1$) sets without a perfect subset.

Another important classical result is that $\boldsymbol{\Pi}_1^1$ and $\boldsymbol{\Sigma}_2^1$ sets have the so-called *reduction property* and $\boldsymbol{\Sigma}_1^1$ and $\boldsymbol{\Pi}_2^1$ sets do not have it. (A class of sets Γ is said to have the *reduction property* if for all $A, B \in \Gamma$ there exist $A', B' \in \Gamma$ such that $A' \subseteq A$, $B' \subseteq B$, $A' \cup B' = A \cup B$, and $A' \cap B' = \emptyset$.) One might expect $\boldsymbol{\Pi}_3^1$, $\boldsymbol{\Sigma}_4^1$, $\boldsymbol{\Pi}_5^1$, etc., sets to have the reduction property, and $\boldsymbol{\Sigma}_3^1$, $\boldsymbol{\Pi}_4^1$, $\boldsymbol{\Sigma}_5^1$, etc., to fail to have it, but the Axiom of Constructibility leaves the case $n = 1$ an odd exception and proves instead that all $\boldsymbol{\Sigma}_n^1$ sets for $n \geq 2$ have the reduction property and all $\boldsymbol{\Pi}_n^1$ sets, $n \geq 2$, fail to have it.

Surprisingly, a more satisfactory answer can be obtained by using some of the so-called *large cardinal axioms*. The inaccessible cardinal numbers defined in Chapter 9 provide the simplest examples of large cardinals. We remind the reader that a cardinal number $\kappa > \aleph_0$ is called *inaccessible* if it is regular and a limit of smaller cardinal numbers. It is shown in Section 2 of Chapter 9 that the first inaccessible cardinal (assuming it exists) must be greater than $\aleph_1, \aleph_2, \ldots, \aleph_\omega, \ldots, \aleph_{\aleph_1}, \ldots, \aleph_{\aleph_{\aleph_1}}, \ldots, \aleph_{\aleph_{\aleph_\omega}}, \ldots$, etc., hence the adjective "large." It is known that the existence of inaccessible cardinals cannot be proved in ZFC; in fact, the construction of a model for ZFC in which there are no inaccessible cardinals is rather simple. If the constructible universe L contains no inaccessible cardinals (in the sense of L), then it is such a model. Otherwise, we take the least inaccessible cardinal ϑ in L and consider a model, sets of which are precisely the elements of L_ϑ, and the membership relation is the usual one. It is quite easy to use the inaccessibility of ϑ and prove that all axioms of ZFC are satisfied in this model. The fact that ϑ is the least inaccessible cardinal in L guarantees that there are no inaccessible cardinals in this model. So existence of inaccessible cardinals is independent of ZFC. Unlike the Continuum Hypothesis or the Suslin's problem, however, it is not possible to prove that inaccessible cardinals are consistent with ZFC by way of constructing an appropriate model (doing so would contradict the celebrated Second Incompleteness Theorem of Gödel). This means that a set theory with an inaccessible cardinal is essentially stronger than plain ZFC. The assumption of existence of inaccessible cardinals requires a "leap of faith" (similar to that required for acceptance of the Axiom of Infinity), but some intuitive justification for the plausibility of making it can be given. It goes roughly as follows.

Mathematicians ordinarily consider infinite collections, such as the set of natu ral numbers or the set of real numbers, as finished, completed totalities. On th e other hand, it is not in the power of an ordinary mathematician to regard the collection of all sets as a finished totality, i.e., a set — it would lead to a contradiction. We call the collections which mathematicians ordinarily view as finished *the sets of the first order*; they are the sets we have been con cerned with until now. Let us now place ourselves in a position of an extraordin ary, more abstract, mathematician, who has examined the universe of the first-order sets and then collected all those sets into a new finished totality, a *second-order set* V. Having gotten V, he can, of course, use the methods of Chapters 1–14 to form various other second-order sets, such as $V - \omega$, $\aleph_1 \times V$, $\mathcal{P}(\mathcal{P}(V))$, $O = \{\alpha \in V \mid \alpha \text{ is an ordinal}\}$, $O + 1$, $O + O + \omega$, etc. [Notice that O is the (second-order) set of all first-order ordinals.] The intuitive arguments which justified the axioms of ZFC for the first-order mathematician also convince the second-order mathematician of the validity of these axioms in his "second-order" universe of set theory. (Notice that Russell's Paradox is avoided by this viewpoint: The second-order mathematician can form the set R of all first-order sets which are not elements of themselves, but R is not a first-order set, so clearly $R \notin R$, and no contradictions appear. Of course, the second-order mathematician still cannot form the "set" of all second-order sets which are not elements of themselves; this is a task for a third-order mathematician.) We now claim that O, the least second-order ordinal, is an inaccessible cardinal in the second-order universe. Certainly $O > \aleph_0$ and O is a limit of (the first-order) cardinal numbers. If $\langle \kappa_\iota \mid \iota < \alpha \rangle$ is a sequence of ordinals less than O having length $\alpha < O$, then both α and all κ_ι are first-order ordinals, and the arguments we used to justify the Axiom Schema of Replacement again convince us that a first-order mathematician should be able to collect $\{\kappa_\iota \mid \iota < \alpha\}$ into a first-order set. But then $\sup_{\iota<\alpha} \kappa_\iota$ is a first-order ordinal, so $\sup_{\iota<\alpha} \kappa_\iota < O$, and O is regular. We can conclude that the set-theoretic universe of the second-order mathematician satisfies the axiom "There exists an inaccessible cardinal."

Let us now return to descriptive set theory, in particular to the question whether all countable coanalytic sets have a perfect subset. Robert Solovay discovered a profound connection between this and inaccessible cardinals: If every uncountable coanalytic set has a perfect subset, then $\aleph_1$ is an inaccessible cardinal in the constructible universe L. (We pointed out in Section 2 that the notion of cardinality is not absolute, so the "real" cardinal $\aleph_1$ need not be "$\aleph_1$ in the sense of the model L" if not all sets belong to L.) In fact, more than this is true:

(*) For any real number a, $\aleph_1$ is inaccessible in $L[a]$,

where $L[a]$ is a model constructed just as L, but starting with $L_0[a] = \omega \cup \{a\}$ (it is the least model of set theory containing all ordinals and the real number a). Conversely, from the "large cardinal axiom" (*) it follows that all uncountable $\mathbf{\Pi}^1_1$ sets have a perfect subset. In addition, (*) further implies that all uncountable $\mathbf{\Sigma}^1_2$ sets have a perfect subset (and thus cardinality $2^{\aleph_0}$) and that all $\mathbf{\Sigma}^1_2$ and $\mathbf{\Pi}^1_2$ sets are Lebesgue measurable (but it does not imply the existence of perfect subsets of uncountable $\mathbf{\Pi}^1_2$ sets, nor measurability of $\mathbf{\Sigma}^1_3$ sets). Overall, the axiom (*) produces a mild improvement upon the results obtainable from the Axiom of Constructibility. To get better results, we have to assume existence of cardinals much larger than mere inaccessibles. But before outlining some of them, we have to make another digression.

A very important role in modern descriptive set theory is played by a certain kind of infinitary game. Let us consider a game between two players, I and II, described by the following rules. A finite set M of possible moves is given. The players choose moves, taking turns and each moving n times. That is, player I begins by making a move $p_1 \in M$; player II responds by making a move $q_1 \in M$. Then it is I's turn to make a move $p_2 \in M$, to which II responds by playing $q_2 \in M$, etc. The resulting sequence of moves $\langle p_1, q_1, p_2, q_2, \dots, p_n, q_n \rangle$ is called a *play*. A set of plays S is specified in advance and known to both players. Player I wins if $\langle p_1, \dots, q_n \rangle \in S$; player II wins if $\langle p_1, \dots, q_n \rangle \notin S$. Many intellectual games between two players, including checkers, chess, and go, can be mathematically represented in this abstract form by choosing suitable M, n, and S (some artificial conventions have to be adopted, e.g., in order to allow for draws).

A fundamental notion in game theory is that of a strategy. A *strategy* for player I is a rule which tells him what move to make at each of his turns, depending on the moves played by both players previously. If a strategy for player I has the property that, by following it, I always wins, it is called a *winning strategy* for I. Similarly, one can talk about a winning strategy for II.

The basic fact about games of this type is that one of the players always has a winning strategy; in other words, the game is *determined*. The reason is rather simple:

If there is a move p_1 such that for every q_1
there is a move p_2 such that for every q_2 ...
there is a move p_n such that for every q_n the play $\langle p_1, q_1, \dots, p_n, q_n \rangle \in S$,

then obviously player I has a winning strategy. If the opposite holds, it means that

For every p_1 there is q_1 such that
for every p_2 there is q_2 such that ...
for every p_n there is q_n such that the play $\langle p_1, q_1, \dots, p_n, q_n \rangle \notin S$.

But in this case, II is the player with a winning strategy.

We can now modify the game by allowing each player infinitely many turns. The play now is an infinite sequence of moves: $\langle p_1, q_1, p_2, q_2, \dots \rangle$. The payoff set S is a set of infinite sequences of elements of M, and I is the winner if and only if $\langle p_1, q_1, p_2, q_2, \dots \rangle \in S$. The game played according to these rules is referred to as G_S. It is no longer obvious that such games are determined, and as a matter of fact they need not be. Using the Axiom of Choice one can construct a payoff set $S \subseteq M^{\boldsymbol{N}}$ for any M with $|M| \geq 2$ such that neither player has a winning strategy in the game G_S. (The argument is quite similar to the one we used to construct an uncountable set without a perfect subset in Example 4.11 of Chapter 10.) The interesting question is whether the game G_S is determined for "simple" sets S. In order to study this question, we take the finite set M to be a natural number m; infinite sequences of elements of m can then be viewed as expansions of real numbers in base m, and the payoff set S can be viewed as a subset of $\boldsymbol{R}$. In this way we can talk about payoff sets being Borel, analytic, etc.

A profound theorem of D. Anthony Martin states that all games with Borel payoff sets are determined. The situation at higher levels of the projective hierarchy is similar to that discussed in connection with the question of existence of perfect subsets, but the large cardinals involved are much larger. To be specific, the Axiom of Constructibility can be used to construct games with analytic ($\boldsymbol{\Sigma}_1^1$) payoff sets that are not determined. The work of Martin and Leo Harrington showed that determinacy of all games with $\boldsymbol{\Sigma}_1^1$ (or, equivalently, $\boldsymbol{\Pi}_1^1$) payoff sets is equivalent to a certain large cardinal axiom. It would take us too far to try to state this axiom here (it is known technically as "For all real numbers x, $x^\#$ exists."); for our purposes it suffices to say that this axiom implies Solovay's assumption (*), and much more (for example, it implies the existence of Mahlo cardinals and weakly compact cardinals in the constructible universe L), and is itself a consequence of the existence of measurable cardinals. So if a measurable cardinal exists then all analytic and coanalytic games are determined. The existence of measurable cardinals does not suffice to prove determinacy of all games with $\mathbf{\Sigma}^1_2$ (or $\mathbf{\Pi}^1_2$) payoff sets. The work of Martin, John Steel, and Hugh Woodin in the early nineties showed that there are large cardinals (much larger than the first measurable one) whose existence implies the *Axiom of Projective Determinacy*: all games with projective payoff sets are determined. Conversely, the Axiom of Projective Determinacy implies existence of models of set theory with these large cardinals.

What reasons are there for the belief that something like the Axiom of Projective Determinacy is true? Besides its own intrinsic plausibility, it stems from the fact that determinacy of infinitary games at a certain level in the projective hierarchy implies all the nice descriptive set-theoretic properties that the sets at or near that level are expected to have. For example, determinacy of $\mathbf{\Sigma}^1_1$ games implies that all uncountable $\mathbf{\Pi}^1_1$ and $\mathbf{\Sigma}^1_2$ sets have a perfect subset, and that all $\mathbf{\Sigma}^1_2$ and $\mathbf{\Pi}^1_2$ sets are Lebesgue measurable. The determinacy of $\mathbf{\Sigma}^1_2$ games implies that all uncountable $\mathbf{\Pi}^1_2$ and $\mathbf{\Sigma}^1_3$ sets have a perfect subset, and that all $\mathbf{\Sigma}^1_3$ and $\mathbf{\Pi}^1_3$ sets are Lebesgue measurable. Moreover, it also generalizes the reduction property correctly to the third level of the projective hierarchy; i.e., it implies that $\mathbf{\Pi}^1_3$ sets have the reduction property and $\mathbf{\Sigma}^1_3$ sets do not. The full Axiom of Projective Determinacy has as its consequences the Lebesgue measurability of all projective sets, the fact that every uncountable projective set has a perfect subset, and the "correct" behavior of the reduction property ($\mathbf{\Pi}^1_1$, $\mathbf{\Sigma}^1_2$, $\mathbf{\Pi}^1_3$, $\mathbf{\Sigma}^1_4$, ... sets have it, $\mathbf{\Sigma}^1_1$, $\mathbf{\Pi}^1_2$, $\mathbf{\Sigma}^1_3$, $\mathbf{\Pi}^1_4$, ... do not), among many others not stated here. It is the results like these that are convincing workers in descriptive set theory that PD must be true.

What emerges from the above considerations is a hierarchy postulating the existence of larger and larger cardinals and providing better and better approximations to the ultimate truth about the universe of sets. Further support for this general picture has come from the work on undecidable statements in arithmetic. By arithmetic we mean the theory of *Peano arithmetic* whose axioms were given in Chapter 3; it is easy to establish that Peano arithmetic is equivalent to the *theory of finite sets*; i.e., the theory that is obtained from ZFC by omitting the Axiom of Infinity (in the resulting theory, only finite sets can be proved to exist). It has been known since the fundamental work of Kurt Gödel in 1931 that there are true statements about natural numbers (or finite sets) that cannot be decided (either proved or disproved) from the axioms of Peano arithmetic (or the theory of finite sets). The statements are true because they can be proved in ZFC, but the use of at least some infinite sets is essential for the proofs. However, Gödel's examples of such statements are based on logical considerations (similar to Russell's Paradox) and have no transparent mathematical meaning. In 1977, Jeffrey Paris discovered the first simple mathematical examples of such statements, and others have been found subsequently. One of the most interesting is Theorem 6.7 in Chapter 6, the claim that every Goodstein sequence terminates with a value of 0 after a finite number of steps. Our proof there uses infinite ordinals; the work of Paris showed that some use of infinity is necessary. Another example is a version of the Finite Ramsey's Theorem discussed in Section 1 of Chapter 12. It is possible to determine the size of infinite sets needed to prove a particular statement, and it turns out that here, too, there is a hierarchy of theories postulating the existence of larger and larger infinite sets and providing better and better approximations to the truth about natural numbers (or finite sets). In most cases these theories are subtheories of ZFC (so few mathematicians doubt that they are true), but there are examples of (rather complicated) statements of arithmetic that cannot be decided even in ZFC (but can be decided, for example, in ZFC with an inaccessible cardinal), so the hierarchy obtained from the study of the strength of arithmetic statements merges into the hierarchy of large cardinals needed for the study of the strength of statements about real numbers arising in descriptive set theory, with ZFC being just one of the stages. Even the techniques used to prove the results about arithmetic are closely related to those used to study large cardinals; they rely heavily on such concepts as partitions, trees, and games.

Most of the work discussed in this section is relatively recent, and by no means complete. Both the theory of large cardinals and the study of the undecidable statements of arithmetic are very active research areas where new insights and interconnections continue to be discovered. As Gödel assures us in his Incompleteness Theorem, no axiomatic theory can decide all statements of arithmetic or set theory. We can thus feel confident that the enterprise of getting closer and closer to the ultimate truth about the mathematical universe will continue indefinitely.

Bibliography

[1] Peter Aczel. *Non-well-founded sets*. Stanford University Center for the Study of Language and Information, Stanford, CA, 1988. With a foreword by Jon Barwise [K. Jon Barwise].

[2] Keith Devlin. *The joy of sets*. Springer-Verlag, New York, second edition, 1993. Fundamentals of contemporary set theory.

[3] F. R. Drake and D. Singh. *Intermediate set theory*. John Wiley & Sons Ltd., Chichester, 1996.

[4] Herbert B. Enderton. *Elements of set theory*. Academic Press [Harcourt Brace Jovanovich Publishers], New York, 1977.

[5] Paul Erdős, András Hajnal, Attila Máté, and Richard Rado. *Combinatorial set theory: partition relations for cardinals*. North-Holland Publishing Co., Amsterdam, 1984.

[6] Ronald L. Graham, Bruce L. Rothschild, and Joel H. Spencer. *Ramsey theory*. John Wiley & Sons Inc., New York, second edition, 1990. A Wiley-Interscience Publication.

[7] Paul R. Halmos. *Naive set theory*. Springer-Verlag, New York, 1974. Reprint of the 1960 edition, Undergraduate Texts in Mathematics.

[8] James M. Henle. *An outline of set theory*. Springer-Verlag, New York, 1986.

[9] Thomas Jech. *Set theory*. Springer-Verlag, Berlin, second edition, 1997.

[10] Winfried Just and Martin Weese. *Discovering modern set theory. I*. American Mathematical Society, Providence, RI, 1996. The basics.

[11] Akihiro Kanamori. *The higher infinite*. Springer-Verlag, Berlin, 1994. Large cardinals in set theory from their beginnings.

[12] Alexander S. Kechris. *Classical descriptive set theory*. Springer-Verlag, New York, 1995.

[13] Kenneth Kunen. *Set theory.* North-Holland Publishing Co., Amsterdam, 1983. An introduction to independence proofs, Reprint of the 1980 original.

[14] Yiannis N. Moschovakis. *Descriptive set theory.* North-Holland Publishing Co., Amsterdam, 1980.

[15] Yiannis N. Moschovakis. *Notes on set theory.* Springer-Verlag, New York, 1994.

[16] Judith Roitman. *Introduction to modern set theory.* John Wiley & Sons Inc., New York, 1990. A Wiley-Interscience Publication.

[17] Robert L. Vaught. *Set theory.* Birkhäuser Boston Inc., Boston, MA, second edition, 1995. An introduction.

Index